D1030355

Temperature

MONOGRAPHS IN PHYSICAL MEASUREMENT

Series Editor

A. H. COOK F.R.S., F.R,S.E.

Jacksonian Professor of Natural Philosophy,
The University of Cambridge.

P. KARTASCHOFF: Frequency and Time (1978)

T. J. QUINN: Temperature (1983)

Temperature

T. J. Quinn

*Bureau International
des Poids et Mesures
Sèvres, France*

1983

ACADEMIC PRESS

A Subsidiary of Harcourt Brace Jovanovich, Publishers

London   New York
Paris   San Diego   San Francisco
São Paulo   Sydney   Tokyo   Toronto

ACADEMIC PRESS INC. (LONDON) LTD.
24/28 Oval Road
London NW1

United States Edition published by
ACADEMIC PRESS INC.
111 Fifth Avenue
New York, New York 10003

British Library Cataloguing in Publication Data

Quinn, T. J.
Temperature — (Monographs in physical measurement)
1. Thermometers and thermometry I. Title
536'.5   QC271

ISBN 0-12-569680-9

Photoset in Great Britain by
Mid County Press, London, SW15
and printed by Thomson Litho Ltd,
East Kilbride, Scotland

Preface

The purpose of this book is to give a comprehensive account of the principles of thermometry over the range 0.5 K to about 3000 K. Although the subject has evolved in a steady rather than spectacular way over the past 25 years, the time is ripe for a book in which the considerable progress that has been made is reviewed and collated. Much of the new work in thermometry has been at low-temperatures, that is to say below about 30 K, the results of which provided the basis for the 1976 Provisional 0.5 K to 30 K Temperature Scale. A temperature of 0.5 K is thus a convenient lower limit to the temperature domain considered in this book. The upper limit is much less well defined since, in principle, radiation thermometry which is treated in Chapter 7, can be used to arbitrarily high temperatures provided that something can be said about the state of thermal equilibrium of the system whose temperature is being measured. Despite the wide range of conditions encountered in thermometry between, say, liquid-helium temperatures and those near the melting point of platinum, there exist the common requirements for thermal equilibrium and for thermal contact with the thermometer. These requirements are present in any thermometric measurement at any temperature and provide a constant theme running throughout the book. It is difficult to enter in any detail into the principles of the various methods of thermometry, their accuracy, range and ultimate limitations, without a clear idea of the underlying physics which is the basis of each method. For this reason each of the main Chapters begins with a brief summary of the physical principles of the method insofar as these are relevant to the theory and practice of thermometry.

The book opens with an Introductory Chapter which includes a brief discussion on the meaning of "temperature", an outline of the history of thermometry and a section on the important distinction between primary and secondary thermometry. In Chapter Two the origins of the present international agreement on thermometry are described followed by a discussion on the evolution and present status of the International Practical Temperature Scale. In Chapter Three the main methods of determining thermodynamic temperature are discussed, these include gas thermometry, acoustic thermometry and noise thermometry. In Chapter Four the establishment of fixed points of temperature is described; these include triple and boiling points of gases, freezing points and superconducting-transition points of metals. Uniform-temperature enclosures for the comparison of thermometers are also treated in this chapter. The ducceeding three chapters then deal with the principal practical methods of thermometry: resistance thermometry, thermocouple thermometry and radiation thermometry. In each chapter as well as the introductory section on the physical principles and a detailed description of the use of the methods to the highest accuracy in the standards laboratory, some examples are also given of industrial applications. The book ends with a short chapter on mercury-in-glass thermometry. Each chapter includes a comprehensive bibliography.

In preparing this book I am very conscious of the debt that I owe to my friends and colleagues who have given of their time in reading and correcting the various parts of the manuscript. Without their helpful criticism there would be many more errors and omissions than there undoubtedly are, and for which the author is, of course, wholly responsible. Special thanks are due to Jan de Boer (University of Amsterdam), Ron Bedford and Hugh Preston-Thomas (National Research Council, Ottawa), Ralph Hudson (formerly of the National Bureau of Standards, Washington, now at BIPM), J. S. Johnson, Rosemount Engineering Co., and my former colleagues of the Temperature Section NPL, Keith Berry, Peter Coates, Tony Colclough, Maurice Chattle, and Richard Rusby. I would also like to thank Jacqueline Monprofit, helped in the later stages by Monique Petit, for typing the manuscript, Christian Veyradier for preparing the figures and the Director, BIPM, for offering me these facilities. In addition I must acknowledge the continued encouragement given to me by the Series Editor, Alan Cook of Cambridge University during the preparation of the manuscript and finally thank my wife who has been so patient and understanding during the very many evenings and weekends which I have devoted to this book.

BIPM   Terry Quinn
January 1983

Contents

Preface . . . . *v*
List of Abbreviations . . . . *x*
Note regarding References to the Proceedings of the
    Symposia on Temperature . . . . *xi*

1. *The Meaning of Temperature and the Development of Thermometry*
    1-1   Temperature: some basic ideas 3
    1-2   Temperature in classical thermodynamics. 5
    1-3   Temperature in statistical mechanics 10
    1-4   Temperature in non-equilibrium conditions, and negative
          temperatures 13
    1-5   Numerical values of T, k and R 15
    1-6   The origins of thermometry 17
    1-7   Primary and secondary thermometry 22

2. *Thermodynamic and Practical Temperature Scales*
    2-1   Introduction. 25
    2-2   The normal hydrogen scale 27
    2-3   The origins of the International Temperature Scale of 1927 . 29
    2-4   The principles of a practical temperature scale . . . 31
    2-5   The evolution of ITS-27 and its modifications in 1948. . . 33
    2-6   The new definition of the unit of thermodynamic temperature: the
          kelvin 35
    2-7   The International Practical Temperature Scale of 1968 (IPTS-68) 37
    2-8   The uniqueness of IPTS-68 at low temperatures . . . 43
    2-9   The departures of IPTS-68 from thermodynamic temperatures . 46
    2-10   The 1976 Provisional 0.5 K to 30 K Temperature Scale, EPT-76 50
    2-11   The vapour pressure scales of ^{3}He and ^{4}He. . . . 53

3. *The Measurement of Thermodynamic Temperature*
    3-1   Introduction. 61
    3-2   Gas thermometry. 61
    3-3   Acoustic thermometry 84
    3-4   Noise thermometry 98

3-5   Magnetic thermometry.   107
3-6   Dielectric-constant and refractive-index gas thermometry   .   .   113

4.   Fixed Points and Comparison Baths
4-1   Introduction.   121
4-2   Comparison baths and furnaces   122
4-3   Heat pipes   128
4-4   Boiling points of water and sulphur   132
4-5   The low-temperature boiling and triple points   134
4-6   Superconducting-transition points   148
4-7   The melting and freezing of metals   130
4-8   The triple point of water   160
4-9   The triple point of gallium   162

5.   Resistance Thermometry
5-1   The electrical resistance of metals, alloys and semiconductors   .   167
5-2   The resistivity of a pure metal as a function of temperature   .   174
5-3   The resistivity of a semiconductor as a function of temperature   .   177
5-4   The high-precision platinum resistance thermometer   .   .   .   182
5-5   Industrial platinum resistance thermometry.   200
5-6   The rhodium/0.5 % iron resistance thermometer   209
5-7   The germanium resistance thermometer   213
5-8   Thermistors   220
5-9   Carbon and carbon–glass thermometers   223
5-10   The effects of magnetic fields on resistance thermometers   .   .   226
5-11   Resistance measurement in thermometry   233

6.   Thermocouples
6-1   Introduction.   241
6-2   Elementary theory of thermoelectricity.   243
6-3   Thermocouple types   249
6-4   Extension and compensating wires   272
6-5   The calibration of thermocouples.   274
6-6   Reference junctions   279
6-7   The pressure dependence of the emf of thermocouples.   .   .   280

7.   Radiation Thermometry
7-1   Introduction.   284
7-2   The properties of thermal radiation   286
7-3   The calculation of the emissivity of practical blackbody cavities   .   300
7-4   Practical blackbody cavities.   316
7-5   Tungsten ribbon lamps as reproducible sources for radiation
       pyrometry   320
7-6   Radiation pyrometers   334
7-7   Radiation thermometry for the determination of thermodynamic
       temperature   349
7-8   Practical radiation thermometry aimed at overcoming the
       emittance problem   351
7-9   Radiation thermometry of semi-transparent media   .   .   .   359

8. **Mercury-in-glass Thermometry**
    8-1   Introduction.   368
    8-3   Thermometric glasses 373
    8-4   Secular change and temporary depression of zero   .   .   . 374
    8-5   Thermometers for special applications   375

Appendixes
    I   International Temperature Scales: dates and editions .   .   . 380
    II   IPTS-68: Extracts from the text. Parts I (Introduction) and II
          (Definition) 381
    III   IEC Draft Table of Resistance-Ratio/Temperature for industrial
          platinum resistance thermometers   389
    IV   Thermocouple Reference Tables: skeleton tables for Thermocouple
          Types B, E, J, K, R, S and T   389
    V   Polynomial   expressions   (forward   and   reverse)   for   the
          Thermocouple Reference Tables of Appendix IV .   .   . 398
    VI   Nicrosil-Nisil thermocouple, skeleton reference table and reference
          functions   407

Index.   411

List of Abbreviations

In the text frequent reference is made to certain laboratories, institutions, committees and organizations and for convenience abbreviations are often used, the full title being given only once. The following is an index of such abbreviations.

ASTM   American Society for Testing and Materials
BIPM   Bureau International des Poids et Mesures, Sèvres
CCT   Comité Consultatif de Thermométrie
CGPM   Conférence Générale des Poids et Mesures
CIPM   Comité International des Poids et Mesures
CSIRO Commonwealth Scientific and Industrial Research Organization, Division of
       Physics, Sydney, formerly National Measurement Laboratory (NML)
IEC   International Electrotechnical Commission
INM   Institut National de Métrologie, Paris
IMGC   Istituto di Metrologia "G. Colonnetti", Turin
KOL   Kamerlingh Onnes Laboratory, Leiden
MIM   Mendeleev Institute of Metrology, Leningrad
NBS   National Bureau of Standards, Washington
NIM   National Institute of Metrology, Beijing
NPL   National Physical Laboratory, Teddington
NRC   National Research Council, Ottawa
NRLM   National Research Laboratory of Metrology, Tsukuba
PRMI   Physico-Radio-Technical Measurement Institute, Moscow
PSU   Pennsylvania State University
PTB   Physikalisch-Technische Bundesanstalt, Braunschweig

Note regarding References to the Proceedings of the Symposia on Temperature

Six symposia have been held under the general title of "Temperature, its Measurement and Control in Science and Industry". The proceedings of the first were not published; those of the second were published in 1941 by Reinhold Publishing Co. and are now known as: "Temperature, its Measurement and Control in Science and Industry, Volume 1, 1941" (Reinhold).

The third symposium was held in 1954 and its proceedings were published as: "Temperature, its Measurement and Control in Science and Industry, Volume 2, 1955" published by Reinhold (New York) and Chapman and Hall (London), edited by H. C. Wolfe.

The fourth symposium was held in 1961 and its proceedings were published as: "Temperature, its Measurement and Control in Science and Industry Volume 3, parts 1, 2 and 3, 1962" published by Reinhold (New York) and Chapman and Hall (London), edited by C. M. Herzfeld.

The fifth symposium was held in 1971 and its proceedings were published as: "Temperature, its Measurement and Control in Science and Industry, Volume 4, parts 1, 2 and 3, 1972" published by Instrument Society of America, edited by H. H. Plumb.

The sixth symposium was held in 1982 and its proceedings were published as: "Temperature, its Measurement and Control in Science and Industry, Volume 5, parts 1 and 2, 1982" published by the American Institute of Physics, edited by J. F. Schooley.

Since these symposia have brought together so much important work in thermometry, references to them are very frequent in this book. Such references are given in the following abbreviated form:

Bedford, R. E. (1972). (Title). *TMCSI*, **4**, 15–25.

Similarly, references to the European Conference on Temperature Measurement held at the National Physical Laboratory, Teddington in April 1975 and published by the Institute of Physics (London) as Conference Series No 26, Temperature Measurement 1975, are referred to in the following way:

Berry, K. H. (1975). (Title). Temperature-75, 27–32.

1

The Meaning of Temperature and the Development of Thermometry

---

1-1   Temperature: some basic ideas

Temperature is a quantity which takes the same value in two systems that are brought into thermal contact with one another and allowed to come to thermal equilibrium. Although this statement on the meaning of temperature requires a little explanation and the definition of some terms, which we shall come to below, it already provides the basis for thermometry. The essential ideas of thermal contact and thermal equilibrium are present; what is lacking is any indication of how a numerical value may be associated with the quantity temperature. Before coming to this, however, we must first say something more about the nature of the quantity itself.

   Among the seven base quantities of the International System of Units*, four: Mass, Length, Time and Temperature are so intimately connected with human existence that it is, at first sight, surprising that there existed almost no comprehension concerning one of them, temperature, until the eighteenth century. Even then, such ideas as there were took a further century to crystallize into anything that we would recognize as a proper definition of temperature. On closer examination, however, it becomes very much less surprising that these ideas took so long to develop. Indeed, even today, only a small proportion of those people who use thermometers have an intuitive understanding of what it is that is being measured. Most of the difficulty associated with the concept of temperature stems from its being not only an intensive quantity, but an intensive quantity which is not directly related to any easily perceived extensive quantity. It is probably here that the real stumbling block to the understanding of temperature is to be found. Pressure, an intensive quantity, is easier to understand since it manifests itself as something related to a force. Pressure thereby becomes an example of an intensive quantity about which it is easy to make quantitative statements,

---

* Le Système International d'Unités (SI) 4th Edition, BIPM 1981.

3

since force is a quantity that is directly perceived in a way that allows us to make statements of the sort "this force is twice as large as that one". Density (easily quantified) and colour (quantified only with difficulty) are other examples of intensive quantities that can be perceived directly in a way that can be immediately understood.

Temperature, while it can be perceived directly by the senses, is much more elusive. Subjective sensations of temperature may allow us to say that "this object is hotter or colder than that one". But even this apparently simple statement is fraught with pitfalls for the unwary. For example, take hold in turn of a block of wood, a piece of expanded polystyrene and a rod of copper, all near room temperature but differing slightly in temperature from one another. It is not easy to make any useful statements about which is hotter or colder. This means, of course, that the human hand is a poor thermometer, but the reasons for this being so are by no means straightforward: they are related to the way in which sensations of hotness and coldness are generated in the human body and they lie far outside the scope of this book.

Closely linked to the concept of temperature, and often confused with it, is the concept of heat. While it is a matter of common experience that some substances require more heat to warm them than do others, it is not immediately obvious why this should be so. Nevertheless, given sufficient insight, common experience allows us to make a number of very fundamental statements about the thermal behaviour of matter: these statements comprise the laws of thermodynamics. The zeroth law, so-called because it came to be formulated after the first and second laws, is concerned with the state of bodies brought into thermal contact with one another. In order to understand clearly what this means, we must first of all define a few terms. The following definitions, although not rigorously exact, allow us to make a few general statements about the meaning of temperature and the thermal behaviour of matter that are useful in introducing the subject of thermometry. The reader is referred to the texts on thermodynamics and statistical mechanics cited in the bibliography to this chapter for a more detailed discussion of the fundamentals of thermal physics.

We shall often find ourselves talking about a "system", by which we mean a macroscopic entity extending in space and time and accessible to normal processes of measurement. Such a system is considered to consist of a very large number of material particles or field quantities such as photons, or both. In all cases they are dynamical systems having an extremely large number of degrees of freedom. An "isolated" system is one which has no interaction whatever with its surroundings, while a "closed" system is one which has no material exchange with its surroundings. If an isolated system is left standing, it eventually reaches a configuration or state from which it does not subsequently depart. This final state is called the thermal equilibrium state of the system. Although macroscopically the final state does not change, microscopically the particles which make up the system continue their permanent complex motion. Despite this, we find that only a very few parameters are required to describe the macroscopic state of a system in thermal equilibrium: one of these is, as we shall see, the property called temperature.

If two initially isolated systems are brought into contact with one another via a common separating wall, what happens subsequently depends upon the nature of that wall. If the wall allows thermal but no material exchange, we call it a "diathermal" wall, and in such a situation a new state of thermal equilibrium of the combined system will eventually be reached; subsequent separation of the two original systems will not lead to any change in the new thermal state of either system. The opposite of a diathermal wall is one that is impervious to heat (but which may allow, for example, mechanical work to be done on the system which it encloses); this is known as an adiabatic wall.

It is very important for a proper understanding of thermometry to have a very clear idea of what is meant by thermal equilibrium and thermal contact. We have defined both of these in terms which strictly speaking are applicable only to an idealized world, in which it is possible both to isolate a system and at the same time, to observe its approach to a final state of thermal equilibrium. However in the real world, we can, by taking enough care, approach such idealized conditions as closely as we like, and this is one of the justifications of the use of classical thermodynamics. There can always be devised real systems which approach in one or more respects (but not in all respects) those ideal systems or conditions in which the fundamental laws of thermodynamics are expressed. Whenever this has been done the predictions of classical thermodynamics have, without exception, been confirmed.

1-2   Temperature in classical thermodynamics

Having defined some of our terms we may now come to a statement of the fundamental law for thermometry, *the zeroth law of thermodynamics*:

> If two systems are separately in thermal equilibrium with a third, then they are in thermal equilibrium with each other.

This may also be presented slightly differently

> If three or more systems are in thermal contact with each other and all in equilibrium together, then any two taken separately are in equilibrium with one another.

We can thus imagine one such system, which we call a thermometer, being brought into thermal contact in turn with other systems to find out whether they are in similar or different thermal states.

We mentioned earlier that the thermal equilibrium state of an isolated system is completely described by only a small number of parameters. These physical quantities have a definite value for each thermal state, and in thermodynamics they are called state parameters (or variables) or thermodynamic parameters (or variables). If a set of independent parameters is chosen, so as to be necessary and sufficient to define a thermal state, then the other state parameters are functions of them. The number of independent parameters required to define the equilibrium state of a system is determined empirically.

Among the thermodynamic parameters which describe the thermal state of systems in equilibrium there is one which has the special property of always taking the same value in different systems which are in thermal equilibrium with each other. It is this one that is called "temperature". Thus all systems in thermal equilibrium with one another have the same temperature, and all systems that individually are in thermal equilibrium and that have the same temperature would be found to be already in thermal equilibrium with one another were they to be brought into thermal contact. A slightly different way of stating the zeroth law of thermodynamics is that if two systems in thermal equilibrium each have the same temperature as a third then they also have the same temperature as each other.

We have said that the thermal equilibrium state of a system can be uniquely defined by specifying the values of a few state parameters. For example, the thermal state of an ideal (or "simple" or "perfect") gas is determined by only two such parameters, the pressure P and the molar volume V_m. This, together with the zeroth law of thermodynamics shows that a functional relationship exists between these state parameters and the temperature. We can write

$$f(P, V_m) = \theta$$

where θ is called the "empirical" temperature. This equation is known as the "equation of state". Its functional form cannot be determined from thermodynamics, but must be deduced either empirically or by means of statistical mechanics.

To be able to put numerical values to empirical temperatures we must move on to the first and second laws of thermodynamics. The first law simply states that energy is conserved provided that not only the work done on the system but also the heat exchange with the surroundings through the walls is taken into account. In an otherwise isolated system, we find that the internal energy U, which is an extensive quantity, can only be increased by doing work on the system. If, however, the system is not thermally isolated and it passes in a particular process from a thermodynamic state indicated by A to another state indicated by B, the amount of work done on the system (indicated by W) depends, of course, on the way the system moves from the state A to the state B. On the other hand, the increase in internal energy is equal to $U_B - U_A$, independent of the way in which the work is done. For such a system which is not thermally isolated, the increase in internal energy $U_B - U_A$ will, therefore, be different from W. The difference Q we call the amount of heat, which is thus a measure of the departure from adiabatic conditions. Thus for any thermodynamic process starting in the state A and finishing in the state B, the change in internal energy is given by

$$U_B - U_A = W + Q \qquad (1\text{-}1)$$

which is *the first law of thermodynamics* for an arbitrary process going from state A to state B (the equivalence of heat and work was demonstrated by the famous paddle-wheel experiment of Joule).

For an infinitesimal process the first law reads

$$dU = \dbar W + \dbar Q \qquad (1\text{-}2)$$

where dU is an infinitesimal increase of the internal energy in this infinitesimal process, and $\dbar W$ and $\dbar Q$ indicate the corresponding infinitesimal amounts of work done on, and heat given to the system respectively.

We now establish the convention that heat is transferred from the system having the higher numerical value of empirical temperature to that having the lower. This leads us to the next important generalization of thermodynamics, that heat always flows in the same direction (we can now say from higher to lower temperatures) unless very particular steps are taken to reverse it. These particular steps always include the addition of external work to the system. This is the basis of *the second law of thermodynamics*, which may be stated as follows:

> It is impossible to devise an engine which, working in a cycle, shall produce no effect other than the transfer of heat from a colder to a hotter body.

This is the formulation of the second law due to Clausius.

There are other formulations of the second law: one by Kelvin for example, which is rather similar to that of Clausius but is oriented more to engineering; or one by Caratheodory, which results from an attempt to place thermodynamics on a more axiomatic base than appears to be the case if the basic laws are formulated in terms of the behaviour of heat engines. Kelvin's formulation of the second law is as follows:

> It is impossible to devise an engine which, working in a cycle, shall produce no effect other than the extraction of heat from a reservoir and the performance of an equal amount of mechanical work.

The requirement that the engine works in a cycle, in both the Clausius and Kelvin formulations, is simply to ensure that the system is in exactly the same energy state before and after performing the process so that the amounts of heat and work have to balance each other. Clausius and Kelvin therefore considered only those processes that can, in principle, go on forever. Caratheodory's formulation of the second law is as follows:

> In the neighbourhood of any equilibrium state of a system there exist states that are inaccessible by an adiabatic process.

It has been shown that this formulation leads to exactly the same consequences as those of Kelvin or Clausius. This is not, however, a discussion that we shall go into. The interested reader is referred to texts on thermodynamics, particularly that of Buchdahl, for further information on the work of Caratheodory.

The real *definition of the temperature* as a physical quantity which is basic for the field of thermodynamics is directly related to the basic laws of thermodynamics mentioned above: starting from the first law of thermodynamics and using Kelvin's formulation of the second law, it is a standard proof that for a reversible heat engine operating over a Carnot cycle between two temperatures θ_1 and θ_2, the ratio of the heat Q_1 taken in at the higher temperature θ_1 to that given out Q_2 at the lower temperature θ_2 is proportional simply to the ratio of the same function of each of the two temperatures.

$$\frac{Q_1}{Q_2} = \frac{\Phi(\theta_1)}{\Phi(\theta_2)}$$

where $\Phi(\theta_1)$ is a function only of θ_1, and $\Phi(\theta_2)$ is the same function only of θ_2, independent of the particular properties of the working fluid. William Thomson (later Lord Kelvin) realized, in 1848, that this relation could be used to define the ratio of any two temperatures. The values of the temperatures would depend upon the functional form of $\Phi(\theta)$, but by taking the simplest possible form of the function he defined a temperature, which he called *thermodynamic temperature, T*, by the relation

$$\frac{Q_1}{Q_2} = \frac{T_1}{T_2} \tag{1-3}$$

The relation (1-3) is valid for a reversible Carnot cycle and is independent of the work done W. Thermodynamic temperature thus has the property that ratios of T are defined in terms of the properties of reversible heat engines and are independent of the working substance. The definition of the quantity thermodynamic temperature then has to be completed by assigning a particular numerical value to an arbitrary fixed point of temperature. This we shall come to later. One of the simplest working substances would be a perfect gas, namely a gas for which both the product PV and the internal energy are independent of pressure at constant temperature. The next step in our argument is to show that a temperature definition satisfying the proportionality of equation (1-3) is indeed proportional to the temperature defined by the ideal-gas laws.

NOTE

Before coming to this, however, we must draw another conclusion from the Carnot cycle, which leads to the introduction and definition of another very important physical quantity in thermodynamics, which is closely related to the temperature: *the entropy* of the system. If we apply the reversible Carnot cycle to the case where the two adiabatics of the cycle are very close together, the quantities of heat become infinitesimal and we can write instead of (1-3):

$$\frac{(\text{d}Q)_1}{T_1} = \frac{(\text{d}Q)_2}{T_2} \tag{1-4}$$

showing that the quantity $đQ/T$ along various isotherms connecting two neighbouring adiabatics is independent of the temperature and is typical for the two neighbouring adiabatics. This brings about an ordering of the various adiabatics which can be made more explicit and quantitative by introducing a quantity "entropy" S which is a quantity which is constant along the adiabatic, just as the temperature is constant along the isotherm. For this reason "adiabatic" in a reversible process can now better be called "isentropic". The quantity $đQ/T$ can then be made equal to the difference dS between the values of entropy of neighbouring isentropes.

For a *reversible* process we thus obtain

$$\frac{đQ}{T} = dS \qquad (1\text{-}5)$$

The infinitesimal increase of the entropy dS is equal to the infinitesimal amount of heat dQ added to the system, divided by T.

With equation (1-5) we have thus defined this very important thermodynamic quantity, the entropy of a system; a quantity conjugate to the temperature. Thus S and T are similar to the conjugate pair of quantities P and V. Introducing into the first law, equation (1-2), $đQ = TdS$ and $đW = PdV$ we obtain the well known relation

$$dU = TdS - PdV \qquad (1\text{-}6)$$

which is the basic equation in thermodynamics, really being a combination of the first and second laws. The introduction of this new quantity "entropy" thus is a direct consequence of the first and second laws of thermodynamics upon which the Carnot process was based. If we do not limit ourselves to reversible processes, the second law states that in general the value of $đQ/T$ is smaller than or equal to the increase dS of the two neighbouring adiabatics. In the limit, for *any* infinitesimal process for which $đQ = 0$, the entropy can only increase. This reinforces the ordering of the successive isentropes: any adiabatic process, i.e. any reversible or irreversible process taking place in a thermally isolated system, can *only* result in a zero or positive increase of the entropy. This formulation of the second law is obviously very close to that of Caratheodory. We now return to the proof that the thermodynamic temperature defined by (1-5) is proportional to the temperature defined by a perfect gas. For a real gas in the limit of low pressures, the product PV and the internal energy are independent of pressure at constant temperature so that we may write, using equation (1-6)

$$\left(\frac{\partial U}{\partial V}\right)_T = T\left(\frac{\partial S}{\partial V}\right)_T - P. \qquad (1\text{-}7)$$

Employing one of Maxwell's thermodynamic relations (see Pippard, 1964) this can be written

$$\left(\frac{\partial U}{\partial V}\right)_T = T\left(\frac{\partial P}{\partial T}\right)_V - P. \tag{1-8}$$

For an ideal gas $(\partial U/\partial V)_T = 0$, from Joule's law, so that

$$\left(\frac{\partial P}{\partial T}\right)_V = \frac{P}{T}$$

or

$$P = Tf(V) \tag{1-9}$$

where $f(V)$ is an unknown function of V. But Boyle's law tells us that, at constant temperature, $P \propto 1/V$, so that $f(V) \propto 1/V$ leading to

$$PV \propto T \tag{1-10}$$

but we already have found that $PV \propto \theta$, our empirical temperature. Thus for an ideal gas $T \propto \theta$. Since for a real gas, in the limit of low pressures, the product PV and the internal energy are independent of pressure at constant temperature, temperatures measured using a real gas thermometer are, in the limit of low pressures, proportional to thermodynamic temperatures. We can make T and θ identical by assigning the same numerical values to T and θ at a single arbitrary fixed point of temperature, as we shall see later on.

1-3   Temperature in statistical mechanics

We have, so far, considered the quantity temperature solely in terms of the behaviour of macroscopic systems, excluding from consideration the individual behaviour of the microscopic particles that make up such systems. Soon after the beginning of classical thermodynamics, however, a parallel development took place in the theory of molecular dynamics, and Maxwell in 1859 and Boltzmann in 1869 were able to give expressions for the velocity or energy distribution in a system of molecules in thermal equilibrium.

According to their theory, the probability $p(\varepsilon)$ in such a system of a molecule having an energy between ε and $\varepsilon + d\varepsilon$ is

$$p(\varepsilon)d\varepsilon = g(\varepsilon)e^{-\varepsilon/kT}d\varepsilon \tag{1-11}$$

where $g(\varepsilon)d\varepsilon$ is proportional to the available phase space (i.e. space having co-ordinates of both position and momentum) for a molecule having energy between ε and $\varepsilon + d\varepsilon$. The quantity kT is thus a characteristic energy, in which k is called the Boltzmann constant, which determines the velocity or energy distribution among the molecules. For a molecule of mass m, this leads to a value for the average of the square of the velocity $\bar{v}^2$, equal to $3\,kT/m$. The expression for $\bar{v}^2$ deduced from classical thermodynamics by Clausius, in terms of the molar gas constant R and molar mass M, is $\bar{v}^2 = 3\,RT/M$. Comparison of the two expressions gives the relation $R = N_A k$ where N_A is the number of molecules per mole.

The work of Maxwell and Boltzmann provided a most important step forward in our understanding of thermal quantities. Thenceforth temperature became a quantity definable either in terms of macroscopic thermodynamic quantities such as heat and work or, with equal validity and identical results, in terms of a quantity which characterized the energy distribution among the particles in a system. The limitation of the kinetic theory of Maxwell and Boltzmann was, however, that it applied only to systems of non-interacting particles, i.e. strictly only to ideal gases but in practice to real gases in the limits of low densities or high temperatures. **NOTE**

The kinetic approach to the explanation of the behaviour of gases was generalized by Boltzmann in 1872 and later by Gibbs who, in 1902, introduced statistical mechanics. The fundamental advance made by Gibbs was that he showed how the average values of properties of a system as a whole could be deduced by examining the likely distribution of these properties, at a given moment, among an arbitrary but very large number of identical systems. He called the large number of identical systems an ensemble. The systems of the ensemble are distributed over the various accessible states, an accessible state being any of the possible configurations that the system may take up. The probability of finding the real system in a particular state then corresponds to the probability of finding the systems of the ensemble in that particular state. Averages of the real system in a certain period of time then correspond to ensemble averages over the Gibbs ensemble. Gibbs showed that a system in a closed volume and in thermal equilibrium with a heat reservoir is represented by the so-called canonical ensemble in which the probability $P(E)dE$ of finding a system having an energy between E and $E + dE$ is given by

$$P(E)dE \propto \Omega(E)e^{-E/\theta}dE \qquad (1\text{-}12)$$

where $\Omega(E)dE$ is the number of accessible states between E and $E + dE$, which is, in purely classical-mechanical terms, proportional to the available phase space for the complete system of N particles between E and $E + dE$. θ is a parameter characterizing the canonical ensemble. The factor $\Omega(E)$ is a very rapidly increasing function of E while $e^{-E/\theta}$ is a very rapidly decreasing function of E, the rate of decrease being determined by the parameter θ. The product of these two factors in equation (1-12) is a function having a very narrow and sharp maximum at an energy E_{max} given by

$$\frac{d\Omega(E)}{dE} - \frac{1}{\theta}\Omega(E) = 0 \quad \text{or} \quad \frac{d \ln \Omega(E)}{dE} = \frac{1}{\theta} \tag{1-13}$$

the maximum energy in the canonical ensemble. The correspondence with the various quantities in thermodynamics follows from the following identifications: The energy E_{max}, at which the distribution $P(E)$ is a maximum, should occur at

$$U = E_{max}. \tag{1-14}$$

If the quantities $\ln \Omega(U)$ and θ are taken to be proportional to the entropy $S(U)$ and the temperature T respectively

$$S(U) = k \ln \Omega(U), \qquad T = \theta/k. \tag{1-15}$$

The relation (1-13) then corresponds to the well known thermodynamic relation

$$\left(\frac{\partial S(U)}{\partial U}\right)_V = \frac{1}{T}. \tag{1-16}$$

In the statistical mechanical interpretation of thermodynamics, the parameter θ, which characterizes the distribution, is thus directly proportional to the thermodynamic temperature T. Applying the mechanism of statistical mechanics to a classical system, the velocity distribution turns out to be given by the Maxwellian distribution, equation (1-11) with the same parameter $\theta = kT$. This then identifies the thermodynamic temperature once again with that used in Maxwell's distribution and with that in the perfect gas law.

We have thus outlined briefly how the principal parameters of state, T and S, in classical thermodynamics are related to the equivalent parameters θ and Ω of a statistical-mechanical system. The importance of the Boltzmann constant k is clear; it provides the link between numerical values of mechanical (classical or quantum) quantities and thermal quantities. At this point we can note another definition of the quantity "temperature" expressed in equation (1-16). Temperature is the parameter of state that is inversely proportional to the rate of change of the logarithm of the number of accessible states as a function of energy for a system in thermal equilibrium. Since the number of accessible states increases as an extremely large power of the energy, the quantity temperature so defined will always be positive.

1-4   Temperature in non-equilibrium conditions, and negative temperatures

All of what we have said so far concerns only systems in thermal equilibrium or approaching thermal equilibrium. For a system far from equilibrium, the concept of temperature is ill-defined. In the real world, of course, there is no such thing as a system in perfect "thermal equilibrium", indeed if there were, we would have no way of observing it. We always find, therefore, in trying to make ever more precise measurements of temperature that in the end, the quantity itself becomes elusive. This happens either because, in trying to make the measurements, we disturb the state of thermal equilibrium or, in striving to preserve the state of equilibrium, we cut off our means of making precise measurements. This will be a theme running throughout the book: how to measure the temperature of a system while at the same time preserving its state of near thermal equilibrium. In Chapter 3 on primary thermometry, in which we deal with the direct measurement of thermodynamic temperature, we find that it becomes increasingly difficult to define and maintain a volume of gas at a known pressure, to a very high accuracy, while at the same time allowing the presence of a pressure-sensing tube sufficiently wide to permit accurate measurements of pressure. In Chapter 5 on resistance thermometry, we shall see that one of the limiting factors in the determination of the temperature in a triple-point-of-gallium cell, for example, is the difficulty of measuring the heating effect of the thermometer resulting from the measuring current. In Chapter 7 on optical thermometry, we shall see how we attempt to measure the energy density of thermal radiation in a completely enclosed cavity (the only conditions under which we can relate energy density to temperature) by observing the radiance of a small hole made in the walls of the cavity. We must make careful calculations on the likely departure from thermal equilibrium of the radiation inside the cavity caused by the presence of the small hole. Without the hole, of course, we would have no way of observing the radiation inside the cavity.

The thermal contact necessary between a thermometer and the body the temperature of which we are trying to measure does not have to be made by physical contact. As we remark above, the passage of thermal radiation from one body to another provides a perfectly adequate means of thermal contact. Indeed, good physical contact does not necessarily imply good thermal contact. At very low temperatures, it is possible to have magnetic spin systems that are integral with, but in very poor thermal contact with, the host lattice. This provides a means of achieving extremely low temperatures. Also at very high temperatures, in plasmas, the energy distribution among the electrons can be quite different from that among the ions. In this way it can be said that the "electron temperature" is different from the "ion temperature".

For systems that are not close to overall thermal equilibrium, it may still be possible to make useful statements about the temperature of small regions of the system. The size of the small region, within which we can say that local thermal equilibrium exists, must be decided by the limits within which we wish to specify the temperature. Suppose that we have a large quantity of a fluid in a container and preliminary measurements with a thermometer show that the temperature apparently fluctuates, and is different in different parts of the fluid, within overall limits of, say, ten degrees. Clearly, if we wish to specify the temperature of the fluid to no better than about ten degrees, we are entitled to say that we have thermal equilibrium. If, however, we are looking for a resolution of only one tenth of a degree, we are unable to give such a figure for the whole bath since the quantity itself is not defined at this level. We may, on the other hand, be able to find small regions of the fluid within which the apparently measured quantity does not change and is uniform to one tenth of a degree. We shall, under these circumstances, be entitled to call the measured quantity temperature. We shall not be so entitled if the volume within which the measured quantity appears uniform is comparable to the size of the thermometer. For in this case, the indications of the thermometer represent some average value of the local temperatures in the region of the thermometer and do not, therefore, represent local equilibrium conditions of the fluid. Nor shall we be entitled to call the measured quantity the temperature of the fluid if the rate of change of apparent local temperature be comparable to the rate of response of the thermometer. In all of these examples, it is possible to estimate the temperature of the fluid using the indications of the thermometer as one piece of information. But it is necessary, in addition, to have previously obtained knowledge about the thermal behaviour of the thermometer and its thermal contact with the surroundings, as well as the thermal behaviour of the fluid.

We stated earlier that temperatures are positive provided that $(d\Omega(E)/dE)$ is positive, i.e. the number of accessible states always increases with energy. This is always the case for a freely moving particle or a harmonic oscillator, and thus fluids and crystal lattices will always have positive temperatures. There are, however, certain very special systems in which there is an upper limit to the spectrum of energy states. If the particles in these states are in thermal equilibrium with one another, while at the same time being thermally isolated from those states having no upper limit to the energy, then they can behave as if they had negative temperatures. Because of the absence of any energy level above the upper limiting level, as the internal energy of the system increases, a state is reached at which all the levels are equally populated. According to statistical mechanics, this can only occur as $T \rightarrow \infty$. If yet more energy is added to the system, the population in the higher levels will exceed that in the lower levels. Statistical mechanics can only deal with this by treating the system as one having a decreasing negative temperature as a function of increasing energy. Finally, when all of the energy is concentrated in the topmost energy level, the distribution is just the inverse of that at absolute zero and so we must assign to the system a temperature of -0.

Since negative temperatures correspond to higher energies than do positive temperatures, if a system having a negative temperature is brought into thermal contact with a system having a positive temperature, heat flows in the direction negative temperature→positive temperature. The exchange of energy between a system at a negative temperature and one that has an unbounded energy spectrum (i.e. can only have a positive temperature) will always result in an equilibrium at a positive temperature.

While systems having negative temperatures are of great interest in certain special applications, they fall outside the scope of this book and the reader is referred elsewhere for further information on this subject (Ramsey, 1956).

1-5   Values of T, k and R

The numerical value of the Boltzmann constant, k, is established by assigning an arbitrary value to the temperature of the triple point of water and comparing the equations of state of a system written in classical and statistical mechanical terms. The most straightforward system is, of course, an ideal gas, for which in classical terms

$$PV_m = RT \text{ per mole} \qquad (1\text{-}17)$$

where R is the molar gas constant and V_m is the molar volume. The equivalent statistical mechanical result is

$$PV_m = N_A kT \qquad (1\text{-}18)$$

where N_A is the Avogadro constant, the number of particles per mole. The Avogadro constant is known from independent measurements of the lattice constant and density of silicon (Deslattes, 1974; Seyfried 1982), and has an estimated uncertainty of only a few parts in 10^6. Thus by choosing a value of 273.16 K for the temperature of the triple point of water (see Chapter 2), we fix the values of k and R and they become quantities to be determined experimentally and linked by the Avogadro constant

$$k = R/N_A. \qquad (1\text{-}19)$$

The experimental determination of k or R requires, in principle, that we measure the parameters of state of a system, in equilibrium at 273.16 K, for which we can write down the equation of state explicitly, and in which the only unknown is k or R. Such a system is a real gas in the limit of low pressures. Until recently, the method that led to the most accurate experimental values for k and R was indeed the method of limiting density of a gas.

For an ideal gas we write the equation of state

$$P_0 V_m = R T_0 \qquad (1\text{-}20)$$

where P_0 is one standard atmosphere, defined as 101 325 Pa, and T_0 is 273.15 K. Note that the reference temperature is taken as 273.15 K and not 273.16 K. This is because early measurements of R were based upon a reference temperature of 0 °C for the freezing point of water and to maintain continuity in numerical values of V_m and other quantities we continue to define a temperature T_0 which is 0.01 K, exactly, below the temperature of the triple point of water.

The molar volume and molar mass of an ideal gas are related by

$$M = L_0 V_m \qquad (1\text{-}21)$$

where L_0 is the density of the gas at P_0 and T_0. For a given mass m of a real gas occupying a volume v, the density $\rho(p)$ at a pressure p is given by

$$\rho(p) = \frac{m}{v}. \qquad (1\text{-}22)$$

We may define a quantity L_p by

$$L_p = \frac{P_0}{p} \rho(p) = \frac{P_0}{p} \frac{m}{v} \qquad (1\text{-}23)$$

so that when $p = P_0$ we may write

$$L_{P_0} = \frac{m}{v}. \qquad (1\text{-}24)$$

Since the departure from ideal behaviour tends to zero as p tends to zero, it is apparent that in the limit of low pressures

$$L_p \rightarrow L_0$$

for all gases. Experimental measurements of $\rho(p)$ at successively lower pressures can, therefore, be used to arrive at a value for L_0 and, for a given value of M, V_m and hence R (Batuecas, 1972).

More recently, other methods have been used for the determination of R, in particular one based upon the speed of sound in a gas (Colclough et al., 1979). The speed of sound, c_0, in an ideal gas at a temperature T_0 is given by

$$c_0^2 = \frac{\gamma R T_0}{M} \qquad (1\text{-}25)$$

where γ is the ratio of the specific heats of the gas. Once again, experimental measurements of the speed of sound in a real gas, as a function of pressure, allow the value in the limit of low pressures, namely c_0, to be determined. Since γ may be calculated and M determined independently, equation (1-25) thus permits us to determine R.

Quite a different method has recently been used to determine a value for k. The total thermal radiation $E(T_0)$ emitted by a blackbody at a temperature T_0 is given (see Chapter 7) by

$$E(T_0) = \frac{2\pi^5 k^4}{15c^2 h^3} \, T_0^4 = \sigma T_0^4 \tag{1-26}$$

where c is the speed of light and h is Planck's constant; σ is known as the Stefan–Boltzmann constant. By measuring $E(T_0)$, Quinn and Martin (1982) have deduced a value for σ and hence k, since c and h are both known to a much higher accuracy than is needed for a determination of k. In carrying out the experiment, it was necessary to measure a geometrical factor introduced into equation (1-26) since it is not experimentally feasible to measure the radiation emitted over a hemisphere.

Another possible method of obtaining k is based upon a measurement of the electrical noise in a resistor. The mean square noise voltage $\overline{V^2}$ in a resistor of Ω ohms is given (see Chapter 3) by

$$\overline{V^2} = 4k T_0 \Omega \Delta f \tag{1-27}$$

where Δf is the bandwidth. In principle, this equation may be used to determine k, but there remain serious difficulties in determining Δf with sufficient accuracy.

It must be clear by now that each of these methods of determining R or k could be transformed into methods of determining T, once values of k or R have been established. This is indeed the case, and this leads us to the concepts of primary and secondary thermometry and how we ensure that, no matter what thermometer we are using, the numerical values obtained for T/K are coherent.

We have now finished our discussion on the meaning of the quantity *temperature* and come to the beginning of the principal subject of the book, namely *thermometry*, the measurement of temperature.

1-6   The origins of thermometry

At the beginning of this chapter, we remarked that it appeared, at first sight, strange that ideas concerning the meaning of temperature took so long to develop. By now it must be evident that it could have hardly been otherwise. The concept of temperature is so intimately linked to thermodynamics and statistical mechanics that before the development of these branches of science it was not possible to have any clear idea of what it meant. This is the reason why it was only in Europe that the beginnings of thermometry were to be found, during the rise of physical science in the seventeenth century. Nothing of this nature appears to have taken place, for example, in China where no independent development of physical science occurred.

At the opening of the seventeenth century, very little was known about heat and temperature, most opinions at that time still being based upon the medical writings of Galen (A.D. 130–200). His clinical thermometry was founded on the ideas of Aristotle, and he assumed that people differed in their proportions of heat, cold, moisture and dryness. It is interesting to note that he proposed a standard of "neutral" temperature which was to be made up of equal quantities of boiling water and ice, either side of which he put four degrees of heat and four degrees of cold. Nothing appears to have survived telling us of the use to which such a standard could have been put. (A temperature of about 10 °C would have been obtained by his procedure.) More than a millenium after Galen, in 1578, we find another medical writer, Hasler of Berne in his "De logistica medica", following Galen in ascribing various degrees of heat and cold to mixtures of drugs. To assist in their prescription, he set up a temperature scale in which there were Galen's four degrees of heat and four degrees of cold with a zero in the middle. Against this he set a scale of latitude, postulating that inhabitants of the equatorial regions have the fourth degree of heat and those of the polar regions the fourth degree of cold. Using these scales the appropriate mixtures of drugs could then be calculated depending upon where the patient lived!

At this time there was no instrument in existence which could be called a thermometer. Admittedly, instruments had been made which had demonstrated the expansion of air on heating as early as the second century B.C. (Philo of Byzantium), but they had not been used or designed to give indications of temperature. The first recorded description of an instrument designed to measure temperature was published by the physiologist Santorio of Padua in his "Commentaries on Galen" in 1612. The credit as the inventor of the thermometer is generally given, however, to Galileo who is thought to have invented the air thermometer in about 1592. (Sherwood Taylor, 1942 and Middleton, 1966 give a comprehensive discussion of early thermometry.) Regardless of who first invented the air thermometer, we find that by the second decade of the seventeenth century the air thermometer was a well-known device. Among letters to Galileo we find one dated May 9, 1613 from his friend Sagredo (quoted by Middleton), in which he claims to have discovered that the north wind is sometimes colder than ice or snow and that in winter small bodies of water are colder than large ones. In addition, Sagredo later mentions measurements he made covering the differences in temperature found in various parts of the body. At this time no mention is made of a serious attempt to produce a scale.

The next major advance, that of using a liquid rather than air as the thermometric medium, was made in 1632 by another physician, Jean Rey, who used an open-ended water-in-glass thermometer. This was not a very permanent device and it was Ferdinand II, Grand Duke of Tuscany, who is credited with the invention of what we would recognize as a real thermometer. This was the sealed alcohol-in-glass thermometer probably first made successfully in 1641. The stems of these thermometers were marked in equal intervals of fractions of the volume of the bulb. By the year 1654 a number of such thermometers, having 50 "degree" markings on the stem, had been sent to various observers at Parma, Milan and Bologna. The fame of these new spirit thermometers spread rapidly as they were evidently far superior to anything previously available. The art of glass blowing was at that time very advanced in Northern Italy, and the skill of the Florentine glass blowers allowed the members of the famous Accademia del Cimento to indulge their fancies in producing thermometers with extraordinarily long coiled stems. These were sufficiently sensitive to respond to a warm breath and, because of the skill of the now forgotten artisan who made them, were astonishingly uniform. In 1657, according to the records of the Accademia, the academicians experimented with mercury-in-glass thermometers, but eventually concluded that the mercury thermometer was less suitable than that containing spirit. This was a pity, because with their skill in glass blowing they would probably have developed the precision thermometer some sixty years before its eventual appearance, at the hands of Fahrenheit, in about 1713.

We thus find that in the middle of the seventeenth century, sensitive thermometers were available but still no serious attempt had been made to produce a worthwhile universal scale. In 1661 however, Sir Robert Southwell, later to become President of the Royal Society, returned from the Grand Tour with a Florentine spirit thermometer. Robert Hooke, at the time Curator of the Royal Society, modified the Italian design, introduced a suitable red dye into the spirit and set about devising a scale. Hooke published his method in his *Micrographia* of 1664. In it he showed how, starting from first principles, comparable thermometers could be made without making the dimensions of the thermometers exactly the same as the Florentines had tried to do. He devised a method based upon equal increments of volume starting from the freezing point of water. As an indication of the difficulty of building up information concerning fixed points of temperature when starting from a position of almost complete ignorance, Hooke apparently tried at one time to use two fixed points at the freezing point of water. He thought that the temperature at which the surface of a water bath began to freeze was different from that at which the whole bath became solid. It is possible that he was misled by the density maximum of water near 4 °C leading to the lower region of an unstirred water bath being warmer than the surface when the latter is starting to freeze. Nevertheless, he produced a scale in which every degree was equivalent to a change of about 1/500 part of the volume of his thermometer liquid (the equivalent of about 2.4 °C). His scale ranged from −7 degrees, at extreme cold, to +13 degrees for the greatest summer heat. This scale was disseminated on various thermometers which were calibrated against the original held by the Royal Society which itself had been calibrated using Hooke's method. This original thermometer, described by Hooke at a meeting of the Royal Society in January 1665, became known as the standard of Gresham College, and was used by the Society until 1709. The influence of the scale thus made available, by means of calibrations against the Gresham College thermometer, was very great during the latter part of the seventeenth century. The first intelligible meteorological records began to be kept using thermometers carrying this Royal Society scale. These are of great interest and are discussed at some length together with the provenance of the thermometers used by the various observers, by Patterson (1951). In the Journal of Robert Hooke, kept at Gresham College from March 1672 until April 1673, and in that of John Locke from December 1669 to January 1675, we find the temperatures measured at various times during these periods. On the eleven days when they both give the temperature, the differences between them never exceeded about 4°C, and on average agree to within a little over $1\frac{1}{2}$°C. This occurred before the birth of Fahrenheit, Reaumur and Celsius, and only some ten years after the first sealed spirit thermometer had been brought to England!

In a space of only about fifty years, thermometry had advanced from a state of almost complete ignorance to one in which meaningful meteorological records had begun to be kept. The concept of a scale had appeared, but as yet there was little appreciation of the dependence of the scale upon the properties of the thermometric fluid. For this we have to wait until the time of Réaumur, who realized in 1739 that the scales of spirit thermometers and mercury thermometers differed because these fluids did not expand in the same way with temperature. It is not clear whether or not he was led to think that there might be some ideal fluid which could be used to make a thermometer which would give a temperature which was in some sense more "absolute" than spirit or mercury.

At the beginning of the eighteenth century we come to the work of Fahrenheit and Amontons, both of whom made important advances in thermometry but in quite different directions. Each one laid the foundations of what are two, almost independent, sides of thermometry which have remained separate, in many respects, right up to the present day and which we shall shortly distinguish as primary and secondary thermometry. Fahrenheit seems to have been the first person to have learned how to make reliable mercury-in-glass thermometers. In addition he developed, between 1708 and 1724 after discussions with the Danish astronomer Römer, the method of making a scale by taking two fixed points and dividing the interval between them into a convenient number of degrees. He eventually produced a scale by taking the temperature of the human body as one fixed point, which he called 96 degrees, and that of freezing water as the other, at 32 degrees. Using this scale, published in 1724 (see Middleton for a fuller discussion), he made measurements of the boiling points of liquids at temperatures up to 600 degrees.

The importance of Fahrenheit's work stems from his ability to produce both stable thermometers and a reproducible scale. He was not the first to suggest a scale in terms of two fixed points, but he was the first to exploit the method using good thermometers.

At about the same time, the French scientist Amontons developed the constant volume gas thermometer. He used air as the thermometric medium and found that the ratio of the greatest summer heat to greatest winter cold was, in Paris, approximately in the ratio of six to five. He then went further and concluded that the lowest temperature which could exist would correspond to a zero gas pressure. This must have been the first step on the way to an understanding of the concept of temperature. According to Amontons we could define the temperature as being simply proportional to the pressure of a gas, and thus we would need only one fixed point to define a scale. Despite the earlier work of Boyle and Marriotte, this suggestion was not taken up, probably for the very good reason that the gas thermometer was too cumbersome an instrument. It was not appreciated, at the time, that a scale set up in this way was of much greater physical significance than that of Fahrenheit.

The history of thermometry since the early eighteenth century can be followed along the two paths initiated by Fahrenheit and Amontons. On the one hand we have the development of more refined practical scales based upon arbitrary fixed points, such as those of Fahrenheit, Celsius and Réaumur, together with the development of better practical thermometers, while on the other hand we have the parallel developments of gas thermometry and thermodynamics. The former path led, via mercury thermometers, to the introduction of the platinum resistance thermometer and the work of Callendar and the platinum/rhodium thermocouple of Chatelier toward the end of the nineteenth century. As we shall see in Chapter 2, the culmination of the work on practical thermometry was the adoption of the International Temperature Scale of 1927 (ITS-27). Along the gas thermometry path we find the work of Charles, Dalton, Gay-Lussac and Regnault on the properties of gases, leading to the important conclusion that all gases have very nearly the same thermal expansion coefficient. This provided the clue to the eventual understanding of how a gas approximates to the ideal working fluid for a thermometer and how a scale could be made which would be independent of the properties of the fluid. The conclusion that the expansion coefficients of all gases are very nearly the same could have led straightaway to a temperature scale based upon a single fixed point. This had already been pointed out by Amontons in 1702, but was not taken up. Instead, gas thermometry came to be based upon two fixed points using an equation of the form

$$p(t) = p(0)[1 + \alpha t]. \tag{1-28}$$

One fixed point was required to establish $p(0)$ and the other to establish α. As we shall see in Chapter 2, this situation was not reversed until the unit of thermodynamic temperature "the kelvin" was re-defined in 1960 when, finally, the unit of temperature was defined by assigning a numerical value to just one fixed point.

1-7 Primary and secondary thermometry

In distinguishing between primary and secondary thermometry we first of all define what we mean by primary thermometry. Primary thermometry we define as thermometry carried out using a thermometer for which the equation of state can be written down explicitly without having to introduce unknown, temperature-dependent, constants. Earlier in this chapter we showed how the Boltzmann constant provides the essential link between numerical values of

NOTE

mechanical and thermal quantities and how its numerical value is fixed by assigning a temperature of 273.16 K to the triple point of water. It also followed that the numerical value of the gas constant was fixed in the same way. We thus have three mutually consistent constants, T (t.p. of water) or T_0 (ice point), k and R. In principle, we may now write down the equation of state of any system and use it as a thermometer, in the confident expectation that the temperatures thereby obtained will be in thermodynamic and numerical agreement with those obtained using another system and a different equation of state. Examples of such systems, suitable for thermometry, are those we have already mentioned when we were considering ways of determining k and R, namely the gas thermometer, acoustic thermometer, noise thermometer and total radiation thermometer. The presence of temperature-independent constants, such as the geometrical factor in the total radiation thermometer, can be accommodated by carrying out a single measurement at T_0. Subsequent measurements of $E(T)$ lead to a value for T by ratio to $E(T_0)$. Suppose that

$$E'(T) = \varepsilon A g \sigma T^4 \tag{1-29}$$

where $E(T)'$ is the measured total radiation from a near-blackbody of emissivity ε, effective aperture A and via an optical system of throughput g. It may well be that although neither ε, A nor g can be measured to high accuracy, we may be confident that they are not functions of T. In this case, by carrying out a measurement of $E(T_0)'$, we may write

$$\frac{E'(T)}{E'(T_0)} = \frac{T^4}{T_0^4} \qquad\qquad (1\text{-}30)$$

from which T may be found. This procedure is also used in noise thermometry, where it is very difficult to determine the bandwidth, Δf, of equation (1-27). In acoustic thermometry, on the other hand, a measurement at T_0 is not required, since there are no temperature-independent quantities that cannot be measured. The presence of R, in equation (1-25), ensures that the numerical values of measured temperature are properly assigned. These are all examples of primary thermometers.

Secondary thermometers are all those that are not primary. An obvious example of a secondary thermometer is a platinum resistance thermometer. We cannot use a platinum resistance thermometer as a primary thermometer because we do not know enough about the behaviour of platinum (see Chapter 5) to be able to write down an explicit equation of state. Any expression that we are at present able to write down, describing the electrical resistance of a platinum wire as a function of temperature, contains unknown, temperature-dependent terms that we do not know how to calculate from first principles. In order to calibrate such a thermometer, therefore, we must compare it with a primary thermometer, either directly or indirectly, at as many temperatures as are necessary to determine the form of the unknown temperature-dependent terms.

The platinum resistance thermometer is a clear example of a secondary thermometer. It is not always so obvious, however, whether or not a particular thermometer should properly be called primary or secondary. The difficulty arises if we take, for example, a gas thermometer and use it in such a way that we need to know the numerical values of the virial coefficients, as indeed is the case in one of the methods of gas thermometry described in Chapter 3. Strictly speaking, such gas thermometry should not be considered primary thermometry if the values of the virial coefficients (which are temperature-dependent quantities) are based upon experimental measurements which themselves relied upon thermometry. However, if the size of the correction stemming from the virial coefficients is small, and since we know a good deal about the behaviour of virial coefficients from theory, the dependence upon the other thermometry may turn out to be very slight. Such thermometry could probably best be called semi-primary thermometry.

An example of a secondary thermometer that, nevertheless, provides very useful information concerning the results of primary thermometry, is magnetic thermometry. Magnetic thermometry is so closely linked to primary thermometry that we deal with it in Chapter 3, which is otherwise mostly concerned with primary thermometry. Magnetic thermometry is not primary thermometry because there are up to four constants in the equation of state that must be evaluated *in situ*. But having evaluated these constants by comparison with another thermometer over a range of temperature, magnetic thermometry is sufficiently well understood for it to give very valuable information on the smoothness of the results of primary thermometry. The sense in which we use "smoothness" in this context is explained in Chapter 2.

References

Batuecas, T. (1972). Molecular volume of an ideal gas. *In* "Atomic Masses and Fundamental Constants 4", (J.-H. Sanders and A. H. Wapstra, eds) Plenum Press, New York.

de Boer, J. (1965). Temperature as a basic physical quantity. *Metrologia*, **1**, 158–169.

Buchdahl, H. A. (1975). "Twenty Lectures on Thermodynamics", Pergamon Press, Oxford.

Colclough, A. R., Quinn, T. J. and Chandler, T. R. D. (1979). An acoustic redetermination of the gas constant. *Proc. Roy. Soc.*, **A368**, 125–139.

Deslattes, R. D., Hennins, A., Bowman, H. A., Schoonover, P. H., Carroll, C. L., Barnes, I. L., Machlan, L. A., Moore, L. J. and Shields, W. R. (1974). Determination of the Avogadro constant, *Phys. Rev. Letters*, **33**, 436.

Middleton, W. E. K. (1966). "A History of the Thermometer and its Use in Meteorology", Johns Hopkins Press, Baltimore.

Patterson (1951). Thermometers of the Royal Society 1663–1768. *Amer. J. Phys.* **19**, 523–535.

Quinn, T. J. and Martin, J. E. (1983). A New Determination of the Stefan–Boltzmann Constant. Proceedings of PMFC II, NBS Special Publication 617.

Ramsey, N. F. (1956). Thermodynamics and Statistical Mechanics at Negative Absolute Temperatures. *Phys. Rev.* **103**, 20–28.

Seyfried, P. (1983). Work Related to the Determination of the Avogadro Constant at PTB, Proceedings of PMFC II, NBS Special Publication 617.

Sherwood-Taylor, F. (1942). "The Origin of the Thermometer", *Ann. Sci.* **5**, 129–156.

Bibliography of texts on classical thermodynamics and statistical mechanics

Guggenheim, E. A. (1977). "Thermodynamics", (sixth edition), North-Holland, Amsterdam.

Kittel, C. (1969). "Thermal Physics", (an elementary statistical mechanics text), John Wiley, New York.

Kubo, R. (1968). "Thermodynamics, an Advanced Course with Problems and Solutions", North-Holland, Amsterdam.

Kubo, R. (1965) "Statistical Mechanics, an Advanced Course with Problems and Solutions", North-Holland, Amsterdam.

Pippard, A. B. (1964). "The Elements of Classical Thermodynamics", Cambridge University Press.

Reif, F. (1965). "Fundamentals of Statistical and Thermal Physics", McGraw-Hill, New York.

2

Thermodynamic and Practical Temperature Scales

2-1   Introduction

In Chapter 1 we traced the development of the concept of temperature from its primitive beginnings, more than two thousand years ago, to the sophisticated ideas of present day thermodynamics and statistical mechanics. In this chapter we shall be looking at the way in which these theoretical ideas were translated into practical temperature standards and temperature scales. Before coming to this, however, we shall first look briefly at the developments which led to the establishment of a framework within which international agreements on such matters could be made.

Starting about the middle of the nineteenth century, the rapid growth of world trade, combined with the increasing complexity of technology, led to a general acknowledgement of the need for some sort of international agreement on weights and measures and units of measurement. Both in Great Britain and in continental European countries, moves were made to try and establish such uniformity of measurement. The British Association for the Advancement of Science (BAAS) took early initiatives in the field of electrical measurement, while the International Geodesic Association at its 2nd General Conference held in Berlin in 1867 made definite proposals aimed at unifying length measurement throughout the whole of Europe. One of the proposals was for the creation of a European Bureau of Weights and Measures. The metric system had by then become firmly established in many European countries, and by the time the need for a universal system of weights and measures had become pressing, it was clear that the metric system was the only serious candidate. Already at the Great Universal Exhibitions, held in London in 1851 and 1862 and in Paris in 1855 and 1867, various resolutions had been put forward urging some sort of international co-operation on weights and measures. Finally in 1869, following the proposals of the International Geodesic Association, supported by the Academies of Science of St.

Petersburg and Paris and the French Bureau des Longitudes, the French Government proposed the creation of a Commission charged with preparing the ground for making the metric system international. Twenty-four countries accepted the original invitation to send delegates to Paris, but eventually, after an interruption caused by the Franco–Prussian war of 1870, a total of thirty countries participated in the discussions. The result was the Convention du Mètre, signed by 17 nations in 1875 (and now adhered to by 46 nations) under which members undertook to establish, in Paris at their common expense, a Bureau International des Poids et Mesures (BIPM) where would be deposited new international standards of the metre and the kilogram (BIPM, 1975; Moreau, 1975). Among the original tasks of the Bureau were to be the comparisons of national prototype standards of length and mass with the international prototypes. The Convention du Mètre, as well as founding the BIPM, set up an organizational structure within which international agreements on units could be formalized and the administrative and financial affairs of the BIPM be maintained. The Comité International des Poids et Mesures (CIPM) was established, made up of fourteen scientists (increased to eighteen in 1921) each from a different member nation, to administer the affairs of the BIPM and to make Recommendations to periodic Conférences Générales des Poids et Mesures (CGPM), to which all member nations are entitled to send delegates. Under the Convention du Mètre, it is the CGPM alone that has authority to adopt proposals concerned with new units or re-definitions of old units. The CGPM now meets once every four years in Paris, the most recent having been the 16th in 1979.

The origins of international agreement on thermometry can now be put in perspective. It was foreseen at the beginning of the activities of the BIPM that thermometry would be one of its subsidiary, but essential, tasks because of the need to measure the temperature and expansion coefficient of all the newly-made platinum–iridium metre bars. It had already been decided that each national prototype metre should be furnished with two mercury-in-glass thermometers, calibrated at the BIPM. In order to fulfil this requirement, a series of special thermometers was made, to the order of BIPM, by an instrument maker in Paris called Tonnelot. The thermometers were made of "verre dur", a particularly good glass from the point of view of stability. Their performance turned out to be beyond expectation and it soon became apparent that, with these thermometers, a reproducibility of measurement of a few thousandths of a degree was possible. Three types of thermometer were made: the first type, (a), carried the whole scale between 0 °C and 100 °C and was divided at one tenth of degree intervals, the length of each degree being 5 mm. The second type, (b), carried a scale up to 50 °C, then a bulb took up the expanding mercury until 95 °C, when a further length of scale occurred up to 100 °C. The length of each degree was about 7 mm. The third type, (c), carried a scale up as far as 39 °C followed by a bulb; a short length of scale near 66 °C; another bulb and finally a section of scale from 97 °C to 100 °C. The length of the degree in this case was about 8 mm. With thermometers of this sort available, it became urgent to establish a uniform scale of temperature against which they could be calibrated. This was taken up by Chappuis, at the BIPM, who set out to compare the various scales to be obtained by means of constant volume and constant pressure gas thermometry, using hydrogen, nitrogen and carbon dioxide as the working gases.

2-2   The normal hydrogen scale

The gas thermometry of Chappuis can be considered as the origin of modern thermometry. Although lacking most of what would be considered normal in a modern laboratory, he had what was for him the essential prerequisite: a purpose-built laboratory having excellent temperature stability by virtue of its construction. The overall aim of the work of Chappuis was to relate the readings of the very best mercury-in-glass thermometers to absolute (i.e. thermodynamic) temperatures. The first part consisted of a detailed study of the constant-volume gas thermometer using, in turn, hydrogen, nitrogen and carbon dioxide as the working fluids. The results were recorded in terms of the readings of a set of eight Tonnelot mercury-in-glass thermometers, of which four were of type "a" and four of a modified type "b", having a scale extending down to -39 °C. Figure 2-1 shows the results obtained by Chappuis (Chappuis, 1888), for the three gases, between 1885 and 1887. The

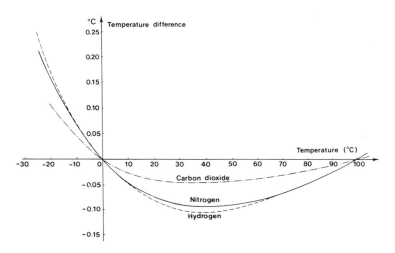

Fig. 2-1. The results of Chappuis, showing the differences between the temperature scales obtained using a gas thermometer employing in turn CO_2, N_2 or H_2 as working substance, plotted against the Tonnelot mercury-in-glass scale. All the thermometers are based upon a fundamental interval of 100 °C between the ice and steam points.

combination of the excellent performance of the Tonnelot thermometers and the extreme care with which the gas thermometry had been carried out, resulted in an estimated accuracy better than one hundredth of a degree over most of the range: a truly remarkable achievement.

In 1887 the CIPM adopted (a decision ratified by the 1st CGPM in 1889) the constant-volume hydrogen scale (called the normal hydrogen scale), based upon fixed points at the ice point (0 °C) and the steam point (100 °C), as the practical scale for international metrology. In describing the scale, it was necessary to specify the initial filling pressure (1 m of mercury at 0 °C), since no corrections had been made to take account of departures from ideal behaviour of hydrogen. It was for this reason that the scale was called a "practical" scale: it was not a thermodynamic scale since there remained the small dependence upon the particular properties of the working fluid. In Chapter 3 we discuss in more detail how the departures from ideal behaviour affect the results of gas thermometry. Here we can just note that in using a constant volume gas thermometer in the way that Chappuis did, i.e. calibrating it at two fixed points and interpolating between them, errors due to departure from ideal behaviour arise only in respect of changes in the non-ideal behaviour between the fixed points. Since for hydrogen such changes between 0 °C and 100 °C are small (the second virial coefficient increases only from about 14 cm^3 mol^{-1} to 15 cm^3 mol^{-1}), the departures from an ideal gas scale of the normal hydrogen scale were indeed very small, and amounted to no more than a few millidegrees near 50 °C (see Hall, 1929).

Chappuis continued his work at BIPM and went on to investigate the constant pressure gas thermometer using the same three gases, concluding that the constant volume thermometer provided a more convenient practical standard than did the constant pressure thermometer. The work was pursued, using a constant volume thermometer, extending the temperature range to higher temperatures. This was done at the instigation of Griffiths, of Kew Observatory, who had been collaborating with Callendar on the development of platinum resistance thermometers. Having successfully made thermometers which were stable up to at least 600 °C, they had decided to use the boiling point of sulphur, near 440 °C, as a third fixed point for calibration, along with the ice and steam points. Callendar and Griffiths had made measurements of the sulphur point using a constant pressure air thermometer (Callendar and Griffiths, 1891) and had deduced a temperature of 444.53 °C. The decision of the CIPM to adopt the normal hydrogen scale had led Callendar and Griffiths to propose to the BIPM that a comparison be made between their platinum resistance thermometers and the constant volume gas thermometer of Chappuis. This comparison was carried out by Chappuis in collaboration with Harker of the Kew Observatory. It involved the establishment of a constant-volume nitrogen scale up to the boiling point of sulphur. Their measurement of the sulphur point led to a value of 444.70 °C, in very close agreement with the earlier result of Callendar and Griffiths.

2-3   The origins of the International Temperature Scale
       of 1927

During the last two decades of the nineteenth century a great deal of gas
thermometry was carried out, much of it at temperatures above 600 °C. Many
freezing and boiling points were measured, mainly using constant-pressure
nitrogen gas thermometers. A detailed review of all this was given by
Callendar (Callendar, 1899), at the 1899 meeting of the BAAS, when he made a
proposal for a practical temperature scale. He proposed that a platinum
resistance thermometer be adopted as the defining instrument of the scale, and
that it be calibrated at the freezing point of water and at the boiling points of
steam and sulphur. He also proposed that a particular batch of platinum wire
be selected from which the thermometers defining the scale be manufactured.
It was his intention that such a scale be called the British Association Scale of
Temperature and that it be related to the ideal temperature scale through
chosen gas thermometer measurements of the sulphur point. He based this
proposal on the verification of the quadratic difference formula, between so-
called platinum temperatures and gas thermometer temperatures, previously
obtained by Chappuis and Harker at the BIPM (Harker, 1907; Chree, 1900).
Callendar also presented a list of secondary fixed points based upon his
evaluation of gas thermometry measurements. His values are shown in
Table 2-1 together with the presently accepted best values on IPTS-68.

Some of the differences between Callendar's proposed values and those of
IPTS-68 are probably due more to differences in purity of the substance than
to errors in gas thermometry; aluminium is clearly a case in point.

It is not clear why the British Association did not take up Callendar's
proposals, since it was not until more than ten years had passed before another
suggestion appeared for an internationally agreed scale. In 1911, the

Table 2-1.

Substance	$t/°C$ (Callendar 1899)	$t_{68}/°C$
tin f.p.	231.9	231.97
bismuth f.p.	269.2	271.44
cadmium f.p.	320.7	321.1
lead, f.p.	327.7	327.5
mercury b.p.	356.7	356.7
zinc f.p.	419.0	419.58
sulphur b.p.	444.5	444.7
antimony f.p.	629.5	630.76
aluminium f.p.	645.5	660.5

Physikalisch-Technische Reichsanstalt (PTR) Berlin (becoming the Physikalisch-Technische Bundesanstalt (PTB) in 1949) addressed a circular letter to the BIPM, the National Physical Laboratory (NPL), Teddington and the Bureau of Standards (BS) in Washington (becoming the National Bureau of Standards (NBS) in 1934), suggesting that the thermodynamic scale be adopted as the International Temperature Scale, and that a practical realization of it be the 1899 proposal of Callendar. Both the NPL and the BS agreed with this proposal and went further, specifying the constants of the platinum, and proposing that above the upper limit (set by increasingly severe problems of contamination) of the platinum resistance thermometer (1100 °C), the scale be defined in terms of the optical pyrometer. At the 5th CGPM, held in Paris in 1913, every encouragement was given to this initiative and a Resolution was adopted asking the Directors of the three laboratories to meet, under the auspices of the BIPM, with the aim of coming to a firm agreement on such a scale. Although the planned meeting did not take place, owing to the outbreak of the First World War, the intention to establish such a scale was maintained. The next time discussions took place on this subject (Hall, 1967) was in 1923, by which time the three national laboratories had put into operation a platinum resistance thermometer scale covering the range from -38 °C, the freezing point of mercury, to 444.5 °C, the boiling point of sulphur, using a quadratic interpolation formula. The minimum resistance ratio R_{100}/R_0 chosen by NPL was 1.385 while that chosen by the BS was 1.388; both laboratories had agreed upon a δ value of 1.52 (see Chapter 5 for the significance of δ). During the course of a visit to NPL and PTR by a representative of the BS, the basis of a scale was agreed upon. It was to consist of a platinum resistance thermometer, to cover the range -38.81 up to 650 °C, calibrated at 0 °C, 100 °C and the boiling point of sulphur at 444.5 °C, and with a minimum value of R_{100}/R_0 of 1.390 and a maximum value of δ of 1.50. Between 650 °C and 1100 °C, the scale was to be defined by a Pt-10 % Rh/Pt thermocouple calibrated at the freezing points of zinc, antimony, silver and gold and using a cubic interpolation formula. Above the gold point, 1063 °C, an optical pyrometer was proposed using the Wien equation for extrapolation with a specified value for the radiation constant, c_2.

The idea of including a thermocouple to define the scale between 650 °C and 1100 °C originated at the BS, where it was felt that the upper limit of the platinum resistance thermometer should not exceed 650°C. This was at first resisted by NPL, where the original proposal of Callendar was preferred in which a platinum resistance thermometer would define the scale up to the gold point. The NPL position may well have been influenced by the problems encountered in the early 1920s (see Section 3.3 in Chapter 6), resulting from the difficulties in manufacturing Pt/Rh alloys of sufficient purity. It is a matter of considerable regret to present day thermometrists that the BS proposal was finally adopted by the three laboratories. As we shall see later in this chapter, considerable effort has been, and continues to be, expended at the national laboratories, notably at NBS, in trying to remove the thermocouple as a defining instrument of the IPTS!

The informal agreement obtained in 1923 between NPL, PTR and the BS, was followed by wider discussions, in which BIPM and the University of Leiden participated, in preparation for a formal proposal to be put to the CIPM in 1927. The draft finally agreed upon in 1925 differed slightly from that mentioned above, in that the range of the platinum resistance thermometer had been extended down to -193 °C, and the cubic equation of the thermocouple had been replaced by a quadratic equation with calibration points at the freezing points of antimony (630 °C), silver (960 °C) and gold. During these discussions it became apparent that the BIPM and the University of Leiden had certain reservations about adopting an International Temperature Scale for which the relationship between it and thermodynamic temperatures was not well known. In the event, the scale adopted by the 7th CGPM in 1927 differed very little from the draft of 1925, but it was adopted provisionally, pending further discussions. It was called the International Temperature Scale of 1927 (ITS-27). It was planned to hold an International Thermometry Conference in 1928 at which the question of the status of the International Temperature Scale would be examined in more detail. This Conference did not take place, and in 1937, the CIPM established a Consultative Committee on Thermometry and Calorimetry to advise it on matters concerned with these subjects. Since then, it has been the Consultative Committee on Thermometry (CCT) that has largely taken the initiative in matters concerned with the evolution of the International Temperature Scale. The interest of the Committee in matters concerned with Calorimetry faded away quite early in its existence.

2-4   The principles of a practical temperature scale

Thus it was that in 1927, international agreement was finally reached on a practical temperature scale. It is worth making very clear the reasons why such a scale was proposed and why, for the same reasons, a practical scale continues to exist. Despite the excellent work of Chappuis and others on gas thermometry, such a method could never become widely used for routine, or even for very special, applications because it is too difficult and too time consuming to carry out. Furthermore, the reproducibility of even the best gas thermometers is inferior to that of the platinum resistance thermometer over most of its range. All of this was well understood by Callendar (1899) who wrote "It is impossible for those who have never worked with a gas thermometer to realize the extent of its shortcomings".

Except at very low temperatures, say below 1 K, primary thermometers remain more difficult to use and less reproducible than the best secondary thermometers. For most purposes, ease of use and reproducibility are more important than absolute accuracy. We must add a caveat to this, since there are many measurements of physical quantities which require accurate measurements of temperature differences. Among them are specific heat, thermal conductivity and many other thermodynamic quantities. If the practical temperature scale being used departs from thermodynamic temperatures by a smooth and slowly varying amount as a function of temperature, then no major problems arise. If, on the other hand, the practical scale contains small, but significant, abrupt changes in its departure from thermodynamic temperatures, then measurements of such physical quantities will show apparent sudden variations as a function of temperature which, in reality, are quite spurious since they result only from the temperature scale. To avoid such problems, we say that a practical scale must be smooth with respect to thermodynamic temperatures. This means that the first and second derivatives of the differences between the practical and thermodynamic scales must be continuous as a function of temperature. While it is easy to maintain a smooth practical scale for a particular secondary thermometer, such as a platinum resistance thermometer, it is much more difficult to arrange for a smooth junction between the ranges of two different secondary thermometers. This is because the two parts of the scale would be based upon different physical principles misrepresenting thermodynamic temperatures in slightly different ways. The junction between the platinum resistance thermometer and the platinum/rhodium thermocouple, in the ITS-27 and in both of its revisions, IPTS-48 and IPTS-68, provides a good example of the difficulties encountered. In IPTS-68 there remains a discontinuity in the first derivative of "$t - t_{68}$" amounting to about 0.2 %. Such discontinuities can only be avoided by having accurate measurements of thermodynamic temperature available, either side of the junction, at the time the interpolating equations of the two secondary thermometers are established. These data were not available in 1927 and are only just becoming available today, as we shall see later on.

Although the details of the ITS-27 have been modified in successive versions, the principles remain the same. The scale consists of a set of defining fixed points together with specified instruments with specified equations for interpolating between them. A set of fixed points alone is not sufficient to define a scale, nor do additional fixed points, over and above the set of defining fixed points, add anything to the scale. The platinum resistance thermometer part of ITS-27 above 0 °C is *completely defined* using the ice, steam and sulphur points together with the specified quadratic interpolation equation. The presence of additional fixed points within a particular range may have other uses, but has nothing to do with the realization of the temperature scale itself. These remarks apply equally, of course, to IPTS-68.

An important property of a practical temperature scale is its "uniqueness". This is related to the variations in the properties of individual thermometers used to define the practical scale. In the case of the platinum resistance thermometer, we assume that all samples of perfectly pure, annealed, platinum behave in exactly the same way. Non-uniqueness arises because of the small variations in purity, or state of anneal, invariably found between different samples of platinum. It shows itself in the following way: suppose that a set of three platinum thermometers calibrated at the ice, steam and sulphur points are then placed together in a uniform temperature bath at, say, 250 °C. They will each give a slightly different value of temperature, calculated according to the proper quadratic interpolation equation. Each one is right and each one has quite correctly given the temperature on ITS-27. The differences between them are a measure of the "non-uniqueness" of the definition of ITS-27. Non-uniqueness is thus quite a different quantity from "irreproducibility", which is the lack of reproducibility of successive realizations of the scale by a single thermometer, resulting from changes taking place in the thermometer itself.

In drawing up the definition of a practical scale, it is sometimes suggested that it be defined in terms of the behaviour of a group of thermometers all made from the same batch of material, to reduce the non-uniqueness of the resulting scale. This was part of Callendar's 1899 proposal to the BAAS. There is much to be said for adopting a scale based upon such a principle, often called, in the case of resistance thermometers, "a wire scale". In fact in national laboratories, it is not unusual to find that the IPTS is maintained for many years on a small number of thermometers for which the differences in behaviour are known with respect to similar thermometers held by other laboratories. The argument against a wire scale is, of course, that of non-universality. If, for any reason, all the particular thermometers carrying the scale at a particular laboratory are lost or damaged, along with all those similar thermometers held by other laboratories, there is no way of retrieving the scale, whereas the continued existence of a scale defined in terms of fixed points and specified interpolation functions is not dependent upon any particular thermometer or group of thermometers. The price paid for this security is the non-uniqueness of such a scale together with the excessive complexity of the definition required in order to reduce the non-uniqueness to an acceptable level.

2-5   The evolution of ITS-27 and its modifications in 1948

The defining fixed points of the ITS-27 are shown in Table 2-2 along with the values of the defining fixed points of its successors IPTS-48 and IPTS-68. For the purposes of interpolation, ITS-27 was divided into three ranges, from −190 °C to 660 °C, from 660 °C to 1063 °C and above 1063 °C. The lower range was defined by a platinum resistance thermometer using the Callendar–van Dusen cubic equation below 0 °C and the Callendar quadratic equation above 0 °C (see Chapter 5 for a detailed discussion on interpolation equations for platinum thermometers). Between 660 °C and 1063 °C, the scale was

Table 2.2. The fixed points of ITS-27, IPTS-48 and IPTS-68 (1975 edition).

Fixed point	ITS-27		IPTS-48		IPTS-68	
	T/K††	$t/°C$	T/K††	$t/°C$	T_{68}/K	$t_{68}/°C$
e-H_2 t.p.					13.81	−259.34
e-H_2 b.p. (25/76 atmos.)					17.042	−256.108
e-H_2 n.b.p.					20.28	−252.87
Ne n.b.p.					27.102	−246.048
O_2 t.p.					54.361	−218.789
Ar t.p.*					83.798	−189.352
O_2 n.b.p.†	90.18	−182.97	90.18	−182.97	90.188	−182.962
H_2O f.p.	273.15	0	273.15	0		
H_2O t.p.					273.16	0.01
H_2O n.b.p.	373.15	100	373.15	100	373.15	100
Sn f.p.*					505.1181	231.9681
Zn f.p.					692.73	419.58
S n.b.p.	717.75	444.6	717.75	444.6		
Ag f.p.	1233.65	960.5	1233.95	960.8	1235.08	961.93
Au f.p.	1336.15	1063	1336.15	1063	1337.58	1064.43

* These are alternatives given in the Scale.

† The O_2 b.p. is more closely defined in IPTS-68 as condensation point.

†† These values did not appear in ITS-27 or IPTS-48, and are here based upon $t = T − 273.15$ K.

Rh/Pt thermocouple calibrated at the freezing points of antimony, silver and gold and using a quadratic interpolation formula. Above 1063 °C, the scale was defined by Wien's equation, taking a value for c_2 of 1.432 cm K.

The first revision (Hall, 1954) of the ITS took place in 1948. In this revision, the only change below 0 °C was the disappearance of the extrapolation below the oxygen point, to −190 °C, which had been found to be unreliable. The IPTS-48 extended down only to −182.97 °C. The junction between the resistance thermometer and the thermocouple was changed from 660 °C to the freezing point of antimony, 630.5 °C. This was done to rectify the bizarre situation in which the freezing point of aluminium had become inaccessible on the ITS-27! This had arisen due to the lower calibration point of the thermocouple, the freezing point of antimony at 630.5 °C being below the lower limit of the range defined by the thermocouple. It so happened that a temperature measured with a thermocouple and found to be just below 660 °C and thus not in the thermocouple range, when measured with a platinum resistance thermometer, calibrated at 0 °C, 100 °C and 444.5 °C, was found to be just above 660 °C, and therefore not in the platinum resistance thermometer range either! Improvements in the manufacturing of aluminium had resulted in a small change in the melting point which had shifted it into this small inaccessible temperature range.

Another change that took place in the scale in 1948 was a small increase in the temperature assigned to the silver point, from 960.5 °C to 960.8 °C. This was done to reduce the discontinuity, previously found in ITS-27, at the junction of the resistance thermometer and the thermocouple. In the optical pyrometer range, the value assigned to c_2 was increased to 1.438 cm K following improved estimates of the values of the atomic constants. In addition, Wien's equation was replaced by Planck's equation. The numerical changes between temperatures measured on ITS-27 and IPTS-48 are shown in Fig. 2-2. In 1948 it was also decided to drop the name "degree Centigrade" for the unit and replace it by degree Celsius. This was done partly to remove the possible confusion, in French, between Centigrade, the unit of temperature, and the same word which can mean one hundredth of a right angle. The other reason for changing was to arrive at the satisfactory position in which all the common temperature scales and units would be named after someone closely concerned with their origin; Kelvin, Celsius, Fahrenheit, Réaumur and Rankine.

2-6   The definition of the unit of thermodynamic temperature: the kelvin

The 1948 revision of the International Temperature Scale was followed by a period of some twenty years during which considerable activity in thermometry took place. The growth of science in general was reflected in the increased

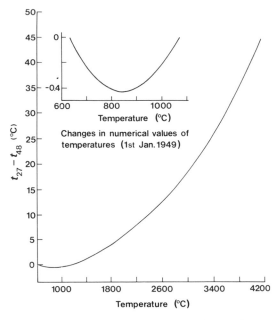

Fig. 2-2. Differences between the International Temperature Scales of 1927 and 1948.

effort devoted to metrology. In thermometry, much of the new work was in the low temperature range, below 90 K, the lower limit of IPTS-48. At higher temperatures, between 0 °C and the gold point, new gas thermometry measurements were made. All of this led eventually to the second revision of the Temperature Scale, that which took place in 1968. In the meantime, however, in 1960 an important change was made in the definition of the unit of thermodynamic temperature. The 1854 proposal of Kelvin was finally adopted, namely that the unit of thermodynamic temperature be defined in terms of the interval between the absolute zero and a single fixed point. The fixed point chosen was the triple point of water. The temperature of the triple point of water was fixed at exactly 0.01 °K above the ice point, which in turn was assigned the thermodynamic temperature of 273.15 °K. This proposal had already been made in 1948 but at that time, there was still a divergence of view as to whether the absolute zero should be assigned a temperature of −273.15 °C or −273.16 °C. It was, however, already recognized that the triple point of water was a much more reproducible fixed point than the ice point. The question was finally resolved in 1954 (CIPM, 1955), and the new definition of the kelvin adopted by the 10th CGPM in 1960. This resulted in the curious situation that thermodynamic temperatures were defined in quite a different way to International Practical Temperatures. It therefore became necessary to distinguish between the degree Kelvin (unit of thermodynamic temperature) and the International Practical degree Kelvin (unit of International Practical

Kelvin Temperature). The two were almost certainly not identically equal since °K(Int-1948) was defined in terms of an interval of exactly 100 °K(Int-1948) between the ice and steam points, while °K(thermodynamic) was defined in terms of an interval of exactly 273.16 °K between the absolute zero and the triple point of water. Since the number 273.16 resulted from experimental measurements using gas thermometers which had been calibrated at 0 °C and 100 °C, the two units °K(thermodynamic) and °K(Int-1948) would be identical if, and only if, these experiments had been exactly right in giving a temperature of -273.15 °C to the absolute zero.

This awkward situation was resolved in the 1968 revision of the Temperature Scale, when both thermodynamic and Practical units were defined to be identical and equal to 1/273.16 of the thermodynamic temperature of the triple point of water. The unit itself was renamed "the kelvin" in place of "degree Kelvin" and designated "K" in place of "°K". This left the interval between the ice and steam points as an experimental quantity to be decided upon the basis of the best measurements of the thermodynamic temperature of the steam point. In fact, in the 1968 revision of the Temperature Scale, the value of the steam point was kept as 100 °C exactly, because there was not sufficient evidence to indicate otherwise. Since 1968, however, good experimental evidence from gas and radiation thermometry has been obtained (see Chapter 3) indicating that a value closer to 99.975 °C is to be preferred. The fact that new primary thermometry, based upon a defined temperature of 273.16 K for the triple point of water, leads to a value of 99.975 °C for the steam point means simply that the previous gas thermometry, based upon a defined interval of 100 °C between ice and steam points, was wrong in assigning a value of -273.15 °C to the absolute zero. The correct value would have been -273.22 °C.

NOTE

2-7 The International Practical Temperature Scale of 1968 (IPTS-68)

The developments that eventually led to the adoption of the IPTS-68 had already begun at the time of the adoption of its predecessor in 1948. The NBS made a proposal (CIPM, 1949) to the CCT in 1948 that the International Temperature Scale could be extended down to the boiling point of hydrogen (≈ 20 K) by the simple expedient of using the boiling point of hydrogen as an additional calibration point together with a so-called Z-function for interpolation, where:

$$Z = \frac{R(T) - R(H_2)}{R(O_2) - R(H_2)} \qquad (2\text{-}1)$$

in which $R(T)$, $R(H_2)$ and $R(O_2)$ are the resistances of the thermometer at a temperature T, and at the boiling points of hydrogen and oxygen respectively. It was suggested that for thermometers having a $W(100\ °C)$, that is to say R_{100}/R_0, greater than 1.392, a scale defined in this way would be unique to within ± 10 mK. Although the proposal was not taken up at the time, discussions took place at subsequent meetings of the CCT on the best values to be assigned to a range of low-temperature fixed points and on various empirical relations for resistance as a function of temperature for pure platinum. It was soon recognized that it would be necessary to include one or more intermediate fixed points between 20 K and 90 K in order to achieve a scale of adequate uniqueness. One of the most promising schemes required simply the triple point of oxygen, at about 54 K, used in conjunction with a Z-function table. This appeared to give a scale unique to within a few millikelvins over the whole range from 20 K to 90 K. At the same time, a number of laboratories were engaged upon gas thermometry in the range from the triple point of hydrogen, 13.8 K, up to 273 K. In 1961 it was agreed that NPL and the Physicotechnical and Radiotechnical Measurements Institute (PRMI), Moscow, would undertake an intercomparison of platinum resistance thermometers calibrated on four of the most important gas thermometer scales. These were the NPL(1961), NBS(1955), PRMI(1954) and Pennsylvania State University PSU(1954) scales. Details of these scales and the way in which the intercomparison was carried out are to be found in Bedford *et al.*, 1969. The NBS-55 scale is of particular interest since it is an example of the way in which a so-called "wire-scale" can be successfully

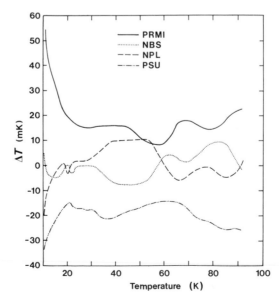

Fig. 2-3. The differences between the various gas thermometer scales upon which the low temperature part of IPTS-68 is based. $\Delta T =$ difference from mean of all scales (after Orlova *et al.*, 1966).

operated. NBS-55 is a scale based upon gas thermometry carried out in 1939 (Hoge and Brickwedde, 1939). It was originally maintained on a group of six platinum resistance thermometers, and was known as NBS-39. In 1955 an arbitrary shift of 10 mK was made over the whole of the scale and the name was changed to NBS-55. As we shall see later on, the successors to these six original NBS-39 thermometers continue to ne used to maintain an NBS version of IPTS-68.

The results of the intercomparison (Orlova et al., 1966) of these four scales, shown in Fig. 2-3, provided the basis for the eventual low temperature part of IPTS-68. A mean table of $W(T)$ vs T was deduced from the results shown in Fig. 2-3 by first normalizing each of the four scales to an oxygen boiling point of 90.170 K and a hydrogen boiling point of 20.267 K. A polynomial was fitted to the mean values of $W(T)$, of the form:

$$T = \sum_{i=0}^{11} A_i (\ln W)^i \qquad (2\text{-}2)$$

The table obtained from equation (2-2) covered the range 12 K to 95 K, and did not depart from the mean experimental data by more than about 3 mK. A complete table was then prepared for the range 12 K to 273.15 K by adding the values calculated from the following polynomial originally prepared for the range 70 K to 273.15 K:

$$T = \sum_{i=0}^{6} B_i (\ln W)^i \qquad (2\text{-}3)$$

This polynomial had previously been fitted to the results of NPL gas thermometry (Barber and Horsford, 1965). The resulting table was published by the CCT (BIPM, 1964) under the title "Provisional reference table CCT-64 of W against T for platinum resistance thermometers in the range 12 K to 273.15 K". Table 2-3 shows the best values of the fixed points consistent with

Table 2-3.

Fixed point	CCT-64* T/K	1966 T/K	IPTS-68 T/K
e-H$_2$ t.p.	13.825	13.809	13.810
e-H$_2$ b.p. at 25/26 atmosphere pressure	17.040		17.042
e-H$_2$ b.p.	20.275	20.267	20.280
Ne b.p.	27.096		27.102
O$_2$ t.p.	54.352	54.354	54.361
O$_2$ b.p.	90.1727	90.180	90.188

* Best values calculated after publication of the CCT-64 Table (Bedford et al., 1969).

the CCT-64 reference table (Bedford *et al.*, 1969). The values given here were arrived at after the CCT-64 Table had been published, and differ slightly from the values recommended in the Table itself.

Soon after CCT-64 was prepared, a CCT Working Group in 1966 proposed a provisional scale (Hall and Barber, 1967), taking into account further gas thermometry results for the oxygen boiling point and hydrogen triple point. The recommended fixed point values of this 1966 Provisional Scale are also given in Table 2-3.

The detailed discussions that took place between 1966 and the adoption of IPTS-68 are too involved to go into here. The interested reader is referred to the lively account by Preston-Thomas (Preston-Thomas, 1972) for an insider's view on how such matters come to be decided. The main point to note is that the CCT, having begun with the intention of simply adding a downward extension to IPTS-48, found itself proposing a completely new Scale which differed significantly from IPTS-48 over the whole of its range. Part of the reason for what happened is to be found in the difficulty of adding to, or extending an existing Scale, without introducing discontinuities of the sort we have already mentioned earlier and which at all costs must be avoided. Other reasons for modifying IPTS-48 above the ice point stemmed from the results of new gas thermometry showing that temperatures previously assigned to the fixed points above 100 °C were in error. It was also decided to take account of the latest values of atomic constants and modify the radiation constant, c_2, of Planck's equation. The text of IPTS-68 (1975 edition) is reproduced in Appendix II. Included is the Table of Differences between IPTS-68 and IPTS-48 which, together with Fig. 2-2, allows differences to be found between IPTS-68 and ITS-27. Between 0 °C and 100 °C, ITS-27 is essentially identical to the 1887 normal hydrogen scale (Hall, 1929).

In outline, IPTS-68 is made up of four parts: (a) between 13.81 K and 273.15 K; (b) 0 °C to 630.74 °C; (c) 630.74 °C to 1064.43 °C, and (d) above 1064.43 °C. In part (a) the Scale is defined in terms of a set of six low-temperature fixed points (Table 2-3) together with a reference function, $W_{CCT-68}(T_{68})$. This reference function is a modified version of CCT-64. Individual thermometers are calibrated at the six low-temperature fixed points and at the ice and steam points and deviation functions $\Delta W(T_{68})$ calculated from these calibrations for the four sub-ranges, as shown schematically in Fig. 2-4. Within each sub-range, $W(T_{68})$ for the thermometer is thus the sum of $W_{CCT-68}(T_{68})$ and $\Delta W(T_{68})$ obtained from the deviation functions. The need for continuity between sub-ranges led to the requirement that the first derivative of $\Delta W(T_{68})$ be matched at the junction with the next higher sub-range. In the 1975 Edition of IPTS-68, while no numerical values of T_{68} were changed, the triple point of argon, at 83.798 K, was added as an alternative to the boiling point of oxygen. This was a consequence of the recognition (see Chapter 4) that triple points generally provided better fixed points than do boiling points, mainly because of the absence of the need to carry out pressure measurement.

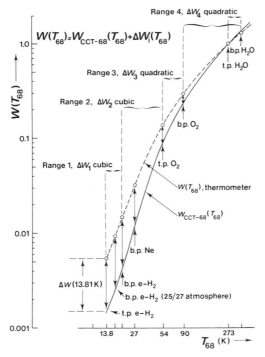

Fig. 2-4. An illustration of the way in which IPTS-68 is defined between 13.81 K and 373 K (see text for a full explanation).

In the range (b), 0 °C to 630.75 °C, the Scale is defined in terms of the old Callendar quadratic equation, but modified to take account of the new gas thermometry values for the fixed points (Terrien and Preston-Thomas, 1967). This modification introduced into the Callendar equation a deviation represented by equation 10 of the IPTS-68 (see Appendix II), which is often referred to as the "Moser wobble" after its originator (Preston-Thomas, 1972). In this part of the Scale, the freezing point of tin (231.9681 °C) is given as an alternative to the steam point as one of the defining fixed points. The temperature assigned to the tin point was chosen with the intention of making it as coherent as possible with the steam point for the purposes of α-coefficient determination (see Chapter 5). Once again this is because the freezing point of a metal allows very high precision to be obtained without the need for pressure measurement. One of the difficulties that has arisen, however, with the widespread adoption of the tin point rather that the steam point, is the problem of making high-precision measurements of $R(T)$ at the tin point for capsule-type thermometers. It is necessary to take great care in order to avoid electrical leakage across the glass bead between the leads of the thermometer (see Chapter 5, Section 4-2). There is also some evidence (Ward and Compton, 1979) that the present value of the tin point is not wholly consistent with the steam point, to the extent of as much as 1 mK. A possible solution to this difficulty may be found in the use of the triple point of gallium at 29.774 °C. Although very much closer to the triple point of water than the tin point, and thus providing a less sensitive measure of the α-coefficient, its reproducibility appears to be extraordinarily good (Mangum and Thornton, 1979).

In the third part of IPTS-68, (c), the defining instrument is the Pt-10 % Rh/Pt thermocouple, calibrated at 630.74 °C and the freezing points of silver and gold, and using a quadratic interpolation formula. Various specifications are given, in terms of the required thermal emfs at the fixed points, which must be satisfied if the thermocouple is to be acceptable as a defining instrument of the Scale. As we shall see in Chapter 6, these specifications are not very useful, since in most cases they are unnecessarily restrictive. Once the condition has been met that one arm of the thermocouple is pure platinum, then provided that the rhodium content of the second arm lies between about 10 % and 13 %, the temperature scales obtained appear to be invariant. The real problem with the thermocouple is that it is not sufficiently reproducible. The reasons for this are discussed in Chapter 6. Although they are well understood, they remain very difficult to overcome. The heart of the problem is the fact that the measured thermal emf originates in the temperature gradient between the hot and cold junctions, and it depends upon both the temperature gradient and the homogeneity of the thermocouple wire in the temperature gradient. If the wire is not perfectly homogeneous, the measured emf will depend not only upon the difference in temperature between the hot and cold junctions but also upon the details of the temperature gradient along the wire. It turns out that the best accuracy that can be obtained using a Pt-10 % Rh/Pt thermocouple, between 630 °C and 1064 °C, is about ±0.2 °C. This is not adequate for present day requirements, and every effort is being made to find a replacement for the thermocouple in time for the next revision of IPTS-68, either by extending upwards the range of the platinum resistance thermometer or by extending downwards the range of the optical pyrometer or by a combination of both.

The fourth part of IPTS-68, (d), that above the gold point, is defined in terms of the radiation emitted by a blackbody and described by Planck's equation. The spectral radiance of the radiation from a blackbody at the freezing point of gold is taken as the reference. The temperature of a blackbody at any higher temperature may be measured in terms of the ratio of its spectral radiance, at the same wavelength, to that of a blackbody at the reference temperature. The value of the second radiation constant is taken as 0.014 388 metre kelvin. It must be remembered that the wavelength must be the vacuum wavelength, otherwise the refractive index of air must be taken into account in the Planck formula (see Chapter 7). The practical instrument used to realize the Scale in this range is the photoelectric optical pyrometer, usually operating at a wavelength of about 650 nm. There are no conditions laid down, in the definition of the Scale, concerning the wavelength but it has been found that the convenience of working in the visible region of the spectrum outweighs the small advantage in sensitivity that would be had by working in the near infrared, near 1 μm. In any case, both the sensitivity and accuracy already obtained are sufficient for all practical purposes. While a photoelectric pyrometer is used for realizing the Scale, its dissemination is often accomplished by means of tungsten-ribbon lamps calibrated in terms of spectral radiance as a function of dc current. Work in a number of national laboratories between 1960 and 1970 resulted in a very substantial improvement in the whole area of optical pyrometry. By 1971 it had become possible to realize IPTS-68 between the gold point and 1700 °C to an accuracy of about ±0.1 °C. This was demonstrated by an intercomparison carried out between four national laboratories (Lee *et al.*, 1972) the results of which are illustrated in Fig. 7-34 of Chapter 7. Since then, steady improvements have been made in the accuracy of photoelectric pyrometers, mainly with a view to carrying out measurements below the gold point. As we shall see below, there is a need for accurate measurements to be made of the differences between thermodynamic temperature and IPTS-68 over the whole range from the ice point to the gold point. The definition of IPTS-68 above the gold point and its realization by means of photoelectric pyrometers is the one section of IPTS-68 that is wholly satisfactory. In any future revision of the Scale, it is unlikely that any changes will be made in this part. The only factor affecting the radiation part of the Scale which is likely to be changed significantly is the temperature assigned to the gold point, but for the time being it is quite uncertain what this change will be.

2-8   The uniqueness of IPTS-68 at low temperatures

Having outlined the definition of IPTS-68, we now come to the question of its uniqueness, particularly in the low-temperature range. This has recently been tested (Ward and Compton, 1979) by an intercomparison of thirty-seven platinum resistance thermometers, from ten laboratories, carried out at NPL over the range 4.2 K to 373.15 K. The thermometers were compared at fifty temperatures in this range in order to show up differences in their calibrations, both at the fixed points and between them. The estimated accuracy of the comparison measurements is shown in Fig. 2-5. This figure can be taken as a fair representation of the best that can currently be achieved in thermometer comparison at low temperatures. The origins of the thermometers were as diverse as could be expected, the majority coming from among the three commercial manufacturers of such thermometers, while others were made at various laboratories. The platinum came from three different sources and the thermometers were made following a variety of designs.

The results of the intercomparison are shown in Figs. 2-6 to 2-9. In each figure, two sets of points are shown, one is for the maximum difference between any pair of thermometers from amongst the sample measured, and the other is

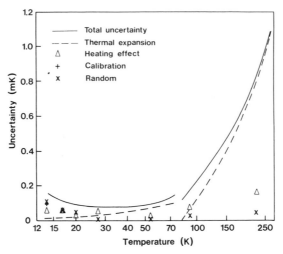

Fig. 2-5. Uncertainties and their sources, in the low temperature intercomparison measurements of Ward and Compton (after Ward and Compton, 1979).

the value below which 68 % of all the differences lie, the so-called semi-intersextile range. From a comparison of Figs 2-6 and 2-7, it is clear that the largest part of the observed differences between thermometers stems from differences in their calibrations at the fixed points. If, as in Fig. 2-7, these differences are removed, the remaining variations are very much smaller. Thus the semi-intersextile range, shown in Fig. 2-7, can be taken as a good estimate of the non-uniqueness of IPTS-68 for present day thermometers. In Fig. 2-8

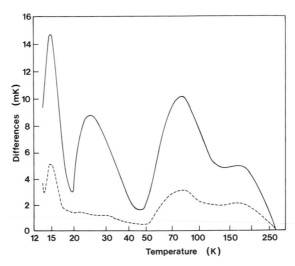

Fig. 2-6. Maximum (—) and semi-intersextile (---) ranges for the differences between 17 thermometers using their original calibrations (after Ward and Compton, 1979).

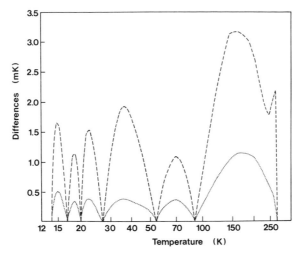

Fig. 2-7. Maximum (- - -) and semi-intersextile (···) ranges for the differences between all 35 thermometers following recalibration (after Ward and Compton, 1979).

are shown the maximum, and semi-intersextile, ranges of the gradients between scales realized by pairs of thermometers carrying their original calibrations. Above about 27 K, these differences are small, but at lower temperatures they become significant with respect to the best specific heat measurements. Care should be taken, therefore, in interpreting specific heat and other such measurements at low temperatures, particularly if they are made over short ranges of temperature. In Fig. 2-9, the gradients are again

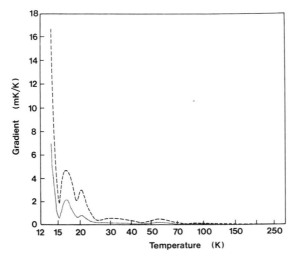

Fig. 2-8. Maximum (- - -) and semi-intersextile (···) ranges for the gradients between 17 thermometers using their original calibrations (after Ward and Compton, 1979).

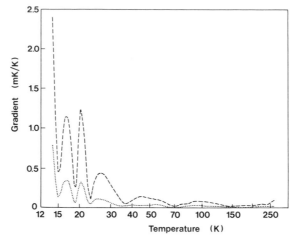

Fig. 2-9. Maximum (---) and semi-intersextile (···) ranges for the gradients between all 35 thermometers following recalibration (after Ward and Compton, 1979).

shown, but after the thermometer scales had been corrected for differences at the calibration points.

The overall conclusions that can be drawn from the work of Ward and Compton is that the intrinsic non-uniqueness of IPTS-68 is of much less significance than are the differences that arise due to inaccuracies in calibration at the fixed points. The rapid development in the design and use of sealed triple-point cells, which has occurred since Ward and Compton carried out their intercomparison, should go a long way in the reduction of this source of uncertainty (Chapter 4). Preliminary results (Pavese, 1982) of an intercomparison of such sealed cells are very encouraging.

2-9   Departure of IPTS-68 from thermodynamic temperatures

So far we have considered the definition of IPTS-68, its reproducibility, smoothness and uniqueness. There remains the important question of how closely IPTS-68 reproduces thermodynamic temperatures. From our discussion earlier, in Section 2-4, we remember that it is not a prerequisite of a practical scale that it reproduce thermodynamic temperatures. Nevertheless if there are significant departures, it becomes very difficult to avoid the presence of discontinuities in slope at the junction of the ranges of different defining instruments. The measurement of thermodynamic temperature is considered in detail in Chapter 3, and also in Chapter 7 in respect of radiation methods.

The thermodynamic temperature of all but one of the fixed points of IPTS-68 was chosen on the basis of gas thermometer measurements alone. The one exception was the boiling point of e-H_2, for which the value was chosen taking into account the results of NBS acoustic thermometry. Our more recent knowledge of numerical values of thermodynamic temperature above 13.81 K is still largely based upon gas thermometry, although some accurate acoustic thermometry results exist up to 20 K, and optical and noise pyrometry measurements of temperature ratios are available above 630 °C with total-radiation data between 327 K and 365 K. Various confirmatory data are available from magnetic thermometry up to about 90 K but, as we shall see in Chapter 3, magnetic thermometry is not primary thermometry and cannot stand alone.

We shall look first at the differences "$t - t_{68}$" above 0 °C, since the differences at low temperatures are closely related to the derivation of the new Provisional Temperature Scale from 0.5 K to 30 K, which we shall come to in Section 2-10. Above 0 °C, the first results have appeared from the monumental gas thermometry work of Guildner and Edsinger at NBS (Guildner and Edsinger, 1976). Taking extreme precautions against every conceivable source of error, they carried out constant volume gas thermometry from 0 °C up to 460 °C, and have determined "$t - t_{68}$" with an estimated uncertainty of only a few millikelvins. Their results are shown in Fig. 2-10. The values of "$t - t_{68}$" shown in this figure are very much larger than the estimated uncertainties (probably estimated at the 3σ level) given in the original 1968 edition of IPTS-68, which are reproduced here in Table 2-4.

In particular, we now find that the steam point no longer has a Celsius temperature of 100 °C, instead it has the measured value of 99.975 °C. That the measured value is not exactly 100 °C is, of course, no surprise (see Section 2-6), but the size of the change is rather unexpected. As we shall see in Chapter 3, the explanation advanced by Guildner and Edsinger is that all previous gas thermometry suffered from hidden errors due to sorption. These results of Guildner and Edsinger have been confirmed by the preliminary results of

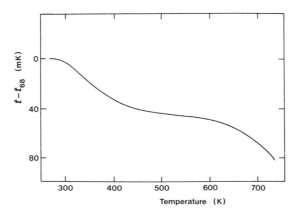

Fig. 2-10. The differences between IPTS-68 and thermodynamic temperature found in recent NBS gas thermometry (after Guildner and Edsinger, 1976).

Table 2-4. Estimated uncertainties, ΔT (1968) of the assigned values of the defining fixed points of IPTS-68, estimated at the time the Scale was defined, together with the author's estimate of the differences $T - T_{68}$.

Fixed point	T_{68}/K	$\Delta T(1968)/\text{K}$	$T - T_{68}/\text{K}$
e-H$_2$ t.p.	13.81	0.01	[see Table 2-5]
17.042 K point	17.042	0.01	[see Table 2-5]
e-H$_2$ b.p.	20.280	0.01	[see Table 2-5]
Ne b.p.	27.102	0.01	[see Table 2-5]
O$_2$ t.p.	54.361	0.01	≈ 0.000
O$_2$ b.p.	90.188	0.01	≈ 0.008
H$_2$O t.p.	273.16	exact by definition	
H$_2$O b.p.	373.15	0.005	-0.025
Sn f.p.	505.118	0.015	-0.044
Zn f.p.	692.73	0.03	-0.066
Ag f.p.	1235.08	0.2	<0.2
Au f.p.	1337.58	0.2	<0.2

Quinn and Martin (1982) obtained from thermal radiation measurements (see Chapter 7, section 7) in the range 327 K to 365 K.

At temperatures above 630 °C, optical pyrometry and noise thermometry have clearly shown that the Scale, defined by the Pt-10 % Rh/Pt thermocouple, departs significantly from thermodynamic temperatures even on the basis of the existing values of the fixed points. This is illustrated in Fig. 2-11. What remains lacking are definitive measurements linking the highest temperature of the NBS gas thermometry, at 460 °C, to the lowest point of the

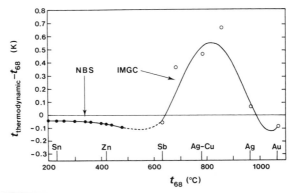

Fig. 2-11. Differences between T and T_{68}; ($\bullet$–$\bullet$) following Guildner and Edsinger from Fig. 2-10; (—) following noise thermometry carried out at IMGC. This curve is made to pass through the gold point value obtained by optical pyrometry using the IMGC noise thermometer value of the silver point as the reference temperature. The dotted line joining the two sets of measurements is based upon Coates and Andrews, 1982.

noise and optical pyrometry measurements at 630 °C. Preliminary results from Coates and Andrews (Coates and Andrews, 1982), shown in Fig. 2-11, are the first indications we have had on how the two sets of measurements are likely to fit together, but further confirmation is needed. It is now clear, however, that the differences $T - T_{68}$ at the gold and silver points are unlikely to exceed 0.2 K.

At low temperatures, the principal new values of thermodynamic temperature come from gas thermometry, carried out by Berry (Berry, 1979) at NPL, over the range 2.6 K to 27.1 K. These have been supported by new results from noise thermometry up to 4.2 K (Klein $et\ al.$, 1979), acoustic thermometry from 4.2 K to 20 K (Colclough, 1979) and by results of a novel method of gas thermometry (Gugan and Michel, 1980; Gugan, 1982) in which the dielectric constant is measured as a function of temperature. Used as an interpolating thermometer, the dielectric constant gas thermometer of Gugan and Michel confirmed the results of Berry over the range 4.2 K to 27 K. The values of the low temperature fixed points on Berry's scale, NPL-75, are shown in Table 2-5.

Before Berry's results were available, however, it had already become clear that IPTS-68 below 27 K and the widely used ^{3}He and ^{4}He vapour pressure scales (see Section 2-11), departed significantly from thermodynamic temperatures. The evidence for this was based largely upon the results of acoustic and magnetic thermometry (for the details of these methods, see Chapter 3). To take account of these differences, a Provisional Scale has been introduced to cover the range 0.5 K to 30 K. Before coming to this, we shall look briefly at the evidence for the departure of IPTS-68 from thermodynamic temperatures in the range just above this, namely between 27 K and 273 K. As we have seen earlier, the reference function $W_{\text{CCT-68}}(T_{68})$ was derived from the mean of gas thermometer measurements, the uncertainties of which were estimated to be about ± 10 mK. The first independent investigation of the departure of IPTS-68 from thermodynamic temperature in the range above 27 K was carried out by Cetas (Cetas, 1976), using a magnetic thermometer. His

$Table\ 2\text{-}5$. Fixed point values on NPL-75 and differences, ΔT, from IPTS-68 or the 1958 ^{4}He vapour pressure scale. (Uncertainties estimated at the 3σ level.)

Fixed point	Uncertainty in fixed point realization/mK	Uncertainty in T_{75}/mK	T_{75}/K	T_{68}/K or T_{58}/K	ΔT/mK
^{4}He b.p.	0.3	0.5	4.2221	4.2150	7.1
e-H_2 t.p.	0.3	0.8	13.803_5	13.8100	-6.5
17.042 K point	0.4	0.9	17.035_6	17.0420	-6.4
e-H_2 b.p.	0.4	0.9	20.271_2	20.280	-8.8
Ne b.p.	0.3	1.4	27.097_9	27.102	-4.1

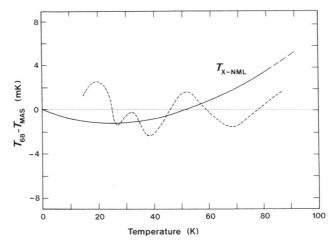

Fig. 2-12. Smoothed differences (- - -) between IPTS-68 and a magnetic temperature scale T_{MAS} based upon a fit of susceptibility data for manganese ammonium sulphate (MAS) to the IPTS-68 from 13.81 K to 83 K. Smoothed differences (—) between $T_{X\text{-}MAS}$ and T_{MAS} where $T_{X\text{-}MAS}$ is another magnetic temperature scale derived by Cetas (after Cetas, 1976).

results are illustrated in Fig. 2-12. On the assumption that the magnetic scale, T_{MAS}, is smooth and its departure from thermodynamic temperatures is likely to be described by a low-order polynomial (see Chapter 3), this figure suggests that IPTS-68 oscillates about thermodynamic temperatures with an amplitude not exceeding about 4 mK. For the range between 90 K and 273 K, no new measurements of thermodynamic temperature have been made. In the absence of any evidence to the contrary, we can assume that the differences "$T - T_{68}$" in this range are of the same order of magnitude as those between 30 K and 83 K, since the gas thermometry upon which the Scale is based was probably of the same quality over the whole of the range 27 K to 273 K. Thus, as a general conclusion, for the range 27 K to 273 K, we can say that IPTS-68 appears to be smooth and not to depart from thermodynamic temperatures by more than about 5 mK. This must be qualified, however, by the comment that since the magnetic thermometry of Cetas made use of the triple and boiling points of oxygen as calibration points, errors in the thermodynamic temperature of these two points would not have been shown up by his work.

2-10   The 1976 Provisional 0.5 K to 30 K Temperature Scale ,
EPT-76

In 1976, the CIPM approved the new low-temperature Scale called the 1976 Provisional 0.5 K to 30 K Temperature Scale, or EPT-76 (the initials of the

French title of the Scale). The purpose of EPT-76 was to provide a unified Scale upon which temperature measurements could be made in this range, pending the revision and downward extension of IPTS-68. The Scale was designed to be thermodynamically smooth, in the sense given in Section 2-4; continuous with IPTS-68 at 27.1 K and to agree with thermodynamic temperatures as closely as these two conditions would allow.

The need for a new Scale below 30 K had become clear following the accumulation of evidence, from acoustic and magnetic thermometry, that the existing Scales in this region, IPTS-68 and the helium vapour pressure scales, departed significantly from thermodynamic temperatures and did so in opposite, and thus inconsistent, directions. The departure from thermodynamic temperatures of the ^{3}He-1962 and ^{4}He-1958 vapour pressure scales, mentioned in the previous Section, had been found by the pioneering work of Plumb and Cataland (Plumb and Cataland, 1966) on acoustic thermometry at NBS in 1965. They found that at 4.2 K, the error in the ^{4}He Scale appeared to be about 10 mK diminishing linearly with temperature at lower temperatures. Above 13.81 K, it was principally the magnetic thermometry of Cetas and Swenson (Cetas and Swenson, 1972) that showed up the errors in IPTS-68. These errors were opposite to, and thus inconsistent with, the errors in the helium vapour pressure scales.

In order to clarify the differences between the various magnetic scales and the NBS 2–20 K acoustic scale, an intercomparison of germanium resistance thermometers was organized and carried out at the NML (Sydney). The results (Besley and Kemp, 1977) showed that it was possible to bring together all of the magnetic scales in a thermodynamically consistent way, to produce a common scale to within about 1 mK. It must be emphasized again that magnetic thermometry is not primary thermometry in this range, since it needs a minimum of four calibration points in order to establish the constants of the equation (Chapter 3).

The magnetic scale chosen to represent the common scale was given the name $T_{XAc'}$ and originated at Iowa State University (ISU). The name indicates that the scale was a modification of a previously existing scale T_{XAc} which itself was based upon a magnetic scale obtained by fitting magnetic data to the NBS 2–20 Acoustic Scale (Swenson, 1973). Having shown that it was possible to establish a consistent magnetic temperature scale below 30 K, the CCT proposed to the CIPM that a provisional temperature scale be set up in this range based upon $T_{XAc'}$. The version of IPTS-68 used to establish the junction at 27.102 K was that realized at NPL (Ward and Compton, 1979) and which formed the basis of the intercomparison of platinum resistance thermometers mentioned earlier. By this time it had become clear that the best measurements of thermodynamic temperature between 2.6 K and 27 K were those of NPL-75, and intercomparisons at NPL and ISU (Anderson and Swenson, 1978) showed that NPL-75 $\cong T_{XAc'}$ below 4.2 K and that at higher temperatures the difference between them is proportional to T^2.

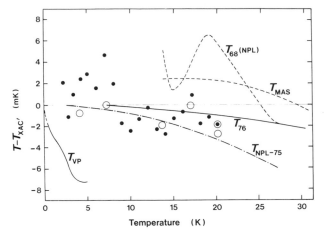

Fig. 2-13. A comparison of low-temperature scales and the definition of the EPT-76 in terms of $T_{XAc'}$. These relations are derived from Tables 2, 3, 4 and 5 of the EPT-76. ●=NBS acoustic thermometry, o=NPL acoustic thermometry.

Therefore T_{76} was chosen (Durieux *et al.*, 1979; EPT-76, 1979) to be defined by:

$$T_{76} = T_{XAc'} - 2.5 \times 10^{-6} \ T^2/K \qquad (2\text{-}4)$$

The differences between T_{76} and the other scales are shown in Fig. 2-13. The differences between T_{76} and NPL-75 have been confirmed by more recent noise thermometry (Klein *et al.*, 1979) by measurements made up to 4.2 K.

The EPT-76 is itself defined in terms of the temperatures assigned to eleven fixed points within the range 0.5 K to 30 K, together with the differences between T_{76} and the following existing scales: IPTS-68; the ^{4}He-1958 and ^{3}He-1962 vapour pressure scales; the NBS 2–20(1975) acoustic scale; various magnetic scales; NPL-75 and the NBS version of IPTS-68 which is defined by difference from NBS-55. In contrast to IPTS-68, the EPT-76 may thus be realized in a number of ways; either by using one of the above scales and the tabulated differences given in the text of EPT-76 or by using a thermodynamic interpolating thermometer, such as a gas thermometer or magnetic thermometer, calibrated at one or more of the specified reference points of EPT-76.

The superconducting transition points of EPT-76 are defined as the temperatures of transition between the superconducting and normal states, in zero magnetic field, of the five elements present in NBS Standard Reference Material 767. A criticism that can be made of EPT-76 is that one of its recommended realizations is thus linked to a particular product for which there is only one supplier, NBS. In due course it may become possible to prepare samples of the five elements in a sufficiently reproducible state so that, regardless of the source of the material, the superconducting transition

Table 2-6. Reference points of EPT-76 (s.c.t. = superconducting transition point).

Reference point	T_{76}/K
cadmium s.c.t.	0.519
zinc s.c.t.	0.851
aluminium s.c.t.	1.1796
indium s.c.t.	3.4145
^{4}He n.b.p.	4.2221
lead s.c.t.	7.1999
e-H_2 t.p.	13.8044
e-H_2 b.p. at 25/76 standard atmosphere	17.0373
e-H_2 n.b.p.	20.2734
neon t.p.	24.5591
neon n.b.p.	27.102

temperature always takes the same value. The temperatures assigned to the superconducting reference points of lead, indium and aluminium are based upon the mean of values determined at a number of different laboratories after correction to $T_{XAc'}$. The uncertainty in these values with respect to $T_{XAc'}$ is estimated to be about ± 0.5 mK. The value assigned to the superconducting transition temperature of zinc is based upon magnetic thermometry at NPL, and that for cadmium upon magnetic measurements at NPL and noise thermometry at NBS. A detailed description of the origin and derivation of EPT-76 is given by Durieux *et al.*, 1979.

2-11   The vapour pressure scales of ^{3}He and ^{4}He

The saturated-vapour pressure/temperature relation for liquid helium provides such a good and reproducible scale that its use as such long pre-dates any internationally agreed scale in the helium range. In fact it even pre-dates ITS-27, since it was in 1924 that Kamerlingh Onnes of the University of Leiden first established a vapour pressure scale of ^{4}He, extending up to the critical point, 5.2 K. Various revisions were made to the scale at Leiden in 1929, 1932 and 1938. The beginnings of an international agreement on a ^{4}He vapour pressure scale came in 1948 when representatives of the Kamerlingh Onnes Laboratory, the Royal Society Mond Laboratory in Cambridge and various cryogenic laboratories in the USA agreed among themselves to adopt a common scale (Van Dijk and Schoenberg, 1949). This scale was to be based upon a thermodynamic formula of Bleaney and Simon (Bleaney and Simon, 1939) below 1.6 K, the vapour pressure measurements of Schmidt and Keesom (Schmidt and Keesom, 1937) between 1.6 K and 4.3 K, and five vapour pressure points between 4.3 K and 5.2 K determined by Kamerlingh Onnes and Weber (Kamerlingh Onnes and Weber, 1915). Although this scale was never formally published, it became very widely known and used by low-temperature researchers up until 1958 when international agreement was formally reached on a ^{4}He vapour pressure scale.

It turned out to be rather more difficult than might have been expected to reach agreement on a ^{4}He vapour pressure scale. The reasons for the difficulty are those that are always encountered in the setting up of a new practical temperature scale. They revolve around the question of whether the equations which describe the scale should be thermodynamic equations, soundly based upon theory, or empirical equations best suited to the fitting of experimental data. Ideally they should be the former, but if the relevant thermodynamic equations contain a number of constants that are difficult to evaluate, and hence uncertain in numerical value, the degrees of freedom in the fitting of experimental data become so large that most of the advantages of using a thermodynamic equation are lost. On the other hand, an empirical equation resulting from the straightforward fitting of experimental data can mask thermodynamic inconsistencies and errors in the data. In the early 1950s, it so happened that the estimated accuracy of the thermodynamic equation, which gives ^{4}He vapour pressure as a function of temperature, was about the same as that of the empirical equation to be obtained by the fitting of experimental data. This was not the case over the whole of the vapour pressure range and was in any case a matter of lively discussion among those concerned (Keller, 1956). In the end a compromise was reached in which simply a Table was published, of vapour pressure against temperature, without any equation being given. This Table was proposed to the CCT in 1958 jointly by those who had previously been the proponents of one sort of equation or the other. Indeed when agreement was very close to being reached, the attractions for each side of the others' method became so strong that for a while it looked as if positions were about to be reversed (Hudson, 1980). Indeed, in the closing stages of the discussions the rival parties did actually exchange roles! The Table adopted by the CIPM in 1958 became known as the 1958 ^{4}He Scale (Brickwedde *et al.*, 1960), and covered the range from 0.5 K to 5.23 K, temperatures measured on it being denoted by T_{58}.

Shortly after the adoption of the 1958 ^{4}He Scale, a further proposal was made in respect of a vapour pressure scale for ^{3}He. This was based mainly upon the work of Sydoriak *et al.* (1964), and was to extend from 0.2 K up to the critical temperature 3.324 K. The proposed Table of vapour pressure was made to fit smoothly and to agree with the 1958 ^{4}He Scale at 2.245 K, just above the λ-point of ^{4}He. The Scale was accepted by the CIPM and became known as the 1962 ^{3}He Scale, temperatures measured on it being denoted by T_{62}.

Although neither the 1958 ^{4}He nor the 1962 ^{3}He Scales included equations from which the Tables could be deduced, the information that went into the calculation of the Tables was quite clear. The 1958 ^{4}He Scale was based upon gas thermometry data smoothed by magnetic thermometry and, below 2.2 K, by thermodynamic calculations. The 1962 ^{3}He Scale was based upon comparison between the vapour pressures of ^{3}He, and the 1958 ^{4}He Scale above 0.9 K and on thermodynamic calculations below 0.9 K.

Complete revisions have now been made of both the ^{3}He and ^{4}He Scales, making them consistent with the results of gas, noise, acoustic and magnetic thermometry carried out since the original versions of the Scales were adopted. A detailed discussion on the origin and derivation of the new Scales is to be found in Durieux and Rusby (1983). The experimental work upon which the Scales are based is largely that which was used to furnish the differences $(T_{(\text{vapour pressure})} - T_{76})$ which appear in Table 3 of EPT-76. In addition, new magnetic thermometry by El Samahy (1979) has provided data on ^{3}He vapour pressures down to 0.5 K, and magnetothermodynamic measurements of ^{3}He vapour pressure have been made by Fisher and Brodale (1981) in the range 0.8 K to 1.5 K. The theoretical basis for the new equations is the thermodynamic equation (Van Dijk and Durieux, 1958):

$$\ln P = -\frac{L_0}{RT} + \frac{5}{2}\ln T + i_0 + \varepsilon(T) - \frac{1}{RT}\int_0^T S_1 dT + \frac{1}{RT}\int_0^P V_1 dP$$

$$(2\text{-}5)$$

In order to evaluate $\ln P$, we need a value for L_0, the latent heat of vaporization at absolute zero; data for $S_1(T)$, the molar entropy and $V_1(T)$, the molar volume of the liquid; a value for $\varepsilon(T)$, which expresses the non-ideality of the vapour in terms of the virial coefficients; and i_0 which is the chemical constant, obtainable from statistical mechanics. It is possible, in principle, to obtain a numerical pressure–temperature relation from equation (2-5) by a process of successive approximations starting from experimental data for $\varepsilon(T)$, $S_1(T)$ and $V_1(T)$, and a value for L_0 obtained from a single empirical P–T point. In practice, however, this method is limited to the low pressure domain because the last three terms of equation (2-5) and their associated uncertainties increase rapidly with T. There exists, therefore, a region at intermediate pressures within which the theoretical Scale, calculated from equation (2-5), and the empirical Scale are of comparable accuracy. The numerical value of L_0 is thus chosen to lead to the best agreement in this region.

The new ^{4}He Scale was calculated between 1.4 K and the λ-point by Durieux et al. (1982) taking $L_0 = 59.83$ J mol^{-1}. This part of the scale agrees closely with the experimental results of Rusby and Swenson (1980) obtained using magnetic thermometry. Above the λ-point, the new Scale is based solely upon experimental measurements while below the λ-point it is based upon theory. The equation chosen to represent the new Scale below the λ-point is:

$$\ln (P/\text{Pa}) = \sum_{k=-1}^{6} a_k T^k \qquad (2\text{-}6)$$

where the coefficients

$$a_{-1} = -7.41816 \text{ K}$$

$$a_0 = 5.421\,28$$

$$a_1 = 9.903\,203 \text{ K}^{-1}$$

$$a_2 = -9.617\,095 \text{ K}^{-2}$$

$$a_3 = 6.804\,602 \text{ K}^{-3}$$

$$a_4 = 3.015\,460\,6 \text{ K}^{-4}$$

$$a_5 = 0.746\,135\,7 \text{ K}^{-5}$$

and $$a_6 = 0.079\,179\,1 \text{ K}^{-6}.$$

The data were weighted by the inverse sensitivity ($dT/d \ln P$), and P and dP/dT were forced to take the values $5\,041.8$ Pa and 2.461 K^{-1}, respectively, at the λ-point (Rusby and Swenson, 1980).

For the range above the λ-point, the equation of McCarty (1973) was chosen:

$$\ln (P/\text{Pa}) = \sum_{k=-1}^{8} a_k t^k + b(1-t)^{1.9} \tag{2-7}$$

where the coefficients

$$a_{-1} = \quad -30.932\,85$$

$$a_0 = \quad 392.473\,61$$

$$a_1 = \quad -2\,328.045\,87$$

$$a_2 = \quad 8\,111.303\,47$$

$$a_3 = -17\,809.809\,01$$

$$a_4 = \quad 25\,766.527\,47$$

$$a_5 = -24\,601.4$$

$$a_6 = \quad 14\,944.651\,42$$

$$a_7 = -5\,240.365\,18$$

$$a_8 = 807.931\,68$$

and
$$b = 14.533\,33$$

$$t = T/T_c$$

where
$$T_c = 5.195\,3 \text{ K}$$

The last term in equation (2-7) ensures that (d^2P/dT^2) diverges at the critical temperature, T_c, as is required by theory. The equation was forced to fit a critical pressure of 227.46 kPa (Kierstead, 1971), a normal boiling point of 4.222 1 K and P and dP/dT at the λ-point as above.

The new ^{3}He Scale is based upon the same equation as was used for the T_{62} Scale, namely:

$$\ln(P/\text{Pa}) = \sum_{k=-1}^{4} a_k T^k + b \ln(T/\text{K}) \qquad (2\text{-}8)$$

where the coefficients

$$a_{-1} = -2.509\,43 \text{ K}$$

$$a_0 = 9.708\,76$$

$$a_1 = -0.304\,433 \text{ K}^{-1}$$

$$a_2 = 0.210\,429 \text{ K}^{-2}$$

$$a_3 = -0.054\,514\,5 \text{ K}^{-3}$$

$$a_4 = 0.005\,606\,7 \text{ K}^{-4}$$

and
$$b = 2.254\,84.$$

The data to be fitted was taken from the experimental results of Rusby and Swenson (1980) down to 0.8 K, and those of El Samahy between 0.5 K and 1.8 K. New calculations were also made by El Samahy of ^{3}He vapour pressures by adjusting the equation of Sydoriak et al. (1964). The results of the theoretical calculations agreed well with the experimental results within their range of overlap. This gave confidence to the authors of the new Scale that an extrapolation of the theoretical equation down to 0.2 K was justified.

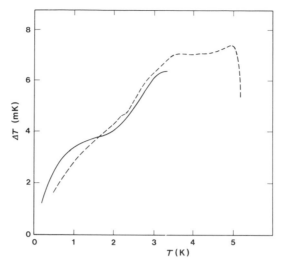

Fig. 2-14. The differences $T_{76}(^4\mathrm{He}) - T_{58}$ (---) and $T_{76}(^3\mathrm{He}) - T_{62}$ (—), where $T_{76}(^4\mathrm{He})$ and $T_{76}(^3\mathrm{He})$ represent temperatures on the new helium vapour-pressure scales (after Durieux and Rusby, 1982).

The differences between T_{58} and T_{62} and the new vapour pressure Scales are shown in Fig. 2-14.

References

Anderson, and Swenson, C. A. (1978). Characteristics of germanium resistance thermometers from 1 to 35 K and the ISU magnetic temperature scale. *Rev. Sci. Inst.*, **49**, 1027–1033.

Barber, C. R. and Horsford, A. (1965). Differences between the thermodynamic scale and the International Practical Scale of Temperature from 0 °C to −183 °C. *Metrologia*, **1**, 75–80.

Bedford, R. E., Durieux, M., Muijlwijk, R. and Barber, C. R. (1969). Relationship between the IPTS-68 and the NBS-55, NPL-61, PRMI-54 and PSU-54 temperature scales in the range from 13.81 to 90.188 K. *Metrologia*, **5**, 47–49.

Berry, K. H. (1979). NPL-75, a low temperature gas thermometry scale from 2.6 K to 27.1 K. *Metrologia*, **15**, 89–115.

Besley, L. M. and Kemp, W. R. G. (1977). An intercomparison of temperature scales in the range 1 to 30 K using germanium resistance thermometry. *Metrologia*, **13**, 35–51.

BIPM (1975). Le Bureau International des Poids et Mesures, 1875–1975, published by the BIPM, Pavillon de Breteuil, F-92310 Sèvres, 1975. Available also in English translation as NBS Special Publication 420. "The International Bureau of Weights and Measures 1875–1975".

BIPM (1964). Provisional reference table CCT-64 of W against T for platinum resistance thermometers in the range 12 K to 273.15 K, published by BIPM.

Bleany, B. and Simon, F. (1939). The vapour pressure curve of liquid helium below the λ-point. *Trans. Farad. Soc.*, **35**, 1205–1214.

Brickwedde, H., van Dijk, H., Durieux, M., Clement, J. R. and Logan, J. K. (1960). The 1958 $^4\mathrm{He}$ scale of temperature. *NBS J. of Res.* **64A**, 1–18.

Callendar, H. L. and Griffiths, E. H. (1891). On the determination of the boiling point of sulphur and on a method of standardizing platinum resistance thermometers by reference to it — experiments made at the Cavendish Laboratory, Cambridge. *Phil. Trans. Roy. Soc.* **A182**, 119–157.

CCT (1982). Report of 14th Session (BIPM).

Callendar, H. L. (1899). On a practical thermometric standard. *Phil. Mag.* **48**, 519–547.

Cetas, T. C. (1976). A magnetic temperature scale from 1 to 83 K. *Metrologia,* **12**, 27–40.

Cetas, T. C. and Swenson, C. A. (1972). A paramagnetic salt temperature scale, 0.9 K to 18 K. *Metrologia,* **8**, 46–64.

Chappuis, M. A. (1888). Etudes sur le Thermomètre à Gaz et Comparaison des Thermomètres à mercure, Travaux et Mémoires du Bureau International des Poids et Mesures, Vol. VI.

Chree, C. (1900). Investigations on platinum thermometry at Kew Observatory, *Proc. Roy. Soc.* **A67**, 3–58.

CIPM (1949). Procès-Verbaux des Séances de 1948, **21**, T84.

CIPM (1955). Procès-Verbaux des Séances de 1954, **24**, T79.

Coates, P. B. and Andrews, J. W. (1982). Measurement of thermodynamic temperatures using the NPL photon-counting pyrometer. *TMSCI,* **5**, 109–114.

Colclough, A. R. (1979). Low frequency acoustic thermometry in the range 4.2–20 K with implications for the value of the gas constant. *Proc. Roy. Soc.* **A365**, 349–370.

Crovini, L. and Actis, A. (1978). Noise thermometry in the range 630–962 °C. *Metrologia,* **14**, 69–78.

Durieux, M., Astrov, D. N., Kemp, W. R. G. and Swenson, C. A. (1979). The derivation and development of the 1976 provisional 0.5 K to 30 K temperature scale *Metrologia,* **15**, 57–63.

Durieux, M. and Rusby, R. L. (1983). Helium vapour pressure equations on the EPT-76. *Metrologia,* **19**.

Durieux, M., Van Dijk, J. E., Rem, P. C., ter Harmsel, H. and Rusby, R. L. (1982). Helium vapour pressure equations on the EPT-76. *TMSCI,* **5**, 145–154.

El Samahy, (1979). Thermometry between 0.5 K and 30 K. Thesis, University of Leiden.

EPT-76 (1979). The 1976 Provisional 0.5 K to 30 K Temperature Scale. *Metrologia,* **15**, 65–68, 1979.

Fisher, R. A. and Brodale, G. E. (1982). Proceedings of LT16 (Los Angeles) *Physica,* **109/ 110 B + C**, 2126–2128.

Gugan, D. and Michel, G. W. (1980). Dielectric-constant gas thermometry from 4.2 K to 27.1 K. *Metrologia,* **16**, 149–168.

Gugan, D. (1982). Surface fitting of helium isotherms: implications for the temperature scale 2.6–27.1 K. *TMSCI,* **5**, 55–64.

Guildner, L. A. and Edsinger, R. E. (1976). Deviation of International Practical Temperatures from Thermodynamic Temperatures in the temperature range from 273.16 K to 730 K. *J. Res. NBS,* **80A**, 70–738.

Hall, J. A. (1929). The International Temperature Scale between 0 °C and 100 °C. *Phil. Trans. Roy Soc.* **A229**, 1–48.

Hall, J. A. (1967). The early history of the International Practical Scale of Temperature. *Metrologia,* **3**, 25–28.

Hall, J. A. (1954). The International Temperature Scale. *TMCSI,* **2**, 115–139.

Hall, J. A. and Barber, C. R. (1967). The evolution of the International Practical Temperature Scale. *Metrologia,* **3**, 78–86.

Harker, J. A. (1907). On the Kew scale of temperature and its relation to the international hydrogen scale. *Proc. Roy. Soc.* **A78**, 225–240.

Hoge, H. J. and Brickwedde, F. G. (1939). Establishment of a temperature scale for the calibration of thermometers between 14 and 83 °K. *J. Res. NBS*, **22**, 351–373.

Hudson, R. P. (1980). Private communication.

Kamerlingh-Onnes, H. and Weber, S. (1915). Leiden Commun., 147b.

Keller, W. E. (1956). The battle of the millidegree. *Nature*, **178**, 883–887.

Kierstead, H. A. (1971). Pressures on the critical isochore of ^{4}He. *Phys. Rev.* **A3**, 329–339.

Klein, H. H., Klempt, G. and Storm, L. (1979). Measurement of the thermodynamic temperature of ^{4}He at various vapour pressures by means of a noise thermometer. *Metrologia*, **15**, 143–154.

Lee, R. D., Kostkowski, H. J., Quinn, T. J., Chandler, T. R., Jones, T. P., Tapping, J. and Kunz, H. (1972). Intercomparison of the IPTS-67 above 1064 °C by four National Laboratories. *TMSCI*, **4**, Part I, 377–393.

Mangum, B. W. and Thornton, D. D. (1979). Determination of the triple point temperature of gallium. *Metrologia*, **15**, 201–215.

McCarty, R. D. (1973). *J. Phys. Chem. Ref. Data*, **2**, 923.

Moreau, H. (1975). "Le Système Métrique", Editions Chiron, Paris.

Orlova, M. P., Sharevskaya, D. I., Astrov, D. N., Krutikova, I. G., Barber, C. R. and Hayes, J. G. (1966). The derivation of the Provisional Reference Table CCT-64, $T = f(W)$ for platinum resistance thermometers for the range from 12 to 273.15 °K. *Metrologia*, **2**, 6–13.

Pavese, F. (1982). On the use of first generation sealed cells in an international comparison of triple point temperatures of gases. *TMSCI*, **5**, 209–216.

Plumb, H. H. and Cataland, G. (1966). An acoustical thermometer and the NBS Provisional Temperature Scale 2–20 (1965). *Metrologia*, **2**, 127–139.

Preston-Thomas, H. (1972). The origin and present status of the IPTS-68. *TMCSI*, **4**, Part 1, 3–14.

Quinn, T. J. and Martin, J. E. (1982). Radiometric measurements of thermodynamic temperature between 327 K and 365 K. *TMSCI*, **5**, 103–108.

Rusby, R. L. and Swenson, C. A. (1980). A new determination of the helium vapour pressure scales using a CMN magnetic thermometer and the NPL-75 gas thermometer scale. *Metrologia*, **16**, 73–87.

Schmidt, G. and Keesom, W. (1937). New measurements of liquid helium temperatures. II The vapour pressure curve of liquid helium. *Physica*, **4**, 971–977.

Swenson, C. A. (1973). Relationship from 1 to 34 K between a paramagnetic salt temperature scale and other scales: an addendum. *Metrologia*, **9**, 99–101.

Sydoriak, S. G., Roberts, T. R. and Sherman, R. H. (1964). The 1962 ^{3}He scale of temperature. I. New vapour pressure comparisons. II. Derivation. III, Evaluation and status. IV. Tables. *NBS J. of Res.*, **68A**, 547–588.

Terrien, J. and Preston-Thomas, H. (1967). Progress in the definition and in the measurement of temperature. *Metrologia*, **3**, 29–30.

Van Dijk, H. and Schoenberg, D. (1949). Tables of vapour pressure of liquid helium. *Nature*, **164**, 151.

Van Dijk, H. and Durieux, M. (1958). Thermodynamic temperature scale (T_{L55}) in the liquid helium region. *Physica*, **2**, 1–19.

Van Dijk, J. E., Rem, P. C., ter Harmsel, H., Durieux, M. and Rusby, R. L. (1982). Helium vapour pressure equations on the EPT-76. *TMSCI*, **5**.

Ward, S. and Compton, J. P. (1979). Intercomparison of platinum resistance thermometers and T_{68} calibrations. *Metrologia*, **15**, 31–46.

3

The Measurement of Thermodynamic Temperature

━━━━━━━━━━━━━━━━━━━━━━━━━

3-1 Introduction

In this introductory chapter on the practical aspects of temperature measurement we shall first of all consider the three principal methods of primary thermometry. These are: classical gas thermometry, acoustic gas thermometry and noise thermometry. This is followed by a discussion on magnetic thermometry. Although not used here as a method of primary thermometry, it has very close connections with primary thermometry and so we shall consider it in this chapter. This is also the case for two other methods: refractive-index gas thermometry and dielectric-constant gas thermometry, both of which show promise as interpolating methods. The measurement of temperature using methods based upon thermal radiation measurement is treated separately and is to be found in Chapter 7. In this chapter, we shall concentrate more on the principles of the methods rather than on the results of measurements since these were covered in Chapter 2 as part of our discussion on Temperature Scales.

3-2 Gas thermometry

3-2-1 *Gas thermometry and the virial equation of state*

The importance of the gas thermometer as a primary thermometer stems from the nearly-ideal behaviour exhibited by many gases and the fact that such small departures from ideal behaviour as are observed are fairly simple functions of density and temperature and have a form soundly based upon theory.

For an ideal gas we have:

$$PV = NRT \qquad (3\text{-}1)$$

where P is the pressure, V the volume, N the quantity of gas present expressed in moles, and R the gas constant.

Equation (3-1) represents the equation of state of an ideal gas, that of a real gas is usually written in the form of a virial expansion, either in terms of the density:

$$PV = NRT\left(1 + B(T)\frac{N}{V} + C(T)\left(\frac{N}{V}\right)^2 + \ldots\right) \qquad (3\text{-}2)$$

where $B(T)$ and $C(T)$ are called the second and third virial coefficients respectively, or in terms of the pressure:

$$PV = NRT(1 + B_p(T)P + C_p(T)P^2 + \ldots). \qquad (3\text{-}3)$$

To avoid confusion, the coefficients $B_p(T)$ and $C_p(T)$ of the pressure expansion are not usually referred to as virial coefficients. They are however related to the virial coefficients in the following way:

$$B_p(T) = \frac{B(T)}{RT} \qquad (3\text{-}4)$$

$$C_p(T) = \frac{C(T) - B(T)^2}{(RT)^2} \qquad (3\text{-}5)$$

$$D_p(T) = \frac{D(T) - 3B(T)C(T) + 2B(T)^2}{(RT)^3}. \qquad (3\text{-}6)$$

Similar but more complex relations exist between the higher-order virial coefficients and the coefficients of the pressure expansion (Mason and Spurling, 1969).

The justification for expressing the departure from ideal behaviour of a gas by a virial expansion stems from the direct link such an expression has with intermolecular forces. Calculation of the equation of state of a gas taking into account molecular interactions leads naturally to a polynomial in increasing powers of the density. The second and higher coefficients of the polynomial represent the effects of intermolecular collisions. The second coefficient takes into account collisions between pairs of molecules, the third between three molecules, the fourth between four and so on. Clearly the calculation of the coefficients becomes very difficult for collisions involving more than two molecules. For the purposes of thermometry, the contributions of the third

Table 3-1. Values of PV_m/RT for argon at 25 °C as calculated from the density series and the pressure series.[a]

P/MPa	PV/RT	
	Density series	Pressure series
0.1	$1-0.00064+0.00000+\ldots (+0.00000)$	$1-0.00064+0.00000+\ldots (+0.00000)$
1	$1-0.00648+0.00020+\ldots (-0.00007)$	$1-0.00644+0.00015+\ldots (-0.00006)$
10	$1-0.06754+0.02127+\ldots (-0.00036)$	$1-0.06439+0.01519+\ldots (+0.00257)$
100	$1-0.38404+0.68788+\ldots (+0.37272)$	$1-0.64387+1.51895+\ldots (-0.19852)$

a After Mason and Spurling, 1969.

and higher-order terms in the virial expansion are very small and except at very low temperatures, can be ignored.

Although the virial equation of state satisfies all the requirements of thermometry, it should be remembered that this is so only because the virial expansion converges very rapidly in those ranges of density and temperature within which we wish to use it. This is not the case at very high densities or at very high or low temperatures. Table 3-1 shows the first few terms of the virial expansion of argon at 25 °C over a range of pressures from one atmosphere up to one thousand atmospheres. It is evident that at the highest pressures the virial equation of state is much less satisfactory, and it is thought (Mason and Spurling, 1969) that convergence depends very much upon the postulated form of the forces operating between molecules undergoing various orders of collision. At high temperatures, the virial equation fails when ionization begins to take place. Below the critical temperature, experiments show that the series remains convergent up to the density of the saturated vapour but diverges for liquid densities.

In using the virial equation it must be remembered that it is an infinite series and that if a truncated series is used, as is always the case in practice, the coefficients of the truncated series are not, strictly, the virial coefficients. The difference between the second coefficient of a low order polynomial and the real second virial coefficient, in our range of interest, is usually quite small since it represents the dominant term of the series. This is not necessarily the case, however, for the third and fourth terms of a truncated series; they may be very different from the third and fourth virial coefficients.

The importance for thermometry of the virial equation of state being soundly based upon theory lies in the confidence that this gives to our knowledge of the variation of the virial coefficients with temperature. If the virial equation were an empirical relation between PV and T, gas thermometry would be much more difficult since it would be necessary at every temperature to determine the coefficients of the equation of state. A prior knowledge of the form of $B(T)$ as a function of temperature allows interpolation to be carried out, thereby saving a large amount of experimental work. This will become clear later in this chapter when we discuss in more detail the practical aspects of gas thermometry. We shall first look in a little more detail at the way in which intermolecular forces lead to departures from ideal behaviour in a gas.

Figure 3-1 shows schematically the potential energy $u(r)$ between two molecules as a function of the distance r between them. A measure of the hard-core diameter of the molecule is given by σ and the depth of the attractive well by ε. Such a potential is fairly well described by a function of the form

$$u(r) = \frac{K_n}{r^n} - \frac{K_m}{r^m} \tag{3-7}$$

known as a bireciprocal potential, where $n > m > 3$, and K_n and K_m are positive

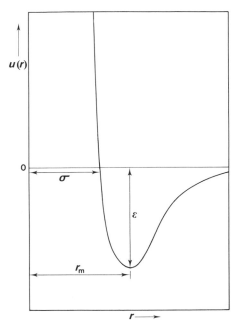

Fig. 3-1. Schematic diagram of the potential energy of the interaction $u(r)$ between two identical molecules separated by a distance r. σ is a measure of the hard-core diameter of the molecules and ε the depth of the potential well.

constants related to σ and ε. Potentials of this type are often called Lennard-Jones potentials, the most successful of which is probably the 6–12 potential in which $n=6$ and $m-12$. Using such a potential it is possible to calculate, without too much difficulty, the second and third virial coefficients and, with rather more difficulty, the fourth and fifth virial coefficients. The great advantage of the 6–12 Lennard-Jones potential, when originally proposed, was that integration leading to the second virial coefficient could be carried out analytically to give a rapidly converging power series. For this reason it has remained widely used even since the advent of computers and is still the basis of modern more refined potentials. For a more detailed discussion on intermolecular potentials, the reader is referred to the monograph by Mason and Spurling (1969). For our purposes it is sufficient to know that there exist well established intermolecular potentials that adequately explain observed kinetic and transport properties of the gases used for thermometry.

The particular properties that we need to know for primary thermometry in a given temperature range are mostly dependent upon only a part of the bireciprocal potential. At low temperatures, interactions between molecules are dependent mostly upon the long-range attractive forces. As the temperature falls, the molecules spend an increasing proportion of their time in association, i.e. in pairs. As a result, the pressure is below that which would

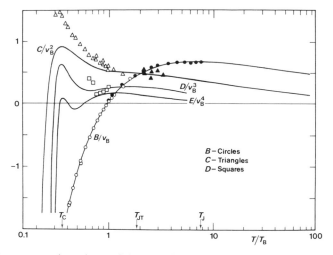

Fig. 3-2. Temperature dependence of the second (B), third (C), fourth (D) and fifth (E) virial coefficients in reduced units. T_B is the Boyle temperature, T_c the critical temperature, T_{JT} the zero pressure Joule–Thomson temperature and T_J the low-pressure Joule inversion temperature. These curves are approximately correct for most ordinary gases, shown also are some experimental points (after Mason and Spurling, 1969).

be expected for an ideal gas and thus the second virial coefficient, $B(T)$, must be negative and increasingly so as the temperature falls still further. At high temperatures, on the other hand, the collisions become more energetic, and the repulsive hard core becomes the dominant factor. This has the effect of producing an excluded volume which leads to an increase in pressure above the perfect gas value and hence to a positive $B(T)$. At still higher temperatures, the collisions become even more energetic and the molecules penetrate even further into their hard cores, thereby reducing the excluded volume and thus reducing $B(T)$. This is shown in Fig. 3-2 together with the calculated behaviour of $C(T)$, $D(T)$ and $E(T)$. The figure is plotted in reduced units following the principle of corresponding states (see for example McGlashan, 1979). The curves are plotted in terms of $B(T)/V_B$ and $C(T)/V_B{}^2$ where

$$V_B = T \left[\frac{dB(T)}{dT} \right]_{T=T_B}$$

$$(3\text{-}8)$$

where T_B is the Boyle temperature, defined as the temperature at which $B(T)=0$. The classical principle of corresponding states does not apply among those gases exhibiting different quantum effects and thus at low temperatures, the second virial coefficients of ^{4}He and ^{3}He differ somewhat from each other and from that shown in Fig. 3-2.

The derivation of the expression for the second virial coefficient of a gas obeying Boltzmann statistics is lengthy but straightforward. The result is the same whether we start from the virial theorem of Clausius, using either classical mechanics or the quantum mechanical time-dependent Schrodinger equation, or whether we start from the canonical ensemble. Using classical mechanics we find that

$$B(T) = -\frac{2\pi N_A}{3kT} \int_0^\infty \exp\left[-u(r)/kT\right] \frac{du(r)}{dr} r^3 dr \qquad (3\text{-}9)$$

where N_A is the Avogadro constant. The term $\exp[-u(r)/kT]$ is obtained by expanding the general expression for the potential energy of the system, ignoring all interactions other than those between pairs of molecules and taking the first term only of the resulting series (see Mason and Spurling, 1969, p. 18 et seq.). Equation (3-9) applies to molecules having no internal degrees of freedom and interacting with central forces only, i.e. the molecules having spherical symmetry.

For thermometry at low temperatures, where we must use helium as our thermometric gas, equation (3-9) is not adequate as it stands because it does not allow us to take into account quantum effects. Considerable effort has been devoted to the question of the second virial coefficients of ^{3}He and ^{4}He in the quantum region, namely below about 8 K, and in the intermediate region between about 8 K and 30 K. The first successful calculations were those of de Boer and Michels in 1939 (de Boer and Michels, 1939). Recent improved calculations are those of Kilpatrick et al. (1954) and Boyd et al. (1969). The full quantum mechanical expression given by Boyd et al. for $B(T)$ is the sum of two parts, $B(T)_{direct}$ and $B(T)_{exchange}$. The former refers to interactions of two particles which obey Boltzmann statistics and the latter to interactions of particles which obey either Bose–Einstein statistics (^{4}He) or Fermi–Dirac statistics (^{3}He). The expressions for $B(T)_{direct}$ and $B(T)_{exchange}$ are written in terms of the direct and exchange phase shift sums and the energies of two body bound states. For ^{4}He, which has zero spin, the two expressions omitting the bound-state terms, are as follows:

$$B(T)_{direct} = -\frac{2^{1/2} N_A \lambda^5}{\pi^2} \int_0^\infty k_m G_+(k_m) \exp\left(\frac{-\lambda^2 k_m^2}{2\pi}\right) dk_m \qquad (3\text{-}10)$$

and

$$B(T)_{exchange} = -\frac{2^{1/2} N_A \lambda^5}{\pi^2} \int_0^\infty k_m \left[G_-(k_m) + \frac{\pi}{8}\right] \exp\left(\frac{-\lambda^2 k_m^2}{2\pi}\right) dk_m \qquad (3\text{-}11)$$

in which $G_+(k_m)$, the direct phase-shift sum, equals $\Sigma_l(2l+1)\eta_l(k_m)$, where l is the angular momentum quantum number and $\eta_l(k_m)$ is the phase shift for momentum k_m and in which $G_-(k_m)$, the exchange-phase shift sum, is equal to $\Sigma_l(-1)^l(2l+1)\eta_l(k_m)$, where λ is the thermal wavelength equal to $(\hbar/mkT)^{1/2}$ for an atom of mass m. The term $\pi/8$ is the contribution of a perfect Bose–Einstein gas. A potential for ^{4}He which describes the interaction is a Lennard-Jones potential

$$u(r) = 4\varepsilon \left[\left(\frac{\sigma}{r} \right)^{12} - \left(\frac{\sigma}{r} \right)^{6} \right] \tag{3-12}$$

in which the potential well depth ε/k and hard core diameter σ have the values given by de Boer and Michels (1939), 10.21 K and 0.255 6 nm respectively.

The magnitude of the direct and exchange phase shift sums are shown in Figs 3-3 and 3-4, where the abscissas are in units of reduced momentum, k_m ($= \sigma k_m$ where $\sigma = 0.2556$ nm). It is clear that the exchange contribution dies away very rapidly at temperatures above the liquid helium range since $G_-(K_m) \to \pi/8$ as k_m increases. Nevertheless if we are interested in the second virial coefficient at temperatures down to 2 K it cannot be ignored. The importance for thermometry of the form of the phase shift sums $G_+(k_m)$ and $G_-(k_m)$, and hence of $B(T)$, arises because of our requirement for interpolating values of $B(T)$ between those temperatures at which experimental

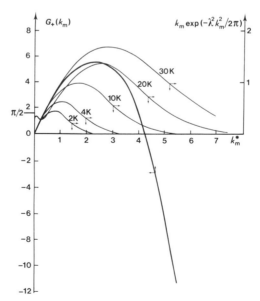

Fig. 3-3. The heavy line is direct phase-shift sum, $G_+(k_m)$ and the light lines are the temperature-dependent integrand factors in the integral of equation (3-10). The abscissa is the reduced momentum $k_m^* = \sigma k_m$ where $\sigma = 2.556 \times 10^{-10}$ m (after Colclough, 1979).

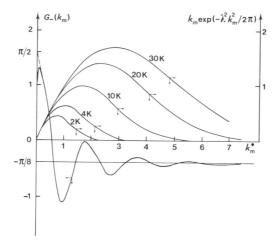

Fig. 3-4. The heavy line is the exchange phase-shift sum $G_-(k_m)$ and the light lines are the temperature dependent factors in the integral of equation (3-11). The abscissa is reduced momentum as in Fig. 3-3 (after Colclough, 1979).

measurements of $B(T)$ are made. Further, we shall find that in acoustic thermometry, the second acoustic virial coefficient depends not only upon $B(T)$ but also upon $dB(T)/dT$ and $d^2B(T)/dT^2$.

Experiments have shown that over a wide range of temperature the second virial coefficient of ^{4}He can be fitted approximately by a simple relation of the form

$$B(T) \approx a + \frac{b}{T} \tag{3-13}$$

As the precision of experimental measurements increased the deficiencies of such a simple relation became evident and a third term, $\delta(T)$, was added, often of the form (C/T^2). Recently, Colclough (1979d) has taken the expressions for $B(T)_{\text{direct}}$ and $B(T)_{\text{exchange}}$, given here as equations (3-10) and (3-11), and has used them to derive an improved interpolation function to describe $B(T)$ as a function of temperature. The range of particular interest is of course that in which quantum effects are significant, that is below about 30 K. This is just the range that has generated much interest in primary thermometry in recent years as we have already seen in Chapter 2. Colclough showed that for k_m^* between 0.5 and 6, a fit of the form

$$G_+(k_m) = a_1 k_m + a_3 k_m^3 + a_2 k_m^2 \tag{3-14}$$

is adequate. This is equivalent to

$$B(T) = a + \frac{b}{T} + \frac{c}{T^{1/2}} \qquad (3\text{-}15a)$$

or, very closely,

$$B(T) = a + b/T \qquad (3\text{-}15b)$$

as we would expect, since the size of the third term in (3-15a) is very small. To take into account the rise in k_m^* below about 0.5, (Fig. 3-3), it is necessary to add a correction term to (3-15), which amounts to about 1% of $B(T)$ for $T \lesssim 4.2$ K.

$$\Delta B(T)_{\text{direct}} = [-25.9(T/K)^{-5/2} + 2.74(T/K)^{-7/2}] \text{ cm}^3 \text{ mol}^{-1}. \qquad (3\text{-}16)$$

This is calculated on the assumption that $G_+ \to \pi/2$ as $k_m^* \to 0$, as is suggested by the fact that modern potentials imply that ^{4}He has a potential well not quite deep enough to permit a bound pair to exist. For this reason, equations (3-10) and (3-11), which do not include the bound-state terms, are valid for ^{4}He.

To take into account the exchange component of $B(T)$ for $T \geqslant 2.5$ K a further term:

$$B(T)_{\text{exchange}} = -427 \ (T/K)^{-6.45} \text{ cm}^3 \text{ mol}^{-1} \qquad (3\text{-}17)$$

is required. Thus the real second virial coefficient will have the form

$$B(T) = a + bT^{-1} - [25.9(T/K)^{-5/2} - 2.74(T/K)^{-7/2}$$
$$+ 427(T/K)^{-6.45}] \text{ cm}^3 \text{ mol}^{-1}. \qquad (3\text{-}18)$$

Note the absence of T^{-2} term, which if it were present would imply a k_m^{-1} term in $G_+(k_m)$ leading to an infinite phase shift for $l=0$ at zero momentum, which would be physically unrealistic. For acoustic thermometry, the second acoustic virial coefficient $A_1(T)$ is given in terms of $B(T)$ by (see section 3-3-1)

$$A_1(T) = \frac{2\gamma}{M}\left[B(T) + \frac{2}{3}T\frac{dB(T)}{dT} + \frac{2}{15}T^2\frac{d^2B(T)}{dT^2} \right] \qquad (3\text{-}19)$$

Substitution of (3-18) in (3-19) leads to

$$A_1 = \frac{2\gamma}{15M}\left(15a + \frac{9b}{T} \right) - [1.08 \times 10^{-2}(T/K)^{-5/2} - 1.75 \times 10^{-3}(T/K)^{-7/2}$$
$$+ 1.1(T/K)^{-6.45}] \text{ m}^2 \text{ s}^{-2} \text{ Pa}^{-1}. \qquad (3\text{-}20)$$

The use of equations (3-18) and (3-20) thus allows smoothing of measured

Table 3-2. Comparison of calculated and measured values of $B(T)$ for ^{4}He.

		$B(T)/\text{cm}^3 \text{ mol}^{-1}$		
T/K	Berry, 1979	Berry (Colclough), 1979[a]	Keesom, 1940 (corrected)[b]	Kilpatrick et al., 1954
2.6014	−142.5	−142.16	−139.4	−136.9
2.7479	−133.2	−133.25	−130.9	−128.5
3.3303	−105.8	−106.1	−104.7	−102.5
4.2201	−79.5	−79.5	−78.6	−76.5
7.1992	−39.0	−39.06	−38.2	−36.4
13.8036	−11.7	−11.81	−10.8	−9.2
20.2712	−2.4	−2.36	−1.3	0.3
20.0979	2.5	2.71	−	5.4

[a] These values result from a fit by Colclough of the data of Berry (1979), Column 2, using equation (3-18) with $a = 17.77 \text{ cm}^3 \text{ mol}^{-1}$ and $b = -407.765 \text{ cm}^3 \text{ K mol}^{-1}$. The differences are in most cases less than the experimental uncertainties in the data of Berry.
[b] These values are from the original data of Keesom corrected by Keller (1955) for errors in the original publication between 2.6 K and 4.2 K.

virial coefficients taking into account the expected theoretical form of $B(T)$.

The best experimental values of $B(T)$ in the low temperature region are probably those of Berry (1979). These are shown in Table 3-2 together with the values of Keesom and Walstra (1940), corrected by Keller (1955) for a numerical error, and the calculated values of Kilpatrick *et al.* (1954) adjusted to fit Keesom's data. Using an equation of the form (3-18), Colclough has fitted the data of Berry, column 2 of Table 3-2, with a standard error of less than

Table 3-3. The second virial coefficient of ^{4}He at high temperature.

$t/°C$	$B(T)/\text{cm}^3 \text{ mol}^{-1}$
0	12.00
25	11.89
50	11.77
75	11.67
100	11.56
125	11.46
150	11.36
300	10.76
400	10.45
500	10.14
600	9.82

0.2 cm^3 mol^{-1}. The calculated values from this fit are shown in column 3 of Table 3-2.

The second virial coefficient of ^{4}He at high temperatures is shown in Table 3-3 calculated by Guildner and Edsinger (1976), and based upon a re-evaluation of the data of Yntema and Schneider (1950) and Gammon and Doushin (1970). The virial coefficient at high temperatures depends upon the behaviour of helium molecules while undergoing energetic collisions and is quite unaffected by quantum effects.

3-2-2   The practice of gas thermometry

Having established that the virial equation of state properly describes the behaviour of helium in the ranges of temperature and density of interest for thermometry, we can now pass on to the practice of gas thermometry. There are two methods widely used, the first is known as absolute PV isotherm thermometry, and the second, somewhat confusingly, constant volume gas thermometry. In the method of absolute PV isotherm thermometry, a gas bulb of known volume V at a constant but unknown temperature T is filled with a series of increasing amounts of gas NR to obtain a series of pressures P. Assuming a value for the gas constant R, we may plot PV/NR as a function of N/V. Thus

$$\frac{PV}{NR} = T\left[1 + B(T)\left(\frac{N}{V}\right) + C(T)\left(\frac{N}{V}\right)^2 + \dots\right].$$ (3-21)

The intercept of the resulting isotherm is the temperature T.

There are a number of ways of measuring the product NR. One method is to weigh a known reference volume before and after it has been connected to the initially-evacuated gas thermometer bulb: the weight difference allows the quantity of gas entering the gas bulb to be established. This method has not been used in recent high precision gas thermometry owing to the purely practical difficulties inherent in the weighing of low density gases. Instead, another method has been used which requires a knowledge of the virial coefficients of the gas at the temperature of the reference volume. For helium at a reference temperature of T_0 (273.15 K), only the second virial coefficient need be taken into account since the total effect of the third and higher virial coefficients at one atmosphere pressure amounts to less than 1 part in 10^6. The virial equation of state for helium at this temperature may therefore be written as

$$PV = NRT_0\left(1 + B(T_0)\left(\frac{N}{V}\right)\right).$$ (3-22)

The procedure is as follows: a known reference volume V_r is filled to a pressure P_i at T_0 and then connected to the evacuated gas thermometer bulb at some temperature T. The resulting pressure in the reference volume is P_f, and we may write:

$$RN = \left[\frac{V_r}{T_0}\right]\left[\frac{P_i}{1+B(T_0)P_i/RT_0} - \frac{P_f}{1+B(T_0)P_f/RT_0}\right]$$

$$+ P_iD_i - P_fD_f \qquad (3\text{-}23)$$

where D_i and D_f are the so-called deadspace corrections arising from the small quantity of gas present in the tube connecting the reference volume to the gas thermometer bulb. We define a deadspace correction D_j by:

$$D_j = \int_0^L \frac{A_j(x)}{T_j(x)}dx \qquad (3\text{-}24)$$

where $T_j(x)$ is the temperature of the jth elemental volume $A_j(x)dx$ in the connecting tube of cross-sectional area $A_j(x)$ and total length L.

The quantity PV/NR may be calculated from

$$\frac{PV}{NR} = \left[\frac{T_0PV}{P_iV_r}\right]\left[\frac{1}{1+B(T_0)P_i/RT_0} - \frac{P_f/P_i}{1+B(T_0)P_f/RT_0}\right.$$

$$\left. + \frac{T_0}{V_r}\left(D_i - \frac{D_fP_f}{P_i}\right)\right]^{-1}. \qquad (3\text{-}25)$$

The second method of gas thermometry, constant volume gas thermometry or CVGT, is based upon the simple relation for a fixed quantity of gas:

$$\frac{P}{P_r} = \frac{T}{T_r}$$

in which P_r is the pressure at a reference temperature T_r. Taking into account the second and third virial coefficients and deadspace corrections this becomes:

$$\frac{T}{T_r} = \frac{P}{P_r}$$

$$\left[\frac{1+B(T_r)(P_r/RT_r)+(C(T_r)-B(T_r)^2)(P_r/T_rR)^2}{[1+B(T)(P/RT)+(C(T)-B(T)^2)(P/RT)^2][1+(T_r/V)(D_r-DP/P_r)]}\right]. \qquad (3\text{-}26)$$

A set of temperatures may be established by measuring the ratio of the pressures P/P_r for each temperature. Provided that the virial coefficients are known at the reference temperature and at each of the unknown temperatures, accurate values of T may be obtained.

These are the two principal methods of gas thermometry and are the methods that have been most highly developed in recent years. There are, of course, modifications that can be made to each method depending upon the temperature range being studied. For example, in carrying out PV isotherm work at low temperatures in the range 2 K to 30 K, it is cumbersome to have to determine the quantity of gas in the gas thermometer bulb always with reference to 273.16 K. Instead, once a single temperature T_r, say, has been established in the low temperature range, the quantity of gas in the bulb at neighbouring temperatures can be established by reference to the known quantity of gas at the reference temperature. For this we can write:

$$NR = \frac{P_r V}{T_r}[1 + B(T_r)(P_r/RT_r) + (C(T_r) - B(T_r)^2)(P_r/RT_r)^2]^{-1}$$
$$+ (P_r D_i - PD_f) \tag{3-27}$$

and

$$\frac{PV}{NR} = T_r \frac{P}{P_r}\Big[[1 + B(T_r)(P_r/RT_r) + (C(T_r) - B(T_r)^2)(P/RT_2)^2]^{-1}$$
$$+ \frac{T_r}{V}\Big(D_i - \frac{D_f P}{P_r}\Big)\Big]^{-1}. \tag{3-28}$$

This requires a knowledge of both $B(T)$ and $C(T)$ at the reference temperature. At high temperatures, for example above room temperature, the third virial coefficient of helium can always be ignored and the method then becomes simpler.

3-2-3   The problem of sorption

Having outlined the basic principles of gas thermometry, we must now turn to the factors which lead to errors. So far we have demanded only that the values of the virial coefficients be known, either at T_0 or T_r for absolute isotherm thermometry and at T for CVGT. As we have seen in Section 3-2-1, virial coefficients are well-behaved quantities, and are not usually subject to experimental determination as part of a thermometry experiment. The uncertainty in T resulting from an uncertainty in $B(T)$ and $C(T)$ is generally one of the small but systematic uncertainties in the measurement. One of the most important sources of error in gas thermometry, and undoubtedly the most important at high temperatures, is sorption of the thermometric gas and other gases at the walls of the gas thermometer bulb. In the gas thermometer equations we have considered so far, it has been an implied condition that the quantity of gas present in the gas thermometer bulb remains unchanged during the various excursions of the temperature and pressure from the reference state. This condition can never be wholly satisfied since there remain always gas molecules adsorbed on the surface of the bulb and gas molecules absorbed in the body of the walls of the bulb. The equilibrium between the rate of adsorption and desorption is a function of both pressure and temperature. Sorption, therefore, directly affects the quantities that we are attempting to measure. The nature of sorption is such that it is very difficult to correct for its effects even when it is known to be present.

Sorption phenomena (Johnson and Messmer, 1974; Redhead et al., 1968) arise through the attractive forces acting between gas molecules and the surface atoms of the adjacent solid. We distinguish between two types of adsorption, physical adsorption and chemical adsorption. In the former the restraining forces are only the relatively weak Van-de-Waals type of intermolecular forces, while in the latter, exchange of electrons takes place and strong chemical bonds form between the adsorbed species and the solid. It often occurs that an initially physisorbed atom is transformed into a chemisorbed atom if the temperature is raised sufficiently to provide the required activation energy for the chemisorption process to take place.

The equilibrium between the adsorption and desorption of molecules from a surface can be represented by an equation (Roth, 1976) giving the fraction of the number of molecules required to form a monolayer $N_0\theta$. N_0 is the total number of molecules required to form a monolayer and θ is the fraction of the possible adsorption sites occupied.

$$N_0\theta = C[P/(MT)^{1/2}]ft' \exp[E_D/(RT_S)](1-\theta), \qquad (3\text{-}29)$$

where M is the molecular weight of the molecule, f the sticking coefficient, t' the period of oscillation of the molecule normal to the surface ($\approx 10^{-13}$ s), E_D the energy of desorption, T_S the temperature of the surface and T the temperature of the gas. The constant, C, has a value of about 3.5×10^{22}. From an equation such as (3-29), adsorption isotherms may be obtained by plotting $N_0\theta$ as a function of P for constant T; or adsorption isobars by plotting $N_0\theta$ as a function of T for constant P. Equation (3-29) serves to show the complexity of the problem of sorption in gas thermometry when both P and T are varied. In addition, it must be remembered that the value of N_0 is a function of the real surface area and not the apparent geometrical area. It is well known (Roth, 1976) that the real surface area, in contrast to the geometrical surface area, depends very much upon the previous treatment of the surface. A mechanically polished surface of nickel, for example, can have a real surface area between 10 and 75 times that of the apparent geometrical surface. For unpolished copper the ratio can be at least 14 (Schram, 1963).

In gas thermometry, the effects of sorption first show themselves in the irreproducibility of the initial pressure at the reference temperature after excursions of the gas thermometer to the temperatures to be measured. Unfortunately, the magnitude of the irreproducibility at the reference temperature is only a very poor guide to the likely error at the temperatures being measured. The form of equation (3-29) shows this. Further complications in high-temperature gas thermometry arise through the possibility, for non-noble gases, of transforming physisorption to chemisorption. In the case of helium, however, the conditions are favourable in respect of the effects of sorption at both high and low temperatures. Helium being a noble gas, chemisorption does not take place; and since it has the lowest desorption energy among the noble gases, the extent of physisorption is minimized. Nevertheless, there remains the problem of outgassing the material of the walls of the bulb and the connecting tubes to remove unwanted gas species. The difficulty that always arises stems from the conflicting requirements of a high pumping speed, for rapid outgassing, and a small connecting tube to minimize the deadspace corrections. The only practical solution lies in careful surface preparation and the cleaning and outgassing of all components before assembly, followed by prolonged pumping of the assembled gas-thermometer bulb at as high a temperature as possible. A typical pumping speed achieved in a gas thermometer bulb is of the order of $10^{-4}\,l\,s^{-1}$ at room temperature. This leads to an outgassing period of several months being required for adequate reduction of the surface-outgassing rate. An experimental check on the effects of sorption can be made if it is possible to increase the internal surface area of the gas-thermometer bulb. If the gas thermometer bulb can be opened, then a series of discs can be introduced having the same surface preparation and finish as the walls of the bulb itself.

To illustrate this and other sources of error, we shall consider briefly two recent examples of gas thermometry. The first was carried out at low temperatures (Berry, 1979) and led to the gas thermometer scale NPL-75 over the range 2.6 K to 27.1 K, and the second was carried out at high temperatures (Guildner and Edsinger, 1976) over the range 0 °C to 460 °C. An outline of the gas thermometer system used to derive the NPL-75 Scale is shown in Figs 3-5 and 3-6 and that used by Guildner for high-temperature gas thermometry is shown in Fig. 3-7. The different requirements for operating at very low and at high temperatures are immediately obvious by comparison of the three figures. At low temperatures the gas bulb is large (about 1 litre), is of substantial wall thickness and is mounted in an evacuated chamber. Rh/Fe resistance thermometers are attached directly to the outside of the bulb. Temperature control is achieved by means of a heater on the radiation shield, using a germanium resistance thermometer as sensor. The thermal diffusivity of the oxygen-free high-conductivity (OFHC) copper is such that at low

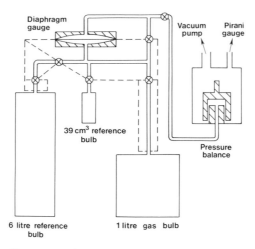

Fig. 3-5. The gas-handling system of the NPL-75 gas thermometer (after Berry, 1979).

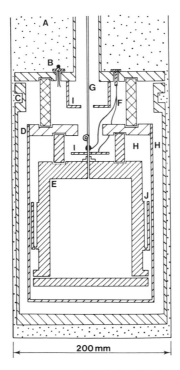

Fig. 3-6. The NPL-75 gas thermometer cryostat. A, liquid helium reservoir; B, Stycast seal for leads; C, stainless-steel vacuum jacket; D, copper isothermal shield; E, copper gas-thermometer bulb; F, heat links to A; G, 1 mm bore stainless-steel tube; H, vacuum space; I, radiation shields; and J, resistance-thermometer wells (after Berry, 1979).

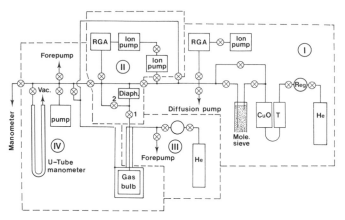

Fig. 3-7. Schematic of NBS high temperature gas thermometer showing the gas handling system. The broken lines show boundaries of parts that can be isolated for pumping (after Guildner and Edsinger, 1976).

temperatures, the presence of significant temperature gradients can safely be discounted. The high-temperature gas-thermometer bulb, on the other hand, is quite different. The gas bulb is made from Pt/20 % Rh alloy to allow a clean chemically-inert surface to be maintained at high temperatures. The walls are only about 1 mm thick and constancy of volume in the presence of pressure changes is ensured by maintaining an equal pressure in an annular space around the gas bulb. This annular space between the bulb and an inconel shield is filled with helium. Temperature control is much more difficult at high temperatures. The whole of the gas bulb and inconel assembly is immersed in a stirred liquid bath having outside temperature control. Figure 3-7 shows the complicated gas handling system required in a high temperature gas thermometer in order to ensure purity of the thermometric gas. At low temperatures, in contrast, practically all of the gaseous impurities condense out and have a negligible vapour pressure. Furthermore, outgassing of the walls of the low-temperature gas bulb is negligible. The only way in which sorption can be troublesome is through physisorption of helium onto the walls of the bulb. Experiments by Berry (1979) in which he increased the surface area of his gas bulb by a factor of about five showed that at a temperature of 2.6 K, the effect on CVGT measurements was no greater than 0.1 mK. For accounts of other modern gas thermometers see Sakurai (1982) and Kemp *et al.* (1982).

3-2-4 The deadspace, aerostatic head and thermomolecular pressure corrections

The principal corrections that must be made to the measured quantities P and V are the following: the deadspace correction, the aerostatic-head correction

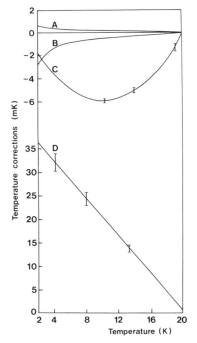

Fig. 3-8. Corrections to gas thermometry due to: A, hydrostatic head; B, thermomolecular pressures; C, dead spaces; D, departures from ideal behaviour in the gas. The uncertainty in any of the corrections is indicated by the error bars (after Berry, 1979).

arising from the difference in the density of the gas at different levels in the pressure-sensing tube and the thermomolecular-pressure correction. The last of these is the result of thermal transpiration along the pressure-sensing tube and is a function of the pressure, the temperature difference between the two ends and the surface condition of the walls. The relative magnitude of these three corrections in the case of the low-temperature gas thermometer of Berry is shown in Fig. 3-8. In a high-temperature gas thermometer, the most important correction is that for the deadspace. This is because the deadspace correction, calculated from equation (3-24), leads to those sections of the pressure sensing tube in which the gas is densest having the most influence. In high-temperature gas thermometry, these are the parts of the pressure-sensing tube that are at room temperature. Very great care, therefore, must be taken in ensuring that the temperature of the connecting tube at room temperature remains constant. In addition, it is necessary to ensure that all the valves open and close without introducing significant volume changes. The measurement of the temperature and volume of the connecting tube and valves to the required accuracy calls for very elaborate methods and is one of the limiting sources of uncertainty in high temperature gas thermometry. In low temperature gas thermometry, the densest gas in the pressure-sensing tube is that adjacent to the gas bulb. Although the determination of the deadspace correction at low temperatures requires care, it is not nearly as difficult as in high-temperature gas thermometry. A direct method (Berry, 1979) of determining the deadspace correction is to detach the gas thermometer bulb from the connecting tube, close the end of the connecting tube, and directly measure the quantity of helium in the whole of the deadspace by expansion from a small known room-temperature volume.

The aerostatic-head correction, arising as it does from the difference in density of the gas in different parts of the pressure-sensing tube, is of minor significance in high-temperature gas thermometry. This is not the case in low-temperature gas thermometry since the ratios of density between gas at room temperature and gas below 10 K become very large. The aerostatic-head correction is very closely related to the deadspace correction and is given by:

$$\Delta P = \int_0^h \rho(h)g\,\mathrm{d}h \tag{3-30}$$

$$\Delta P = \frac{MPg}{R} \int_0^h \frac{\mathrm{d}h}{T(h)}. \tag{3-31}$$

Now

$$\int_0^h \frac{\mathrm{d}h}{T(h)} \cong \int_0^L \frac{\mathrm{d}x}{T(x)},$$

as in the deadspace correction, provided that the pressure-sensing tube is vertical. If, as is usually the case (see Fig. 3-6), the pressure-sensing tube is looped to avoid direct thermal radiation entering the tube, the two integrals differ. The use of a pressure-sensing tube which is horizontal within the temperature gradient would reduce the aerostatic-head correction practically to zero. A further difference between the aerostatic-head correction and the deadspace correction is that the former must, in principle, be applied within the gas thermometer bulb itself. In practice, this only becomes necessary at very low temperatures, below 4 K, where $\Delta P/P$ can amount to as much as 2×10^{-4} within the bulb.

The thermomolecular pressure correction is important in both low temperature and high temperature gas thermometry. A thermomolecular pressure difference will occur between two containers of gas at different temperatures if the orifice connecting them has a diameter comparable to, or smaller than, the mean free path of the gas molecules. At equilibrium, the number of molecules passing from the hot container to the cold container must equal the number passing in the opposite direction. In the limit of low pressures this condition becomes simply, for an orifice or specularly reflecting tube,

$$\frac{P_1}{\sqrt{T_1}} = \frac{P_2}{\sqrt{T_2}} \tag{3-32}$$

where P_1 and P_2, T_1 and T_2 represent the equilibrium pressures and temperatures, respectively, in the two containers.

In the practical situations encountered in gas thermometry, rarely is the mean free path large compared with the diameter of the connecting orifice (which in the practical case is a tube), and thus the simple condition of equation (3-32) does not apply. Instead, a very much more complicated expression is required which must take into account the diameter of the tube, the momentum accommodation coefficient for molecular collisions at the wall of the tube, the molecular weight of the gas and its viscosity. A general expression for the resulting pressure difference was first given by Weber and Schmidt (1936). Many investigations, both theoretical and experimental (Roberts and Sydoriak, 1956; McGonville, 1972), have since been carried out, and it is now clear that the thermomolecular pressure difference is critically dependent upon the state of the surface of the tube. For this reason, it is considered essential to measure the thermomolecular-pressure correction for the particular tube being used rather than to rely upon a calculation using the Weber–Schmidt equation. The measurement may usually be carried out without undue difficulty, since all that is required is a second tube of much larger diameter placed alongside the pressure-sensing tube just for the purposes of this measurement. Figure 3-9 shows the differences obtained between the thermomolecular pressure difference, ΔP, calculated from the Weber–Schmidt equation and that measured for a stainless steel tube in a temperature gradient between 4.2 K and 293 K. The differences amount to about 10 % of ΔP. Thermomolecular pressure differences at high temperatures have been measured by Guildner and Edsinger (1976), and fitted over a range of pressures from 2.5×10^3 Pa to 1×10^5 Pa and at temperatures from 140 °C to 457 °C using an equation of the form:

$$\Delta P = \frac{C}{\bar{P}d^2} \left[\left(\frac{T_2}{T_0} \right)^{2n+2} - \left(\frac{T_1}{T_0} \right)^{2n+2} \right] \tag{3-33}$$

where $\bar{P}$ is the average pressure in the connecting tube, T_1 and T_2 the lower and upper temperatures respectively, d is the diameter of the connecting tube and C is a constant. The reference temperature T_0 enters only because n is defined by

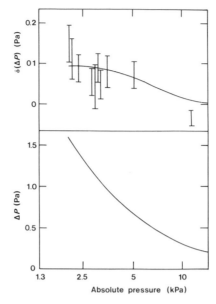

Fig. 3-9. The thermomolecular pressure difference ΔP, calculated from the Weber–Schmidt equation for the NPL-75 gas-thermometer tube in a temperature gradient between 3.2 K and 293 K. $\delta(\Delta P)$ is the difference found by Berry (1979) between this calculated value and his experimentally determined one (after Berry, 1979).

the assumption that the viscosity of helium η is given by:

$$\frac{\eta}{\eta_0} = \left(\frac{T}{T_0}\right)^{n+1/2}.$$

Experimental measurements for a stainless steel tube gave a value for $n = 0.145$ and $C = 2\,532$ Pa^2 mm^2.

3-2-5 Pressure measurement

The measurement of pressure is central to the practice of gas thermometry. However, absolute measurements to high accuracy are not required, instead, we need to be able to make very accurate measurements of pressure ratios. The best absolute measurements of pressure continue to be made using a mercury manometer (Kaneda *et al.*, 1964; Bonhoure and Terrien, 1968; Guildner *et al.*, 1970) and the ultimate accuracy is limited to about 1 part in 10^6 by the uncertainty in the density of mercury (Peggs *et al.*, 1979). In principle, pressure ratios could be measured very much better than this, but in practice other uncertainties intervene and an uncertainty of about 5 parts in 10^7 is the best that can be achieved. For gas thermometry the important components in the overall uncertainty of pressure measurement are the pressure-independent ones since these limit the lowest pressures that can usefully be measured. In CVGT, it is almost always advantageous to work at a low pressure to bring the influence of uncertainties in the virial coefficients down to the level of those stemming from the thermomolecular pressure and other effects. Similarly, in PV isotherm thermometry, the lower the pressure to which the isotherm extends, the smaller are the uncertainties in the extrapolation to zero pressure. The lowest pressure attainable by Guildner and Edsinger (1976) while still maintaining an uncertainty of 2 parts in 10^6, was 3200 Pa. Taking this as the reference pressure at 273.16 K, the total correction arising from the second virial coefficient at 373 K amounted to only 0.5 mK. At such low pressures, the thermomolecular pressure correction becomes as important as that due to $B(T)$.

In the low-temperature gas thermometry of Berry a gas-operated-piston gauge was used for measurement of the pressure ratios. Although such a gauge is not as accurate as a mercury manometer for absolute pressure measurements due to uncertainties in its effective area, it is a very satisfactory device for pressure–ratio measurement (Peggs, 1977). The effective area is only a slowly-varying function of pressure, and pressure ratios can be measured without difficulty to an uncertainty of some 5 parts in 10^6 in the range 2 kPa to 100 kPa.

3-2-6   Conclusions

In spite of all its problems and difficulties, gas thermometry remains the method of thermometry which leads to the most accurate measurements of thermodynamic temperature from 2.6 K up to at least 700 K. In this section we have looked in some detail at the principles of gas thermometry, and rather briefly at some recent practical applications which illustrate these principles. There seems at present very little likelihood that gas thermometry will be displaced as the most important technique of primary thermometry over this temperature range. Indeed, recent developments concerning the possibility of measuring the pressure directly inside the gas bulb hold out the possibility of an extension in the range of applicability of gas thermometry. It appears practicable (Van Degrift et al., 1978) to use a tunnel diode to measure minute deformations of the wall of the gas bulb, thereby transforming pressure changes in the bulb to changes in frequency. Although still at the early development stage, such a device could have very wide application in gas thermometry, although it has yet to be demonstrated that such a device could be used as a primary thermometer.

3-3 Acoustic thermometry

3-3-1 The propagation of sound in gases

We remarked at the beginning of the previous section that PV isotherm thermometry or constant volume gas thermometry based upon Boyle's Law requires a knowledge of the number of moles of gas present in the gas bulb or the value of the second and perhaps third virial coefficients. It has long been realized that considerable advantages might accrue if a method of gas thermometry could be developed that was based upon the variation with temperature of an intensive property of a gas. Such a property might be the speed of sound, the refractive index or the dielectric constant. From Chapter 1 we know that for a method to be considered primary we must be able to write down a relation between the measured quantity and thermodynamic temperature that includes only T_0, R or k together with other constants that can be determined without a prior knowledge of thermodynamic temperature. Acoustic thermometry is the most highly developed of these three methods, and we shall discuss it first.

Sound waves are propagated in a gas by adiabatic compression and rarefaction producing longitudinal waves having a speed, c, given by:

$$c^2 = \frac{B_s}{\rho} \tag{3-34}$$

where B_s is the adiabatic bulk modulus of the gas and ρ is its density. It is worth remarking that the transmission of sound in a gas is an adiabatic rather than an isothermal process because the frequency is always much too *low* for it to be isothermal. This is contrary to what is often asserted in textbooks, but arises because the time needed for temperature equilibrium to occur is proportional to the square of the wavelength, whereas the time available for this to occur is proportional to the wavelength. Thus temperature equilibrium occurs only if the wavelength is less than some critical length, which for a gas at room temperature and atmospheric pressure turns out to be of the order of 100 nm (Tabor, 1979). This implies that the frequency must be greater than 10^9 Hz if isothermal transmission is to occur. This can never happen since at such frequencies, the wavelength would be less than the mean free path and so acoustic propagation would not occur in the same way.

Returning to equation (3-34), since

$$\rho = \frac{NM}{V} = \frac{MP}{RT} \tag{3-35}$$

and

$$B_s = -V \left(\frac{\delta P}{\delta V} \right)_s = \gamma P \tag{3-36}$$

we can write for a perfect gas

$$c_0^2 = -\frac{VRT}{MP}\left(\frac{\delta P}{\delta V}\right)_s \tag{3-37}$$

$$c_0^2 = \gamma\frac{RT}{M} \tag{3-38}$$

where c_0 is the speed of sound in an unbounded ideal gas.

Leaving aside departures from ideal behaviour, equation (3-37) shows that, in principle, acoustic thermometry could be used as a method of primary thermometry. A measurement of c_0 together with a knowledge of γ, R and M would lead directly to the thermodynamic temperature. In order to introduce the effects of non-ideal behaviour we may express equation (3-38) as a virial expansion in pressure:

$$c^2 = A_0(T) + A_1(T)P + A_2(T)P^2 + \dots \tag{3-39}$$

where c is the speed of sound in a real unbounded gas and $A_0(T)$, $A_1(T)$ and $A_2(T)$ are known as the zeroth, first and second acoustic virial coefficients, respectively. Although a similar expansion to that of equation (3-39) could be written in terms of density, such an expansion is not used for acoustic thermometry since pressure is so much easier to measure than is density or molar volume. This is in contrast to PV isotherm gas thermometry in which the density expansion is used since N, the number of moles of gas present, is required in any case to evaluate the density independent term in the virial expansion. This highlights a further important difference between the practice of PV isotherm thermometry and acoustic thermometry. To obtain a value for N, it is usual, as we have seen, to assume a value for the second virial coefficient at a reference temperature, T_0 say, rather than to carry out a direct determination of N by weighing. Acoustic thermometry does not require a measurement of c_0 at T_0, and in this respect is more strictly primary thermometry. Indeed, a measurement of c_0 at T_0 generates a value for the gas constant provided that γ and M are known. Such a measurement has recently been made, and resulted in a value of R in good agreement with the results of previous methods, all of which were based upon weighing (Colclough et al., 1979).

It remains to express the acoustic virial coefficients in terms of the pressure/volume virial coefficients $B(T)$ and $C(T)$ and γ, R and T.

From equations (3-38) and (3-39) it is immediately clear that

$$A_0(T) = \frac{\gamma RT}{M}. \tag{3-40}$$

Starting from equation (3-37) and using standard thermodynamic relations, it can be shown that for a monatomic gas (Quinn, 1969),

$$A_1(T) = \frac{2\gamma}{M}\left[B(T) + \frac{2}{3}T\frac{\mathrm{d}B(T)}{\mathrm{d}T} + \frac{2}{15}T^2\frac{\mathrm{d}^2B(T)}{\mathrm{d}T^2}\right] \tag{3-41}$$

and

$$A_2(T) = \frac{\gamma}{RTM}\left[\frac{13}{5}C(T) + \frac{16}{15}T\frac{\mathrm{d}C(T)}{\mathrm{d}T} + \frac{2}{15}T^2\frac{\mathrm{d}^2C(T)}{\mathrm{d}T^2} - \frac{8}{5}B(T)^2\right.$$

$$+\frac{98}{45}T^2\left(\frac{\mathrm{d}B(T)}{\mathrm{d}T}\right)^2 + \frac{8}{45}T^4\left(\frac{\mathrm{d}^2B(T)}{\mathrm{d}T^2}\right)^2 + \frac{8}{15}TB(T)\frac{\mathrm{d}B(T)}{\mathrm{d}T}$$

$$\left.+\frac{4}{15}T^2B(T)\frac{\mathrm{d}^2B(T)}{\mathrm{d}T^2} + \frac{56}{45}T^3\frac{\mathrm{d}B(T)}{\mathrm{d}T}\cdot\frac{\mathrm{d}^2B(T)}{\mathrm{d}T^2}\right]. \tag{3-42}$$

Fortunately, the departure from ideal behaviour of our thermometric gas, helium, is sufficiently small for $A_2(T)$ to be small except at very low temperatures. In consequence, acoustic isotherms are usually nearly linear and have a slope which is a function mostly of $B(T)$ and its first derivative. We have already seen in Section 3-2-1 that $B(T)$ can be expressed in terms of a polynomial in T, equation (3-18), and that $A_1(T)$ can therefore also be expressed in terms of a polynomial, equation (3-20).

Although $A_2(T)$ is small, care must be taken in interpreting acoustic isotherms. The small curvature at high pressures which results from the $A_2(T)$ term can, either by itself or combined with a curvature in the opposite sense at low pressures due to boundary-layer effects, lead to an apparently linear isotherm. Uncorrected, such an isotherm would have both the wrong slope and the wrong intercept (see Section 3-3-3).

3-3-2 Theory of acoustic interferometry

Up to now we have not mentioned the way in which the speed of sound may be measured. Although we have treated departures from ideal behaviour, we still refer to the speed, c_0, in an unbound gas. Practical considerations, particularly at low temperatures, demand that the speed be measured in a relatively small closed container so that the temperature can be made uniform and can be well controlled. The most accurate measurements of the speed of sound have, up to now, always been made by the method of acoustic interferometry and using a cylindrical resonator. Acoustic waves are propagated in a cylindrical tube by a transducer at one end, and the wavelength is measured in terms of the displacement of a reflecting piston between successive positions of resonance. The presence of standing waves is observed by monitoring the impedance of the transducer. Herein lies the difficulty of acoustic thermometry compared with gas thermometry. In gas thermometry, the measured quantities, pressure and volume, are straightforward static quantities, despite the problems of sorption which we have already mentioned. In acoustic thermometry we are dealing with dynamical processes involving such quantities as the acoustic impedance of a transducer at say 5 kHz, the viscosity and the heat conduction at the walls of the container. These present their own special problems of measurement, and a thorough understanding of the physics of acoustic-wave propagation is required if the results of measurements are to be correctly interpreted.

Space does not allow a full treatment of the theory of acoustic interferometry. Instead a brief outline of the principles will be given illustrating the origin of the main sources of uncertainty. For a full treatment of the subject, the interested reader is referred to the series of papers by Colclough (1970, 1973, 1976, 1979a, b and c) from which the following summary is largely taken. It is interesting to note that the full complexity of acoustic interferometry became apparent only after the method had been taken up with the aim of developing an alternative to gas thermometry less subject to systematic error. It has taken some ten years for the subject to become fully understood despite the fact that all of the principles involved were clearly laid out by Rayleigh in 1877 in "The theory of sound" (see Rayleigh, 1945).

For the purposes of this discussion we shall consider a cylindrical acoustic interferometer having a cross sectional area A, containing a gaseous medium of density ρ in which the speed of sound is c, the acoustic absorption coefficient is α, the wavelength is λ, the wavenumber $k = 2\pi/\lambda$ and R_r and R_t are the reflection coefficients, which may be complex, of the reflector and transducer, respectively. The sum of the mechanical impedance of the transducer, Z_t and of the gas, $Z_L(l)$, is the total impedance $Z(l)$, where l is the length of the cavity, since both transducer and gas load share the same particle velocity at the transducer.

For waves propagating in the tube due to a transducer at $Z = l$, having a displacement $\zeta(l)$ given by

$$\zeta(l) = \zeta(l)e^{j\omega t}$$

we find, by summing the positive and negative going waves:

$$Z_L(l) = A\rho c \left(1 - i\frac{\alpha}{k}\right) R_t^{1/2} \frac{1 + R_r e^{-2\alpha l} e^{-2ikl}}{1 - R_r R_t e^{-2\alpha l} e^{-2ikl}} \tag{3-43}$$

$$= \left(1 - i\frac{\alpha}{k}\right)(R(l) - iX(l)) \tag{3-44}$$

where

$$R(l) = A\rho c R_t^{1/2} \left\{ \frac{1 - R_r^2 R_t e^{-4\alpha l} + R_r(1 - R_t)e^{-2\alpha l} \cos 2kl}{1 + R_r^2 R_t^2 e^{-4\alpha l} - 2R_r R_t e^{-2\alpha l} \cos 2kl} \right\} \tag{3-45}$$

and

$$X(l) = A\rho c R_t^{1/2} \left\{ \frac{R_r(1 + R_t)e^{-2\alpha l} \sin 2kl}{1 + R_r^2 R_t^2 e^{-4\alpha l} - 2R_r R_t e^{-2\alpha l} \cos 2kl} \right\} \tag{3-46}$$

Since $\alpha \leqslant k$, which is a necessary condition for sensitive interferometry stemming from the requirement that the attenuation of the acoustic waves be small over a distance of the order of λ, the factor $(1 - i\alpha/k)$ may be set equal to unity. However since R_t and R_r may be complex of non-negligible magnitude, it is not quite true to say that $R(l)$ and $-X(l)$ are simply the real and imaginary parts of $Z_L(l)$. But since the phase change on reflection is, nevertheless, small we may write:

$$R_t = R_r = R_{r,t} \cong 1 - \beta_{r,t} \cong 1 - \beta. \tag{3-47}$$

For fixed-frequency variable-path interferometry, as opposed to fixed-path variable-frequency interferometry, we may consider that $R_{r,t}$ is real. We may further assume that the second and higher order terms in αl, $\beta_{r,t}$ and, close to the Nth resonance, $2k(l - l_N)$ are negligible, where $l_N = N\lambda/2$.

For convenience we shall write:

$$\Delta N = (\alpha l_N + \beta)/k \tag{3-48}$$

and

$$D_N = \frac{A\rho c}{(\alpha l_N + \beta)}. \tag{3-49}$$

Thus

$$Z(l) = Z_t + R(l) - iX(l) \tag{3-50}$$

where

$$R(l) = \frac{D_N}{1 + [(l_N - l)/\Delta_N]^2} \tag{3-51}$$

and

$$X(l) = \frac{D_N[(l_N - l)/\Delta_N]}{1 + [(l_N - l)/\Delta_N]^2} \tag{3-52}$$

and for the phase φ_N of $Z_L(l)$

$$\tan \varphi_N = -\frac{X(l)}{R(l)} = \frac{l_N - l}{\Delta N}. \tag{3-53}$$

The form of $R(l)$ and $-X(l)$, the real and imaginary parts of the transducer impedance $Z_L(l)$ is easy to find from inspection. D_N and Δ_N are seen to be the maximum and half-width at half height respectively of $R(l)$, and the locus of $Z_L(l)$ is a circle of diameter D_N centred at $Z_t + D_N/2$ in the complex plane since

$$[R(l) - D_N/2]^2 + X(l)^2 = (D_N/2)^2. \tag{3-54}$$

Near anti-resonance, however, this equation may be incorrect since $k(l_N - l)$ is no longer small.

However, $X(l) = 0$ and

$$R(l) = A\rho c(\alpha l + \beta/2) \tag{3-55}$$

which is of second-order smallness compared to D_N. Thus to first order the locus of $Z_L(l)$ is indeed a circle as shown in Fig. 3-10. From the impedance circle of this figure, the values of l_N for which $\varphi_N = 0$ may easily be identified, either

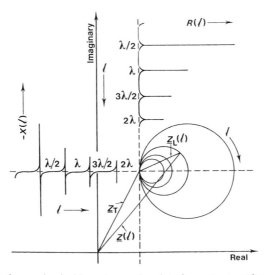

Fig. 3-10. The complex mechanical impedance of an interferometer transducer under load.

directly or indirectly, from a plot of $\tan \varphi_N$ versus l. This yields λ, equal to $2(l_M - l_1)/(M - 1)$, where M is the order of the last resonance, so that:

$$c = \frac{2f(l_M - l_1)}{(M - 1)}. \tag{3-56}$$

Δ_N is given by half of the change in l necessary to traverse the circle from $\varphi_N = +\pi/4$ to $\varphi_N = -\pi/4$. The absorption coefficient α may be calculated from a plot of $k\Delta_N$ against l_N, since from equation (3-48) such a plot has a slope α and intercept β. A rather better use of the data is, however, obtained by plotting D_N^{-1} against l_N to obtain α/β. This may then be used with each measured value of Δ_N to calculate values of α, the mean of which can be taken for a final value.

A more complex analysis taking into account second order terms in αl, β and $2k(l_N - l)$ leads to modifications in equation (3-53) which becomes:

$$\tan \varphi_N'' = \frac{l_N'' - l}{\Delta_N''} \tag{3-57}$$

where $l_N'' \cong l_N - (\alpha/k)\Delta_N$ (Quinn et al., 1976).

3-3-3   The boundary layer correction

The values of c so far obtained refer to the speed of sound in the interferometer tube. To deduce the speed in an unbounded medium we must develop the theory of the acoustic boundary layer. The speed of a plane wave (and we shall consider for the moment only plane waves) in a tube is lower than that in an unbounded medium as a result of viscosity and heat conduction at the walls. The consequent additional absorption coefficient is known as the Kirchhoff–Helmholtz absorption coefficient, α_{K-H}. The total measured absorption coefficient α, is thus the sum of α_{K-H} and the classical absorption coefficient α_0 arising from viscosity and heat conduction in the body of the gas.

$$\alpha_{K-H} = \frac{1}{bc}\left[v^{1/2} + (\gamma - 1)\left(\frac{K}{\rho C_p}\right)^{1/2} \right]\left[\frac{\omega}{2}\right]^{1/2} \tag{3-58}$$

and

$$\alpha_0 = \frac{1}{2c^3}\left\{ \frac{4}{3}v + (\gamma - 1)\frac{K}{\rho C_p} \right\}\omega^2 \tag{3-59}$$

where b is the radius of the interferometer tube, $\omega = 2\pi f$, v the kinematic

Table 3-4. Acoustic losses for ^{4}He at 4.2 K and 273 K and for Ar at 273 K.

Frequency (Hz)	β	$b\alpha_{K\text{-}H}$	α_0 (cm^{-1})
^{4}He at 4.2 K			
10^2	1.4×10^{-4}	1.2×10^{-4}	3.9×10^{-10}
10^3	4.5×10^{-4}	3.8×10^{-4}	3.9×10^{-8}
10^4	1.4×10^{-3}	1.2×10^{-3}	3.9×10^{-6}
10^5	4.5×10^{-3}	3.8×10^{-3}	3.9×10^{-4}

from $Cp = 8002$ J kg^{-1} K^{-1}, $\gamma = 2.453$, $c = 108.2$ ms^{-1}, $\eta = 1.3 \times 10^{-6}$ Pa s, $K = 6.01$ W m^{-1} K^{-1}, $\rho = 14.5$ kg m^{-3}.

Frequency (Hz)	β	$b\alpha_{K\text{-}H}$	α_0 (cm^{-1})
^{4}He at 273 K			
10^2	3×10^{-4}	3.4×10^{-4}	5.2×10^{-10}
10^3	9.6×10^{-4}	1.1×10^{-3}	5.2×10^{-8}
10^4	3×10^{-3}	3.4×10^{-3}	5.2×10^{-6}
10^5	9.6×10^{-3}	1.1×10^{-2}	5.2×10^{-4}

from $Cp = 5193$ J kg^{-1} K^{-1}, $\gamma = 1.667$, $c = 972.6$ m s^{-1}, $\eta = 1.87 \times 10^{-5}$ Pa s, $K = 0.145$ W m^{-1} K^{-1}, $\rho = 0.1787$ kg m^{-3}.

Frequency (Hz)	β	$b\alpha_{K\text{-}H}$	α_0 (cm^{-1})
Ar at 273 K			
10^2	3.2×10^{-4}	3.6×10^{-4}	1.9×10^{-9}
10^3	1.0×10^{-3}	1.1×10^{-3}	1.9×10^{-7}
10^4	3.2×10^{-3}	3.6×10^{-3}	1.9×10^{-5}
10^5	1.0×10^{-2}	1.1×10^{-2}	1.9×10^{3}

from $Cp = 521.8$ J kg^{-1} K^{-1}, $\gamma = 1.6706$, $c = 307.8$ m s^{-1}, $\eta = 2.1$ Pa s, $K = 1.645 \times 10^{-2}$ W m^{-1} K^{-1}, $\rho = 1.784$ kg m^{-3}.

viscosity, K the thermal conductivity and C_p the specific heat at constant pressure. Since α_0 depends upon ω^2, we expect it to become dominant at high frequencies. At audio-frequencies in tubes a few centimetres in diameter, the dominant absorption coefficient is $\alpha_{K\text{-}H}$ with α_0 being very small. Although the form of $\alpha_{K\text{-}H}$, equation (3-58), is well established (Colclough, 1973) its calculation requires a knowledge of the transport properties of the gas. Since these are rarely known to better than 1 %, it is better to measure α as described above. As examples of typical values of $\alpha_{K\text{-}H}$ and α_0 encountered in practical measurements, Table 3-4 lists $\alpha_{K\text{-}H}$ and α_0 over a range of frequencies at 4.2 K and 273 K for a typical interferometer. It may be seen from the table that for frequencies below 10 kHz, α_0 is almost always negligible compared with $\alpha_{K\text{-}H}$.

At sufficiently high frequencies, the acoustic wavelength becomes very small and begins to approach the mean free path of the molecules. In these circumstances, neither the basic equation for c, equation (3-36) nor the equations for $\alpha_{K\text{-}H}$ and α_0 properly describe the real situation since they presuppose that the gas is a continuous fluid. Simple kinetic theory tells us that the mean velocity of the molecules in a gas is of the same order as the speed of sound. Thus, as the wavelength of the sound approaches the mean free path, the sound frequency must approach the mean collision frequency between pairs of molecules. This is an extremely high frequency, $\approx 10^9$ Hz, since the mean free path at room temperature is of the order of 100 nm. Such high frequencies are never encountered in acoustic thermometry, the highest used up to now are about 1 MHz, where such effects are still entirely negligible. The consequences of the boundary-layer thickness at the walls approaching the mean free path can also be ignored. The boundary layer within which the viscosity and heat conduction is assumed to influence the acoustic propagation has a thickness given by

$$d' = \left(\frac{2\eta}{\rho\omega}\right)^{1/2} \tag{3-60}$$

where η is the viscosity and ρ the density. At frequencies of 1 MHz, d' can be as small as 1 μm at 20 K in helium, but the mean free path is still smaller, of the order of 0.01 μm. However in these circumstances, the thickness of the boundary layer may be comparable to surface irregularities, and we must treat such calculated corrections with circumspection.

The correction to be made to the speed due to the boundary layer was originally calculated for plane waves by Kirchhoff in 1868. It was only much more recently, however, that Fritche (1960a, b) developed a more general treatment applicable to non-plane waves (see the following Section for a discussion of non-plane waves). For plane waves the correction is given by

$$c = c_0\left(1 - \frac{\alpha_{K-H}}{k_0}\right) \tag{3-61}$$

from which

$$c_0 - c = \Delta c = \frac{\alpha_{K-H}c_0^2}{\omega} \approx \frac{\alpha_{K-H}c^2}{\omega}\left(1 + \frac{2\alpha_{K-H}c}{\omega}\right) \tag{3-62}$$

and therefore

$$\frac{\Delta c}{c} = \frac{\alpha_{K-H}c}{\omega}\left(1 + \frac{2\alpha_{K-H}c}{\omega}\right). \tag{3-63}$$

Substituting for α_{K-H} using equation (3-58) leads, to first order in α_{K-H}, to

$$\frac{\Delta c}{c} \approx \frac{1}{b}\left\{v^{1/2} + (\gamma - 1)\left(\frac{K}{\rho C_p}\right)^{1/2}\right\}\{2\omega\}^{-1/2} \tag{3-64}$$

which is the equation deduced by Kirchhoff and is identical to that of Fritche for plane waves. Thus we can be confident that by using equation (3-63) with measured values of $\alpha_{\text{K-H}}$ the boundary-layer correction Δc can be properly deduced. There remain some small second order corrections to be applied however if the very highest accuracy is sought. These are discussed by Quinn et al. (1976), but amount to no more than a few parts in 10^6 of the velocity.

3-3-4   Higher modes and non-plane waves

The third and final aspect of acoustic interferometry that must be mentioned concerns the form of the normal modes of propagation of acoustic waves in the cavity. It is straightforward to solve the wave equation for a cylindrical cavity having a transducer executing simple harmonic motion at one end, a reflector at the other end and rigid walls. Following the treatment of Krasnooshkin (1944) modified by Colclough (1970) to take into account the angular dependence of modes as well as their radial dependence, we have for the velocity potential φ

$$\varphi'(r,\,\theta,\,z,\,t) = \varphi(r,\,\theta,\,z)e^{i\omega t} \qquad (3\text{-}65)$$

such that we can write down the wave equation

$$\nabla^2 \varphi(r,\,\theta,\,z) = \frac{1}{c_0^2}\frac{\partial^2 \varphi}{\partial t^2}. \qquad (3\text{-}66)$$

The normal modes, $\varphi_{mn}(r,\,\theta,\,z)$, in the cylindrical cavity may easily be found by separation of variables in the wave equation for the velocity potential, φ, to give (Quinn, 1969; Colclough, 1970)

$$\varphi_{mn}(r,\,\theta,\,z) = J_m\left(\frac{X_{mn}r}{b}\right)(A_{mn}\cos m\theta + B_{mn}\sin m\theta)\cos q_{mn}z \qquad (3\text{-}67)$$

where J_m is a Bessel function of the first kind of order m, where $-m^2$ is the separation constant for the variable θ, and q_{mn} is the wave number, equal to $k_{mn} + i\alpha_{mn}$, of the mnth mode. The constants A_{mn} and B_{mn} give the amplitude of φ_{mn}, which is determined by the amplitude distribution of the displacement $\zeta(r,\,\theta)$ over the face of the transducer. X_{mn} is another real constant, characterizing the mnth mode which is obtained as the $(n+1)$th solution of the radial boundary condition requiring the particle velocity normal to the walls to vanish at the rigid cylindrical walls:

$$\frac{d}{dX}J_m(X)\bigg|_{X = b(q_{00}^2 - q^2)^{1/2}} = 0 \qquad (3\text{-}68)$$

Table 3.5.

	$n=0$	1	2	3
$m=0$	0	3.83	7.01	10.17
1	1.84	5.33	8.54	11.71
2	3.05	6.70	9.97	13.17
3	4.20	8.01	11.37	14.54

Some values of X_{mn} calculated from equation (3-68).

where $-q^2$ is the separation constant for the variable z and $q_{00} = k_{00} - i\alpha_{00}$, the complex wavenumber in the unbounded gas so that α_{00} is the absorption coefficient and $k_{00} = 2\pi/\lambda_{00}$ where λ_{00} is the wavelength. Values of X_{mn} may be tabulated (Table 3-5) to enable the wavenumber for the mnth mode, q_{mn}, to be found from the equation

$$q_{mn} = \left\{ q_{00}^2 - \left(\frac{X_{mn}}{b} \right)^2 \right\}^{1/2}. \tag{3-69}$$

For $m=n=0$, $X_{mn}=0$ and so we have plane waves.

The real and imaginary parts k_{mn} and α_{mn} will be given respectively by:

$$k_{mn} = \frac{1}{\sqrt{2}} \left[k_{00}^2 - \alpha_{00}^2 - \left(\frac{X_{mn}}{b} \right)^2 + \left\{ \left(k_{00}^2 - \alpha_{00}^2 - \left(\frac{X_{mn}}{b} \right)^2 \right)^2 + 4\alpha_{00}^2 k_{00}^2 \right\}^{1/2} \right]^{1/2} \tag{3-70}$$

and
$$\alpha_{mn} = \frac{\alpha_{00} k_{00}}{k_{mn}} \tag{3-71}$$

Figure 3-11 shows k_{mn}/k_{00} and α_{mn}/α_{00} plotted as functions of X_{mn}/bk_{00}. As $X_{mn}/bk_{00} \to 1$ the absorption coefficient rises very rapidly and the wavenumber decreases very rapidly, i.e. the wavelength and the velocity become very large. At this point, the mnth mode ceases to propagate. This occurs for a frequency given by

$$f_{mn} = \frac{X_{mn} c_{00}}{2\pi b} \tag{3-72}$$

known as the cut-off frequency of the mnth mode. Thus for a particular mode to propagate, it is necessary that the cut-off frequency for that particular mode be exceeded. Since f_{mn} is inversely proportional to the diameter of the cylinder,

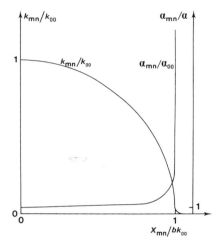

Fig. 3-11. The dependence of the wavenumber and absorption coefficient upon the order of the mode. The cut-off condition for the mnth code is seen to be $k_{00} = X_{mn}/b$.

the smaller the diameter, the higher the frequency below which a given mode will not propagate. Note that the plane wave propagates at all frequencies, since $X_{00} = 0$. It is thus possible to choose a frequency below the cut-off frequency of the lowest of the higher-order modes for which only plane waves propagate. If we take an example of a cylindrical cavity, 2 cm in diameter containing helium at s.t.p., and thus having a value of c_0 of about 900 ms^{-1}, we find that the cut-off frequency of the lowest of the higher-order modes is 26 kHz, while at 4.2 K, it is 3.3 kHz. These are relatively low frequencies, and imply that the boundary layer effects will be relatively large. For example, we find that at 4.2 K at 1 kHz, the Kirchhoff–Helmholtz absorption coefficient at a pressure of 10 kPa reaches a value of about 0.15 m^{-1}. From equation (3-63) we find that this leads to a relative velocity correction of about 3×10^{-3}, which is by no means negligible. Since the boundary layer correction is inversely proportional to the square root of the pressure (see equations (3-58) and (3-63)), the overall effect on an acoustic isotherm is to produce a noticeable curvature at low pressures together with a slight change in overall slope. We have already mentioned the effect of a small $A_2(T)$ term in the acoustic virial expansion in this respect.

At high frequencies, on the other hand (Plumb and Cataland, 1966), the boundary layer correction can be made small, but at the expense of introducing a major uncertainty arising from the possible presence of higher modes. The presence of a higher mode may or may not be evident from the form of the impedance circles or resonance peaks, in the case of a resonating-quartz-crystal transducer. Despite a detailed study of the problem (Colclough, 1970, 1973), it does not appear possible to show unambiguously which, among all of the higher modes that might be present in a high frequency

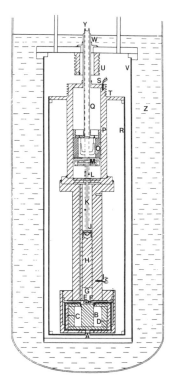

Fig. 3-12. The NPL⁻2–20 K acoustic interferometer. A, Stycast seals; B, permanent magnet assembly; C and D, electrical lead screens; E, p.z.t. accelerometer; F, transducer diaphragm; G, acoustic cavity; H, piston reflector; I, germanium resistance thermometers; J, cube-corner reflector; K, pushrods; L, beam splitter; M, gimbals; N, optical window; O, bearing; P, upper chamber; Q, moving tube; R, radiation shield; S, temperature controlling sensor; T, thermal anchoring grooves (with heater); U, 4.2 K thermal anchoring grooves; V, vacuum can; W, central supporting tube; Y, laser beams; Z, liquid He bath (after Colclough, 1979).

interferometer, actually contribute, and in what proportion, to the measured signal. Whether or not a particular mode propagates depends upon two factors: first, whether or not the cut-off frequency for that particular mode is exceeded and second, whether or not the transducer vibrates in such a way as to excite that particular mode. For a transducer executing perfect piston-like vibrations, the only mode that will be excited will be the plane wave "00" mode, no matter what the frequency happens to be. At high frequencies, detailed information on the way in which the transducer vibrates is not available (since the amplitude of vibration is much too small to be interferometrically visible). It is thus only by inference, based upon an informed guess concerning the behaviour of the transducer and examination of the resonance peaks, that the order of the principal mode being propagated can be found.

The conclusion must be that accurate acoustic interferometry can best be achieved by using frequencies below the cut-off frequency of the first higher order mode. The consequent boundary layer correction, despite being relatively large, is well behaved, and can be measured with an uncertainty that can be properly assessed.

3-3-5   A practical acoustic thermometer

The practical realization of a low frequency acoustic interferometer for thermometry is illustrated in Fig. 3-12. This instrument (Colclough, 1979a) was used for measurements of T in the range 4.2 K to 20 K at about the same time as the gas thermometer of Fig. 3-5. We shall not enter into the details of the design or operation of the equipment, but shall simply outline the main points. Acoustic waves, at a frequency just below the cut-off frequency of the first higher order mode, are excited by an electromagnetically-driven diaphragm to the rear face of which is attached a piezoelectric crystal accelerometer. The piston and cylinder assembly is made from OFHC copper. The movement of the piston is measured by a laser interferometer, one mirror of which is mounted on the piston. Germanium resistance thermometers, mounted in holes drilled in the cylinder block, are used to record the temperature determined acoustically and to compare it subsequently with the temperature determined by other means. Data from the piezoelectric. accelerometer were processed in terms of admittance circles plotted on an $x-y$ recorder after separation of the two components, proportional to $X(1)$ and $R(1)$, by a phase sensitive detector. An acoustic isotherm obtained by measurements over a range of pressure at a temperature of 13.8 K is shown in Fig. 3-13.

The results of primary acoustic thermometry have already been discussed in Chapter 2 along with those of gas and magnetic thermometry. A general conclusion which can be drawn after some twenty years of acoustic thermometry is the following: despite our having come to a good understanding of the systematic uncertainties inherent in the method, and despite the fact that acoustic thermometry overcomes many of the sources of systematic error in gas thermometry, the overall uncertainties remaining in acoustic thermometry still marginally exceed those of gas thermometry. This is because the practice of acoustic thermometry is considerably more complex, due to the nature of the measured quantity, than is gas thermometry. New work by Moldover (1982), however, using a spherical acoustic resonator shows promise of a considerable improvement in accuracy over what has been achieved so far. It may thus turn out that in the end acoustic thermometry will fulfil its early expectations.

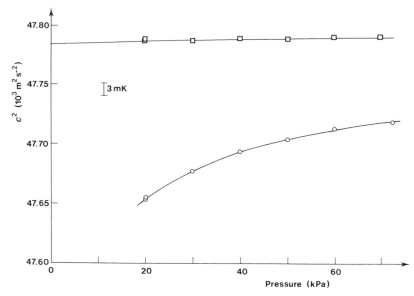

Fig. 3-13. An isotherm obtained with the NPL acoustic thermometer at 13.81 K. o—o, c^2 uncorrected for boundary layer; □-□, c^2 corrected for boundary layer (after Colclough, 1979).

3-4   Noise thermometry

3-4-1   Introduction

The basic equation relating the mean square noise voltage, $\overline{V^2}$, developed in a resistor was originally derived by Nyquist (1928) and modified by Callen and Welton (1951) to include zero-point fluctuations. The complete expression is as follows:

$$\overline{V^2} = \int_{\nu=0}^{\infty} 4h\nu Z(\nu)[\tfrac{1}{2} + (\exp(h\nu/k\tau) - 1)^{-1}]d\nu \qquad (3\text{-}73)$$

where $Z(\nu)$ is a frequency-dependent impedance and ν is the frequency. This resembles, of course, the equation relating the energy density in a blackbody to the temperature of its walls. Both are derived by summing the normal modes in a container. In the case of a blackbody, in Chapter 7 we show how the mode density, or Jeans number, is derived by considering the mode density for electromagnetic radiation in a three-dimensional parallelepiped. Equation (3-73) is derived by considering the mode density of thermal fluctuations in a one-dimensional transmission line linking two resistors and assuming, as we do for electromagnetic radiation, Bose–Einstein statistics for the energy distribution.

Unlike blackbody radiation, however, in noise measurement we are always working in the low-frequency tail of the distribution described by equation (3-73). That is to say, $h\nu/kT \leqslant 1$, and we are in the noise equivalent of the Rayleigh–Jeans domain of the Planck equation. Even at $T = 1$ mK, we have $h\nu/kT \leqslant 5 \times 10^{-3}$ for $\nu = 100$ kHz. Equation (3-73) thus becomes

$$\overline{V^2} \approx 4kTR\Delta\nu \tag{3-74}$$

in which we have replaced $Z(\nu)$ by R, a frequency-independent resistance.

With such a simple relationship between the measured quantity, voltage, and thermodynamic temperature, the question that immediately comes to mind is why is noise thermometry not the most important and widely-used method of primary thermometry? The answer is that it remains just too difficult to make sufficiently accurate measurements of the very small voltages developed (of the order of microvolts) while at the same time avoiding extraneous sources of noise of thermal or non-thermal origin, and maintaining a constant bandwidth and constant gain of the amplifiers. Although considerable progress has been made in recent years, many problems remain and, except at very low temperatures, the accuracies attained do not match those of other primary thermometers. For this reason, we shall not be entering very deeply into the subject of noise thermometry, but shall simply outline the principles of some of the methods that have been tried.

From the form of equation (3-74) it is clear that the proportional accuracy achieved in the measured quantity, $\overline{V^2}$, must reach that required in the final value of T. At present the best proportional accuracy achieved is about 0.01 %, and this only in a very restricted temperature range (Klein *et al.*, 1979). Noise thermometry, therefore, is likely to be of greatest interest either at very low temperatures or at very high temperatures. At helium temperatures, a proportional accuracy of 0.01 % already equals that achieved in the best gas thermometry. At 373 K, on the other hand, such a proportional accuracy is equivalent to about 0.04 K and we can do rather better than this with gas thermometry. At 2000 K, we find that noise thermometry is once again of interest, having particular application in those environments subject to ionizing radiations. This is because it is possible, as we shall see, to devise a method of noise thermometry in which the value of R does not enter into the calculation of T and which is of interest, therefore, in ionising-radiation environments which lead to drifts in the value of R.

An additional constraint present in noise thermometry arises from the statistical nature of the measured quantity. In comparing two noise sources at different temperatures, we find that for a given bandwidth the standard deviation in $\overline{V^2}$, namely $\Delta\overline{V^2}$, for a measurement time t is given by

$$\frac{\Delta\overline{V^2}}{\overline{V^2}} = \frac{K}{\sqrt{t\Delta\nu}} \tag{3-75}$$

where K is a constant to allow for the fraction of the time spent on each noise source. In practice, K is generally equal to about 2.5 so that in terms of T we have

$$\frac{\Delta t}{T} \approx \frac{2.5}{\sqrt{t \Delta v}} \qquad (3\text{-}76)$$

For a typical bandwidth of 100 kHz and for $\Delta T/T \approx 5 \times 10^{-4}$, we require a measurement time of about 5 min. But to improve this by a factor of ten, we require a measurement time one hundred times longer. This is by no means straightforward since it is not easy to maintain the stability of gain of amplifiers for such long periods.

The methods of noise thermometry fall into five main categories which we will now describe in turn.

3-4-2 The method of two unequal resistors

In the first method, two resistors are connected in turn to the input of an amplifier. One resistor is held at a reference temperature, T_0, and the other at the unknown temperature, T. The value of one of the resistors is adjusted until the mean square voltages developed across each resistor are equal, so that $R_1 T_1 = R_2 T_2$ provided that the bandwidth, dv, and the gain and input noise of the amplifier are all independent of the source impedance. This was the method proposed by Garrison and Lawson (1949). Unfortunately, using valve amplifiers it was not possible at that time to achieve these requirements. It was not until the advent of the field-effect transitor, FET, that real advances in noise thermometry took place. The method of Garrison and Lawson was taken up again by Pickup (1975) using FETs at the input stage of his amplifier.

An idea of the difficulty of precise noise thermometry is given by the set of conditions laid down by Pickup to attain accurate measurements of the ratio of thermodynamic temperatures using his equipment, which is illustrated in Fig. 3-14.

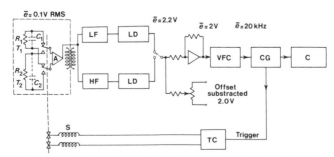

Fig. 3-14. Block diagram of switching noise thermometer. The thermal noise voltages from resistors R_1 and R_2 at temperatures T_1 and T_2 are compared by an alternately switched measuring channel (after Pickup, 1975).

The following is a summary of Pickup's sufficient conditions for accurate noise thermometry:

(1)   The noise resistances R_1 and R_2 should appear as a parallel combination of constant conductance shunted by constant capacitance up to the highest frequency of interest ($\approx 200\,kHz$). Any noise voltages of non-thermal origin generated in the sensors should be negligible compared with the Johnson noise when integrated over the measuring band, the low frequency limit of which is approximately 10 kHz.

(2)   The overall frequency bandwidth of the noise measurement must be the same for each resistance.

(3)   The critical properties of the input stage of the amplifier, including gain, bandwidth and its own noise contribution, should be independent of the noise-source impedance.

(4)   The input conductance of the amplifier must be low enough to avoid errors due to loading of the noise resistance.

(5)   The switch (driven by the solenoid S, in Fig. 3-14) must not introduce additional noise other than an initial transient on contact.

(6)   The low-level parts of the system must be effectively shielded against external electromagnetic interference. In addition, internal feedback from higher level signals must be avoided.

Using the system shown in Fig. 3-14, Pickup obtained measurements of T near 90 K having a standard deviation of about 4 mK. The reference temperature was at room temperature, 293 K.

3-4-4   The method of equal resistors

The second method was devised by Pursey and Pyatt (1959) and subsequently modified by Actis *et al.* (1972) and Crovini and Actis (1978). Instead of using different resistors as in the first method described, they proposed using equal resistors, and measuring the ratio of the noise from the two resistors. This is accomplished by introducing a precision attenuator into the amplifier chain when measuring the noise from the higher temperature source. The method is illustrated in the block diagram of their equipment shown in Fig. 3-15. As is now always the case, the input stage of the amplifier incorporates FET input elements. The method requires a separate determination of the amplifier noise since it enters unequally into the two noise signals. In addition, a high degree of linearity is required in that part of the amplifier preceding the attenuator. A compensating network is introduced into the bypass of the attenuator to ensure equality of bandwidth for the two signals. The switch is a mechanical device operating at about 30 Hz which introduces only an insignificant noise into the amplifier chain. In the input switch the action is "make before break" while in the charging circuit of the memory capacitors it is "break before make" in order to suppress the transients originating in the input switch. Using this system, Crovini and Actis (1978) were able to measure ratios of

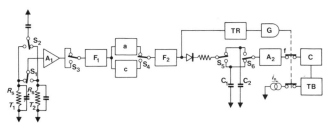

Fig. 3-15. Block diagram of a noise thermometer: using the method of equal resistors; A_1, low-noise amplifier; F_1, F_2, pass-band filters; a, attenuator; c, network to compensate for the high-frequency response of the attenuator; C_1, C_2, memory capacitors; A_2, low-frequency amplifier, demodulator and voltage-to-frequency converter; TB, time–base generator, which does not run in presence of interference; TR, interference detecting circuit and gate trigger; G, gate which disconnects counter and stops time base when TR switches on; S_1 to S_6, motor driven synchronous switches. A further synchronous switch is inside A_2; C, up–down counter (after Crovini and Actis, 1978).

thermodynamic temperature to an accuracy of about 2 parts in 10^4 (at the 3σ level), i.e. 0.25 K at 1000 K.

3-4-4   The correlation method

The third method of noise thermometry is that in which the noise from a single resistor is simultaneously amplified by two separate amplifier chains the outputs of which are compared by a correlator. In this way amplifier noise, which is uncorrelated, is almost completely eliminated and the output of the correlator should be proportional to T. The method, originally suggested by van der Ziel (1970), has been developed by Klein *et al.* (1979) to the stage where measurements of thermodynamic temperature in the liquid helium range have comparable accuracy to those of gas thermometry (Klempt, 1982). In eliminating amplifier noise the correlation method becomes subject to errors of a different sort originating mainly from slow drift in the amplifiers and rectifiers. The circuit of Klein *et al.* is shown in Fig. 3-16. Because measurements

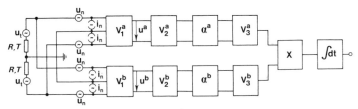

Fig. 3-16. Block diagram of a noise thermometer operating by the correlation method, in which R,T are the temperature-sensing resistors, V_1^a and V_1^b the low noise differential amplifiers, V_2^a and V_2^b differential amplifiers, α^a and α^b symmetrical resistive voltage dividers, V_3^a and V_3^b differential amplifiers and filters, X an analogue multiplier and $\int dt$ an integrating digital voltmeter. The quantities 'u' are noise voltages (after Klein *et al.*, 1979).

different temperatures must be made one after another and the time required at each temperature is long, of the order of 10 hours, extreme stability is required in all the parameters of the measuring system. The method adopted by Klein was to compare the noise from a resistor at the unknown temperature, in the liquid helium range, with that from the same resistor at 273.16 K. The uncertainty in the measured value of T in the range below 4.2 K was ± 0.3 mK of which the major portion, 0.25 mK, was the statistical uncertainty represented by equation (3-76).

3-4-5 The noise power method

The fourth method of noise thermometry is based upon a measurement of the product of the thermal noise voltage and the thermal noise current generated in the same resistor. The principal advantage of the method, developed by Borkowski and Blalock (1974), is that it does not require a knowledge of the value of the resistor (Blalock *et al.*, 1982; Blalock and Shepard, 1982). The method is illustrated in Fig. 3-17, which shows a block diagram of Borkowski and Blalock's apparatus. The noise current generated in a resistor R is given by

$$\overline{i^2} = \frac{4kT}{R}\,dv_i. \tag{3-77}$$

Assuming gain constants K_v and K_i for the voltage and current amplifiers respectively, the product, i.e. the noise power, at the output of the voltage and current amplifiers will be

$$P = (\overline{V^2})^{1/2}(\overline{i^2})^{1/2}$$

$$= 4kTK_vK_i(\Delta v_v \Delta v_i)^{1/2}. \tag{3-78}$$

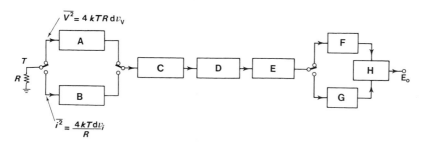

Fig. 3-17. Block diagram of a noise thermometer using the method of thermal noise power in which A is a voltage-sensitive pre-amplifier, B a current-sensitive pre-amplifier, C provides further amplification and filtering, D is a true-RMS meter, E an integrator, F allows the noise voltage to be held, and G the noise current and H is a multiplier (after Borkowski and Blalock, 1974).

Thus, from a measurement of P the value of T may be obtained without a knowledge of R provided that the bandwidth and gain of the noise voltage and noise current amplifiers are known.

In the system shown in Fig. 3-17, a voltage-sensitive preamplifier with a high input impedance and a current-sensitive pre-amplifier with a low input impedance are connected to the same resistor. The effective bandwidth of the system is about 40 kHz centred at a frequency of 45 kHz. The accuracy depends upon the performance of the two pre-amplifiers, in particular their internal noise and their variation of gain with variation of noise-source impedance. The effects of lead-cable impedance shunting the noise source must also be minimized. In common with other methods of noise thermometry, the stability of the amplifiers must be assured and external non-thermal noise sources must be eliminated. The latter condition is often the more difficult to achieve.

The instrument of Borkowski and Blalock was designed for the measurement of high temperatures, up to 1000 K, and employed a 100 Ω platinum resistance element as the noise source. The overall accuracy after correction for pre-amplifier noise was about 0.1 % of the temperature for $T \geqslant 500$ K. Rather larger errors, up to 0.6% of T, occurred at lower temperatures due mainly to noise originating in the pre-amplifiers.

3-4-6   The resistive SQUID method

The fifth and final method of noise thermometry can be used only at low temperatures and employs a Josephson junction in a circuit of very low inductance and resistance to form a resistive SQUID (Superconducting Quantum Interference Device). There are other ways of using a Josephson junction for noise thermometry, for example by using a SQUID magnetometer (Hudson et al., 1975), but the resistive SQUID is the device which leads most directly to the measurement of thermal noise and it is the one we shall consider here. The resistive SQUID noise thermometer was first demonstrated by Kamper and Zimmerman (1971). The principle is illustrated by Figs 3-18 and 3-19. The Josephson junction is connected to a very low resistance, of the order of 10 $\mu\Omega$, through a circuit having an inductance of less than about 10^{-9} H. A direct current of about 1 μA is passed through the resistance from an external source in order to maintain a bias voltage of 10^{-11} V across the Josephson junction causing it to oscillate at about 5 kHz. The oscillation frequency, f, is related to the bias voltage by

$$hf = 2 \text{ eV} \tag{3-79}$$

or
$$f = \frac{V}{\Phi}$$

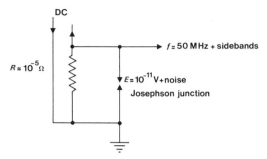

Fig. 3-18. Basic circuit for noise thermometry using the Josephson effect (after Kamper and Zimmerman, 1971).

where $\Phi \equiv h/2e$ and is known as the "flux quantum".

The voltage at any instant across the junction, which is the sum of a steady voltage and a fluctuating one, is

$$V = IR + V_{\text{noise}}. \tag{3-80}$$

Fluctuations in the noise voltage will lead to fluctuations in V which, in turn, will lead to fluctuations in the frequency of oscillation with a bandwidth related to V_{noise}. Since the junction resistance is large compared with the shunt resistance, R, the noise in the former is dominant and thus from a measurement of the noise, the average frequency of oscillation and bias current it is possible to deduce R and T. It has been shown by Kamper (1977) that the components of the noise spectrum that contribute most to the radio-frequency power

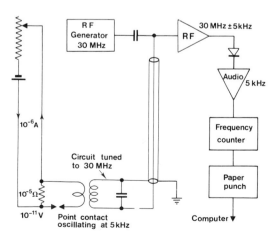

Fig. 3-19. Block diagram of a noise thermometer using the resistive-SQUID method (after Kamper and Zimmerman, 1971).

spectrum of the Josephson oscillation are those of low frequency. For a resistor at 4 K, the upper frequency bound is about 1 kHz, for resistors at still lower temperatures it is even less. The quantity required, namely the mean bandwidth in the frequency spectrum of the Josephson junction, can be determined either by spectrum analysis or by counting. In the spectrum analysis method, the theory of frequency modulation (Silver *et al.*, 1967) leads to the following expression for the bandwidth Δf

$$\Delta f = \frac{4\pi k T R}{\Phi^2}. \tag{3-81}$$

For absolute noise thermometry, a measurement of Δf is not straightforward and, instead, a counting method is to be preferred. For this the signal from the Josephson junction, modulated in frequency by the noise voltage across R, is used to drive a frequency counter. The number of cycles in a given period of time, τ, is measured n times and the variance, σ, of the resulting mean count, $\bar{f}$, is obtained. Assuming that white noise is generated across R we have

$$\sigma^2 = (f - \bar{f})^2 = \frac{2 kT R}{\tau \Phi^2}. \tag{3-82}$$

The measurement of σ is easier and less subject to systematic error due to extraneous noise sources than is the measurement of Δf. One source of systematic error which must be taken into account, however, in the counting method stems from the counting of only complete cycles.

This property of the counter leads to an additional contribution to the variance and hence to an increment in temperature, T_D, given by

$$T_D = \frac{\Phi_0^2}{24 \, Rk \, \tau}. \tag{3-83}$$

Having made the correction for T_D, which may be considered as the limiting noise of the detector, it must be remembered that there remains the statistical uncertainty in the final measured value of T. The uncertainty, δT, in T is given by

$$\delta T = T\sqrt{2/n}. \tag{3-84}$$

The range of application of a SQUID noise thermometer could be extended up to about 20 K. In practice, however, most work has been done at very low temperatures, below 0.5 K down to a few millikelvins (Soulen and

Van Vechten, 1982). No systematic differences greater than 0.3 mK have been observed between the results of such noise thermometers and other primary thermometers in this very low temperature range.

The main problems associated with SQUID noise thermometry still stem from the very long measurement time required to obtain useful accuracy. For example, an accuracy of 1 % in T requires 2×10^4 counts, which leads to $n\tau$ (the total measuring time) being typically about an hour. The longer the measurement time, the more difficult it is to avoid the effects of noise of extraneous origin. However, at very low temperatures, the proportional accuracy sought is not very high, 1 part in 10^3 is more than adequate at 300 mK for example, and the noise thermometer is subject to very few uncertainties. There are no sample-dependent corrections to be applied nor are there any second- or higher-order terms that need to be taken into account. Noise thermometry is therefore one of the best methods of primary thermometry at temperatures below 1 K. At high temperatures, on the other hand, the proportional accuracy sought in primary thermometry remains beyond the reach of noise thermometry. However, the continual advances being made in semi-conductor technology may eventually lead to the necessary improvements in amplifier behaviour that could open the way to useful high-temperature primary noise thermometry.

3-5   Magnetic thermometry

3-5-1   Introduction

The paramagnetic susceptibility, χ, of many materials containing transition and rare-earth elements follows quite closely the Curie Law, which states that χ is inversely proportional to T. The calculation of the paramagnetic susceptibility in real crystals is quite complex, however, and although it is fairly straightforward to give an outline showing the principal factors involved, the details are difficult and in many cases poorly understood. It is mostly for this reason that magnetic thermometry is not primary thermometry, although another contributing factor is that it is very difficult to make absolute measurements of magnetic susceptibility. As we shall see, it is necessary to evaluate some of the constants in the expression for χ as a function of T by calibration against other thermometers. In spite of its not being primary thermometry in the strict sense, magnetic thermometry occupies an important place in primary thermometry because of its special role as an interpolating or, in some cases, extrapolating thermometer. In order to make very clear this special role of magnetic thermometry, we shall begin by looking briefly at some of the factors affecting the temperature dependent paramagnetic susceptibility of certain crystals.

3-5-2   *Paramagnetic susceptibility as a function of T*

In paramagnetic materials, the temperature dependence of the susceptibility is quite easy to calculate on the basis of the individual dipoles being independent of one another and the first excited states being well separated from the ground state. Such a calculation leads to Curie-law behaviour in which

$$\chi = N_A (g\mu_B)^2 \frac{J(J+1)}{3kT} \tag{3-85}$$

$$= \frac{C}{T} \tag{3-86}$$

where N_A is the Avogadro constant; J is the quantum number of the ground state of the atom; g is the Landé factor which relates the applied field, H, to the splitting of the degenerate ground state of the atom into $(2J+1)$ levels separated by energy $g\mu_B H$; μ_B is the Bohr magneton; and C is the so-called "Curie constant".

In a real paramagnetic crystal, the susceptibility follows the simple Curie law over a more or less restricted range of temperature (Hudson, 1972; Durieux *et al.*, 1962). At the upper end of the range, deviations will occur as $T \to E_1/k$, where E_1 is the energy of the first excited state (see below). At low temperatures, deviations will arise due to interactions between the ions and the effects of crystal-field splitting on the ground state if it is any more complex than a doublet. For the purposes of thermometry, the susceptibility is usually written in the following form:

$$\chi = \chi_v + \frac{C}{T - \left(\frac{4\pi}{3} - \varepsilon\right)C - \theta + \delta/T + \ldots} \tag{3-87}$$

$$= \chi_v + \frac{C}{T + \Delta + \delta/T} \tag{3-88}$$

where χ_v is the van Vleck temperature-dependent susceptibility; ε is the demagnetizing factor, depending upon the geometry of the sample, and which is a correction factor to convert the external applied field to the true internal field acting on the ion. $(4\pi/3 - \varepsilon)C$ arises from the local field seen by the dipole and calculated on the basis of first order dipole–dipole coupling in an ellipsoidal sample. For a spherical sample and a lattice of simple cubic symmetry, $\varepsilon = \frac{4}{3}\pi$ and thus $\Delta = 0$; for real materials, especially those favoured for magnetic thermometry, these relations still hold to a very good approximation. θ is the result of exchange interactions, that is electrostatic interactions between the electrons in the partially-filled shells of the magnetic ions. δ arises from crystal field effects and second order dipole–dipole interactions.

The constant C (see equation (3-85)), which is independent of temperature, depends on the ground state of the atom, and for non-cubic lattices it is a function of the orientation of the lattice with respect to the field. For equation (3-88) to be valid, it is necessary that excited states of the ion be unpopulated, otherwise there will be an additional contribution to the term in δ. Fortunately, for most transition metal ions the first excited state is some 10^3 K or 10^4 K above the ground state and can safely be ignored. This is not the case for some rare earth ions. Cerium magnesium nitrate (CMN) for example has a first excited state at only 38 K above the ground state. For this reason, CMN does not exhibit simple Curie law behaviour above about 4 K.

3-5-3   The practice of magnetic thermometry

Before entering further into the properties of individual paramagnetic salts, we should look at the implications for thermometry of equations (3-87) and (3-88). Due to the complexity of the problem, for few salts have C, θ or δ been calculated with sufficient accuracy for equation (3-87) to be used as an explicit relation between χ and T. Nevertheless, quantum mechanics usually gives enough information about the form of these quantities for us to be confident of the manner of their variation over a wide range of temperature and for a wide range of rare-earth and transition-metal salts. Thus by measuring the paramagnetic susceptibility at a small number of known temperatures, the numerical values of the constants of equation (3-88) can be deduced, thereby allowing interpolation, or even extrapolation in certain circumstances, to be carried out.

The anisotropy in the susceptibility of non-cubic crystals can be largely overcome either by careful orientation of a single crystal or by the use of a powdered sample, although packing preferences in a powder may interfere with the complete randomness of the particle orientation and lead to a residual anisotropy. Since it is easier to make thermal contact to a powder however, this, rather than a single crystal, has become the usual form of sample for all magnetic thermometry regardless of crystal symmetry.

Absolute measurements of susceptibility are not made in magnetic thermometry because they are too difficult to carry out. Instead, the mutual inductance of a pair of coils surrounding the sample is measured, which is proportional to the susceptibility of the sample. In earlier magnetic thermometry a Hartshorn mutual-inductance bridge was used (for example, Durieux et al., 1962), but more recently a ratio-transformer version of the Hartshorn bridge has been preferred (for example, Cetas and Swenson, 1972).

In either case, the bridge reading, n, takes the form

$$n = A + \frac{B}{T + \Delta + \delta/T + \dots} \qquad (3\text{-}89)$$

where A is a function of the coil design and the temperature independent susceptibility; B is a function of the coil design and the Curie constant of the salt being used; Δ is a property of the salt and its shape; δ is a property of the salt only. Because of the dependence of the measured quantity upon the shapes of the coil and the sample, the measured values of A, B and Δ cannot be used with another sample, even of the same powder.

Among the salts that have been most widely used for magnetic thermometry are cerium magnesium nitrate (CMN), chromic methyl ammonium alum (CMA) and manganous ammonium sulphate (MAS). The first of these, CMN, $Ce_2Mg_3(NO_3)_{12}24H_2O$, is of use for thermometry only at temperatures below about 4.2 K due to the presence of the first excited state at 38 K and low sensitivity. CMN has a hexagonal structure and its magnetic properties are very anisotropic. Despite this, Δ is still very small: approximately 0.27 mK. The susceptibility in the direction parallel to the hexagonal axis, χ_{11}, is very much smaller than that perpendicular to it, $\chi_\perp$. The susceptibility $\chi_\perp$ is itself small because of a small ionic moment, $J = 1/2$, and also because the ions are very dilute in the lattice. The latter, of course, leads to very good Curie law behaviour which is one of the attractions of this salt for thermometry below 1 K.

The other two salts, CMA, $Cr(CH_3NH_3)(SO_4)_212H_2O$ and MAS, $Mn(NH_4)_2(SO_4)_26H_2O$, can be used to much higher temperatures for thermometry than CMN, since they each have first excited states at very high temperatures. CMA has a cubic-type structure which becomes orthorhombic below a transition temperature of 164 K. The theory of the magnetic behaviour of CMA is well understood (Hudson, 1972) since the ground state is uncomplicated and the ions are well separated in the lattice so that the dipole–dipole interaction is small. Durieux (1962) gave for this salt a value for δ of 0.002 79 K^2 and for θ a value of 12 mK and showed that the $1/T^2$ and higher-order terms could be neglected at temperatures above 1 K. CMA is thus an excellent salt for magnetic thermometry above about 0.3 K. The theory of the magnetic behaviour of MAS, on the other hand, is not well understood since it has a much more complicated ground state than CMA. There are no well established values for θ or δ, each must be determined experimentally for the particular sample used. Nevertheless, individual samples of MAS have been found to follow quite closely the behaviour predicted by equation (3-88) while at the same time having the advantage that the Curie constant is some three times that of CMA.

Most paramagnetic salts contain water of crystallization and are efflorescent, CMN on the other hand is deliquescent and so is much more easily preserved. There is another salt, however, which looks promising for magnetic thermometry above 4 K; this is godolinium metaphosphate, $Gd(PO_3)_3$. It contains no water of crystallization and has a large Curie constant. It can only be prepared in the form of a powder or a glass but it is isotropic.

The practice of magnetic thermometry is, in principle, quite straightforward. Figures 3-20 and 3-21 show, in diagramatic form, the apparatus used by Cetas and Swenson (1972) to establish a magnetic temperature scale from 0.9 K to 18 K. This scale formed the basis upon which EPT-76 was developed (see Chapter 2). The powdered salt sample was held in a non-magnetic nylon capsule supported by a fused-silica rod attached to a temperature-controlled copper block. The copper block contained a reservoir for liquid helium with for pumping, together with germanium and platinum resistance thermometers. Thermal contact between the copper block and the sample was ensured by a number of fine copper wires. The coils were mounted on the outside of the glass vacuum jacket and held at a constant temperature of 4.2 K.

It is important that the relative positions of sample and coils remain

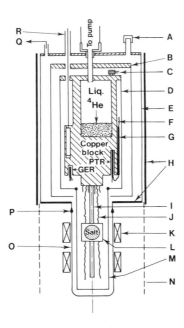

Fig. 3-20. A sketch of the cryostat used by Cetas and Swenson for paramagnetic-salt thermometry, in which A are nylon electrical lead throughs, B an intermediate shield, C the heater sensor, D the block shield, E the brass vacuum jacket, F the thermometer leads, G the thermal anchors, H lead shields, I a Vycor support, J copper wires, K coils, L nylon holder, M "coil foil" shield, N black paper light shield, O pyrex vacuum jacket, P copper-to-pyrex seal, Q high vacuum and R jacketed vapour-pressure line (after Cetas and Swenson, 1972).

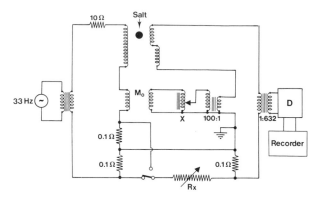

Fig. 3-21. A mutual-inductance bridge circuit used for susceptibility measurements (after Cetas and Swenson, 1972).

unchanged during a series of measurements since some of the constants in equation (3-89) depend upon them. The main difficulties of magnetic thermometry arise from the presence of small temperature gradients, the effects of extraneous temperature-dependent magnetic coupling between various components and relative motion between sample and coils. Careful experimental procedures are also required to ensure purity of the sample and absence of foreign magnetic particles.

The procedure adopted by Cetas and Swenson for obtaining a magnetic temperature scale between 0.9 K and 18 K required independent temperature measurement between 18 K and 34 K to establish the constants A and B, and between 0.9 K and 2.6 K to establish Δ and δ. The values of Δ and δ were established first. This procedure, although simple in outline, is not wholly without difficulty and the original paper should be consulted for a more detailed discussion of the problems involved. Similar work was carried out over a much larger temperature range by Cetas (1976). He used two salts, MAS and an additional rare earth salt gadolinium sulphate octohydrate, $Gd_2(SO_4)_3 8H_2O$.

The reproducibility of the best magnetic thermometry is better than 0.5 mK over a wide range of temperature. The accuracy with respect to thermodynamic temperature depends upon the accuracy with which equation (3-89) properly describes the measured susceptibility, the stability of the constants and the accuracy with which they can be measured. Residual departures from thermodynamic temperature should reflect simply the absence of higher order terms in $(1/T)$ from equation (3-89).

The value of magnetic thermometry for temperature-scale investigations stems from the fact that theory, although not being able to give us exact values of the terms of equation (3-88) or (3-89), can give good estimates of upper limits for the higher order terms ignored by the experimentalist and hence the likely departures from thermodynamic temperature.

3-6   Dielectric-constant and refractive-index gas thermometry

3-6-1   The principles of dielectric-constant gas thermometry

A new method of gas thermometry recently demonstrated by Gugan and Michel (1980) and showing considerable promise both as an interpolating method and even as a primary method is based upon the relations between the dielectric constant of a gas and the density. This is expressed, for a perfect gas, by the Clausius–Mossotti equation

$$\frac{\varepsilon_r - 1}{\varepsilon_r + 2} = \frac{\alpha}{V_m} \qquad (3\text{-}90)$$

where ε_r is the dielectric constant $(\varepsilon_r = \varepsilon/\varepsilon_0)$ and α is the molar polarizability $= 4\pi N_A \alpha_0/3$, and in which α_0 is the atomic polarizability.

Combining equation (3-94) with $PV_m = RT$, we can write

$$P = \frac{R}{\alpha} \left(\frac{\varepsilon_r - 1}{\varepsilon_r + 2} \right) T.$$

Thus, given the molar polarizability and a means of measuring ε_r, equation (3-91) forms the basis of a method of primary thermometry. As in acoustic thermometry, the thermometric parameter is an intensive property of the gas and thus independent of the quantity of gas present. The dielectric constant method differs from the acoustic method, however, in that it involves a knowledge of a quantity, the molar polarizability, that cannot be deduced exactly from theory. A single calibration point is therefore required but, provided that this is at T_0, the method could still be used for primary thermometry. As we shall see, however, the application of the dielectric constant method for primary thermometry is made difficult by the need to know something about the compressibility of the capacitor used in the determination of changes in polarizability.

The departures from ideal behaviour of a real gas enter in two ways. First we have the virial expansion in the density, and second, we have a virial expansion of the Clausius–Mossotti relation. The latter occurs because the polarizability is affected by interactions between atoms in much the same way as is the pressure. Writing the Clausius–Mossotti relation as a virial expansion we have

$$\frac{\varepsilon_r - 1}{\varepsilon_r + 2} = \frac{\alpha}{V_m} \left(1 + \frac{b}{V_m} + \frac{c}{V_m^2} + \ldots \right) \qquad (3\text{-}92)$$

The coefficients b, c, etc. correspond to the second and third virial coefficients of the usual density expansion B and C, in that they refer to pair and triplet collisions, respectively, between molecules or atoms (Hill, 1958).

Although equation (3-92) could be combined directly with the P–V virial expansion of the equation of state, it is best first to convert the dielectric constant to a more-directly-measurable quantity, for which we choose capacitance. The dielectric constant can be found from the ratio of the capacitance of a mechanically stable capacitor with and without the gas between the electrodes. We can write, following Gugan and Michel (1980)

$$\frac{C(P)}{C(O)} = \varepsilon_r(1 + KP) \tag{3-93}$$

where $C(P)$ and $C(O)$ are the capacitances with and without the gas present, and K is an effective linear compressibility for the capacitor assembly. Since the changes in capacitance are small, it is more convenient to write

$$\gamma = (\varepsilon_r - 1) + K\varepsilon_r P \tag{3-94}$$

where

$$\gamma = \frac{C(P) - C(O)}{C(O)}.$$

Combining equations (3-94) and (3-92) with the virial equation of state and defining a polarizability parameter μ by $\mu = \gamma/(\gamma + 3)$, we obtain the following expression

$$P = A_1\mu(1 + A_2\mu + A_3\mu^2 + \ldots) \tag{3-95}$$

in which

$$A_1 = \left(\frac{\alpha}{RT} + \frac{K}{3}\right)^{-1}$$

$$A_2 = \frac{B - b}{\alpha}$$

and

$$A_3 = \frac{C - 2Bb + 2b^2 - c}{\alpha^2}$$

For helium we may write

$$A_3 = \frac{C}{\alpha^2}$$

Small compressibility terms in A_2 and A_3 have been ignored. In theory, $\mu \to 0$ as $P \to 0$, but if there remain any small errors in the measured values of $C(O)$, an additional pressure-independent term must be introduced into (3-95).

Equation (3-95) describes an isotherm having a slope given by A_1. In order to determine T, we need to know R, α and K. Assuming a value for R, we still require two calibration points, one for α and the other for K, unless K can be determined in quite a different way, which at present appears difficult. The alternative would be to obtain a sufficiently accurate value for the polarizability of the thermometric gas, α. In the case of helium, Glover and Weinhold (1976) have recently been able to put upper and lower bounds on the theoretical value for the polarizability of helium-4. They give a "best value", based upon their evaluation of the existing theoretical work, of 517.031×10^{-3} cm^3 mol^{-1}. This is very close to the value of Buckingham and Hibbard (1968) of 517.033×10^{-3} cm^3 mol^{-1}. The experimental value deduced by Gugan and Michel (1980) is $(517.257 \pm 0.025) \times 10^{-3}$ cm^3 mol^{-1}. This value is based upon the NPL-75 temperature scale, and the uncertainty represents a 1 mK uncertainty in NPL-75 together with a 3% uncertainty in the measured value of K. The difference between the experimental and theoretical values remains to be explained.

3-6-2   A practical dielectric-constant gas thermometer

The practical realization of a dielectric-constant gas thermometer is illustrated in Figs 3-22 and 3-23 which show the capacitor cell and cryostat arrangement of Gugan and Michel. The capacitor is made of copper and is in the form of coaxial cylinders spaced about 1.5 mm apart. The capacitance is about 10 pF, the minimum value required to achieve the resolution of 1 part in 10^8 necessary for the measurement. Two identical cells are used so that the capacitance of the gas-filled cell is always measured by comparison with an identical evacuated cell during the determination of an isotherm. The capacitors were designed with the aim of minimizing the mechanical restraints on the individual components so that the measured compressibility would be very closely related to the expansion coefficient of the material from which it was made. In order to minimize the effects of surface films resulting from adsorption, the spacing (1.5 mm) was the maximum consistent with achieving the required minimum value of the capacitance.

The results of the thermometry of Gugan and Michel have already been mentioned in Chapter 2. The interested reader is referred to their paper for a detailed description of their experimental method.

3-6-3   The refractive-index gas thermometer

Very closely related to dielectric-constant gas thermometry is refractive index gas thermometry. At very high frequencies, we may write n^2 instead of ε_r in equation (3-90), where n is the refractive index. The resulting equation is

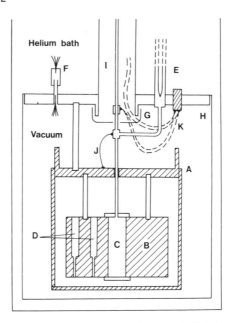

Fig. 3-22. Schematic diagram of the cryostat used by Gugan and Michel for dielectric-constant gas thermometry. A, OFHC copper isothermal shield; B, 10 cm diameter, 10 cm high OFHC copper thermometer block; C, capacitor cell (one of two); D, wells for RhFe, Pt, and Ge thermometers; E, low-temperature needle valve (one for each cell); F, epoxybonded electrical lead-through (one of six); G, radiation trap; H, 17.5 cm diameter stainless-steel vacuum can, indium O-ring sealed to top-plate; I, 1.5 mm diameter stainless steel pressure-sensing tube inside 37.5 mm main pumping; J, thermal link to I; K, AuFe/Chromel thermocouple (one of four along tube I) (after Gugan and Michel, 1980).

sometimes called the Lorentz–Lorenz equation

$$\frac{n^2-1}{n^2+2}=\frac{\alpha}{V_m}.$$ (3-96)

At low densities

$$n-1\propto\rho$$ (3-97)

where ρ is the density. This relation is sometimes called the Gladstone and Dale law. Thus, if a Michelson interferometer is set up having a gas cell in one arm, the optical path length, l, increases as the pressure in the gas cell increases so that

$$\frac{\delta l}{l}=n-1\propto\rho.$$ (3-98)

Although an interpolating thermometer has yet to be demonstrated using the

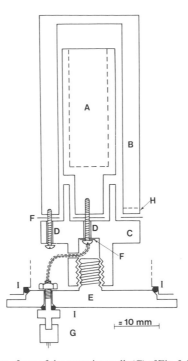

Fig. 3-23. Schematic diagram of one of the capacitor cells (C) of Fig. 3-22. A, low electrode; B, high electrode; C, earthed guard ring; D, insulated screws; E, capacitor support plate and gas-tight cover for end of DCGT cell; F, mica washers; G, co-axial lead (one of two); H, channels for gas flow (one of three); I, indium O-ring seal (after Gugan and Michel, 1980).

principle expressed by equation (3-98), the feasibility has been discussed by Colclough (1974). The difficulties and likely systematic uncertainties of such a thermometer are very similar to those of the dielectric-constant thermometer. The choice will depend upon the relative ease of making and using optical interferometers compared with capacitors. Either one could fulfil the same function of providing an excellent interpolating thermometer soundly based upon thermodynamic principles. The advantage of these methods over magnetic thermometry is that they require the determination of only two constants compared with the four of magnetic thermometry. Indeed, one of the constants, the polarizability, is a well defined physical constant of the thermometric fluid, and in principle could be determined once and for all. If this could be done with sufficient accuracy, then the way would be opened for these methods to be used for primary thermometry.

References

Actis, A., Cibrario, A. and Crovini, L. (1972). Methods of noise thermometry above 400 °C. *TMCSI*, **4**, Part I, 355–364.

Berry, K. H. (1979). NPL-75, A low temperature gas thermometer Scale from 2.6 K to 27.1 K. *Metrologia*, **15**, 89–115.

Blalock, T. V., Horton, J. L. and Shepard, R. L. (1982). Johnson-noise power thermometer in process temperature measurement. *TMSCI*, **5**, 1249–1260.

Blalock, T. V. and Shepard, R. L. (1982). A decade of progress in high temperature Johnson-noise thermometry. *TMSCI*, **5**, 1219–1224.

Bonhoure, J. and Terrien, J. (1968). The new standard manobarometer of the BIPM. *Metrologia*, **4**, 59–68.

Borkowski, C. J. and Blalock, T. V. (1974). A new method of noise thermometry. *Rev. Sci. Instr.* **45**, 151–162.

Boyd, M. E., Larsen, S. Y. and Kilpatrick, J. (1969). Quantum mechanical second virial coefficient of a Lennard–Jones gas, helium. *J. Chem. Phys.*, **50**, 4034–4055.

Buckingham, A. D. and Hibbard, P. G. (1968). Polarizability and hyperpolarizability of the helium atom. Symposium of the Faraday Society, London, **2**, 41–46.

Callen, H. B. and Welton, T. A. (1951). Irreversibility and generalized noise. *Phys. Rev.* **83**, 34–40.

Cetas, T. C. and Swenson, C. A. (1972). A paramagnetic salt temperature scale 0.9 K to 18 K. *Metrologia*, **8**, 46–64.

Cetas, T. C. (1976). A magnetic temperature scale from 1 K to 83 K. *Metrologia*, **12**, 27–40.

Colclough, A. R. (1970). Higher modes in acoustic interferometry. *Acustica*, **23**, 93–99.

Colclough, A. R. (1973). Systematic errors in primary acoustic thermometry in the range 2 K–20 K. *Metrologia*, **9**, 75–98.

Colclough, A. R. (1974). A projected refractive index thermometer for the range 2 K to 20 K. *Metrologia*, **10**, 73–74.

Colclough, A. R. (1976, 1977). Theory and criterion for end-face alignment in the cylindrical acoustic interferometer. *Acustica*, **36**, 259–270.

Colclough, A. R. (1979a). Low frequency acoustic thermometry in the range 4.2 K–20 K with implications for the value of the gas constant. *Proc. Roy. Soc., London*, **A365**, 349–370.

Colclough, A. R. (1979b). Non-linearity errors in acoustic thermometry. *Acustica*, **42**, 28–36.

Colclough, A. R. (1979c). The use of admittance figures in the processing of acoustic interferometry measurements. *Acustica*, **42**, 18–27.

Colclough, A. R. (1979d). The temperature dependence of pressure–volume and acoustic second virial coefficient for ^{4}He in the quantum region. *Metrologia*, **15**, 18–193.

Colclough, A. R., Quinn, T. J. and Chandler, T. R. (1979). An acoustic redetermination of the gas constant. *Proc. Roy. Soc. London*, **A368**, 125–139.

Crovini, L. and Actis, A. (1978). Noise thermometry in the range 630–962 °C. *Metrologia*, **14**, 69–78.

de Boer, J. and Michels, A. (1939). Quantum mechanical calculation of the second virial coefficient of helium at low temperatures. *Physica*, **6**, 409–420.

Durieux, M. (1962). Thermometry at liquid helium and liquid hydrogen temperatures. Thesis, University of Leiden.

Durieux, M., Van Dijk, H., Ter Harmsel, H. and Van Rijn, C. (1962). Some remarks on magnetic thermometry between 1.5 K and 23 K and on the vapour pressure–temperature relation of liquid hydrogen. *TMCSI*, **3**, Part 1, 383–390.

Fritche, L. (1960a). Präsizionsmessung der klassichen Schallabsorption mit Hilf des Zylinderresonators I. *Acustica*, **10**, 189–198.

Fritche, L. (1960b). Theorie des akustischen Zylinderresonators unter Berücksichtigung der Schallanregung II. *Acustica*, **10**, 199–207.

Gammon, B. E. and Douslin, D. R. (1970). Proc. Fifth Symposium on Thermophysical

Properties, p. 107, Am. Soc. Mech. Engr., New York.

Garrison, J. B. and Lawson, A. W. (1949). An absolute noise thermometer for high temperatures and high pressures. *Rev. Sci. Instr.* **20**, 785–794.

Glover, R. M. and Weinhold, F. (1976). Dynamic polarizabilities of two-electron atoms, with rigorous upper and lower bounds, *J. Chem. Phys.* **65**, 4913–4926.

Gugan, D. and Michel, G. W. (1980). Dielectric constant gas thermometry. *Metrologia*, **16**, 149–167.

Guildner, L. A. and Edsinger, R. E. (1976). Deviation of International Practical Temperatures from thermodynamic temperatures in the temperature range from 273.16 K to 730 K. *J. Res. NBS*, **80A**, 703–738.

Guildner, L. A., Stimson, H. F., Edsinger, R. E. and Anderson (1970). An accurate mercury manometer for the NBS gas thermometer. *Metrologia*, **6**, 1–18.

Hill, T. L. (1958). Theory of the dielectric constant of imperfect gases and dilute solutions. *J. Chem. Phys.* **28**, 61–66.

Hudson, R. P. (1972). Principles and applications of magnetic cooling, North-Holland, Amsterdam.

Hudson, R. P., Marshak, H., Soulen, R. J. and Utton, D. B. (1975). Recent advances in thermometry below 300 mK. *J. Low Temp. Phys.*, **20**, 1–102.

Johnson, K. N. and Messmer, R. P. (1974). Clusters, Chemisorption, and catalysis. *J. Vac. Sci. and Technology*, **11**, 236–242.

Kamper, R. A. and Zimmerman, J. E. (1971). Noise thermometry with the Josephson effect. *J. Appl. Phys.*, **42**, 132–136.

Kamper, R. A. (1977). "Superconducting devices for metrology and standards", Chapter 5 in: Superconducting Applications: SQUIDS and Machines, (B. Schwartz and S. Foner, eds), Plenum Press.

Kaneda, R., Sudo, S. and Nisjibata, K. (1964). An interferometric primary standard barometer. *Bull. NRLM*, **9**, 1–28.

Keesom, W. H. and Walstra, W. K. (1940). Isotherms of helium at liquid helium temperatures. *Physica*, **7**, 985–991.

Keesom, W. H. and Walstra, W. K. (1947). *Physica*, **13**, 225.

Keller, W. E. (1955). Pressure–volume isotherms of ^{4}He below 4.2 K. *Phys. Rev.* **97**, 1–12.

Kemp, R. C., Besley, L. M. and Kemp, W. C. (1982). Constant volume gas thermometry between 13.81 K and 83.8 K. *TMSCI*, **5**, 33–38.

Kilpatrick, J., Keller, W., Hammel, E. and Metropolis, N. (1954). Second virial coefficient of ^{3}He and ^{4}He. *Phys. Rev.*, **94**, 1103–1110.

Klein, H. H., Klempt, G. and Storm, L. (1979). Measurement of the thermodynamic temperature of ^{4}He at various vapour pressures by a noise thermometer. *Metrologia*, **15**, 143–154.

Klempt, G. (1982). Errors in Johnson-noise thermometry. *TMSCI*, **5**, 125–128.

Krasnooshkin, P. E. (1944). On supersonic waves in cylindrical tubes and the theory of the acoustic interferometer. *Phys. Rev.* **65**, 190–195.

Mason, E. A. and Spurling, T. H. (1969). "The Virial Equation of State", Pergamon.

McGlashan, M. L. (1979). "Chemical Thermodynamics", Chapter 13, Academic Press, New York and London.

McGonville, G. T. (1969). Thermolecular pressure corrections in helium vapour pressure thermometry: the effect of the tube surface. *Cryogenics*, **9**, 122–127.

McGonville, G. T. (1972). The effect of the measuring tube surface on thermomolecular corrections in vapour pressure thermometry, *TMCSI*, **4**, No. 1, 159–165.

Moldover, R. R. and Mehl, J. B. (1982). Spherical acoustic resonators: promising tools for thermometry and measurement of the gas constant, Precision Measurements and Fundamental Constants 2. NBS.

Nyquist, H. (1928). Thermal agitation of electric charge in conductors. *Phys. Rev.*, **32**,

110–113.

Peggs, G. N., Elliott, K. W. T. and Lewis, S. (1979). An intercomparison between a primary standard mercury barometer and a gas-operated pressure balance standard. *Metrologia*, **15**, 77–85.

Peggs, G. N. (1977). A method for computing the effective area of a pressure balance from diametral measurement. NPL Report No. 23.

Pickup, C. P. (1975). A high resolution noise thermometer for the range 90–100 K. *Metrologia*, **11**, 151–159.

Plumb, H. and Cataland, G. (1966). Acoustical thermometer and the National Bureau of Standards Provisional Temperature Scale 2–20 (1965). *Metrologia*, **2**, 127–139.

Pursey, H. and Pyatt, E. C. (1959). Measurement of equivalent noise resistance of a noise thermometer amplifier. *J. Sci. Instr.* **36**, 260.

Quinn, T. J. (1969). Systematic errors in acoustic thermometry, NPL Report Qu 5.

Quinn, T. J., Colclough, A. R. and Chandler, T. R. (1976). A new determination of the gas constant by an acoustical method. *Phil. Trans. Roy. Soc. London*, **283**, 367–420.

Redhead, P. A., Hobson, J. P., Kornelsen, E. V. (1968). "The Physical Basis of Ultra High Vacuum", Chapman and Hall, London.

Roberts, T. R. and Sydoriak, S. G. (1956). Thermomolecular pressure ratios for ^{3}He and ^{4}He. *Phys. Rev.*, **102**, 304–308.

Roth, A. (1976). "Vacuum Technology", North Holland, page 170 *et seq.*

Sakurai, H. (1982). Constant volume gas thermometry for thermodynamic measurements of the triple point of oxygen. *TMCSI*, **5**, 39–42.

Schram, A. (1963). La désorption sous vide. *Le Vide*, **18**, 55–68.

Silver, A. H., Zimmerman, J. E. and Kamper, R. A. (1967). Contribution of thermal noise to the line-width of Josephson radiation from superconducting point contacts. *Appl. Phys. Letters*, **11**, 209–211.

Soulen, R. J. and Van Vechten, D. (1982). Noise thermometry using a Josephson junction. *TMCSI*, **5**, 115–124.

Tabor, D. (1979). "Gases Liquids and Solids", 2nd Edition, Cambridge University Press.

Van Degrift, C. T., Bowers, W. J., Wildes, D. G. and Pipes, P. B. (1978). In Advances in Instrumentation, **33** Part 4 (Instrument Society of America, Pittsburgh) paper 805, p. 33.

Van der Ziel, A. (1970). "Noise: Sources, Characterisation, Measurement", Prentice Hall, New York, p. 54.

Weber, S. and Schmidt, G. (1936). Leiden Communications, 246C.

Yntema, Y. L. and Schneider, W. G. (1950). Compressibility of gases at high temperatures III. The second virial coefficient of helium in the temperature range 600 °C to 1200 °C. *J. Chem. Phys.*, **18**, 641–646.

4

Fixed Points and Comparison Baths

4-1   Introduction

We have already seen in Chapter 2 how the International Practical Temperature Scale came to be based upon a group of fixed points of temperature together with specified interpolation methods between them. The fixed points of the first International Temperature Scale were the boiling points of oxygen, water and sulphur and the freezing points of water, silver and gold. In the most recent version of the Scale, the boiling points of hydrogen and neon, the triple points of hydrogen, neon, argon, oxygen and water and the freezing points of tin and zinc have all been added, while the boiling point of sulphur has been deleted. In recent years there has been a marked change in emphasis away from boiling points towards triple points or freezing points, for the simple reason that they can be realized without the need for a measurement of pressure. Continued demands for improved precision in boiling point realizations had led to requirements for pressure measurement that were becoming increasingly difficult to meet. For example, at the boiling point of water, a reproducibility of 0.1 mK in temperature requires a reproducibility of pressure measurement of 0.3 Pa, and at the boiling point of sulphur, a pressure change of 0.3 Pa leads to a temperature change of 0.2 mK. In addition to the replacement of boiling points by triple or freezing points, the need to extend the International Practical Temperature Scale below 13.81 K into regions where triple points are not available, has led to the development of fixed points based upon solid state transitions. The most important step in this direction has been the adoption of superconducting transition temperatures as fixed points in the lower part of EPT-76.

In this chapter we shall consider the basic principles to be followed in the high-accuracy realization of fixed points. The question of how to realize a fixed point to a rather lower accuracy is one that is often posed by those who are not interested in, or have no need of thermometry at the highest level of accuracy.

It is rarely easy to give a detailed answer to such a question for the following reason. There are many perturbing factors operating in the realization of a fixed point, any one of which can lead to a reduction in accuracy. To test that a given perturbing factor is small or negligible under a given set of circumstances is usually much easier than to measure its effect if it is large. For example, the melting curve of a nearly-pure metal having, say, no more than 20 ppm of impurity, covers only a very small range of temperature, perhaps less than a millikelvin. The question of how much impurity can be tolerated if the user is only interested in an accuracy of 0.1 K is a complicated one. The answer depends upon the nature of the impurities, whether they raise or lower the melting point, their solubility, the rate of melting and freezing and the previous thermal history of the sample. It, therefore, cannot be given without a great deal of experimental work. Thus the use of fixed points for a so-called rough calibration, when the proper precautions are not taken to ensure a good realization of the fixed point, may lead to errors that can be very much larger than might be imagined. It is mainly for this reason that the quickest, simplest and most economic calibration of a thermometer, for all but the very highest accuracies, can be obtained by comparison with a calibrated thermometer of the same type. All that is necessary, in addition to the calibrated thermometer, is a uniform temperature enclosure to cover the range of temperature in question. Using the comparison method, there are very few errors that are likely to be significant, since even non-uniformities in the temperature of the enclosure, inadequate thermometer immersion or thermometer lag tend to cancel, provided that the two thermometers are of the same type and are mounted in close proximity to each other. Before entering into a discussion on the realization of fixed points, therefore, we shall first consider the various types of uniform temperature enclosures that are useful for the comparison of thermometers.

4-2   Comparison baths and furnaces

The simplest and most widely used comparison bath is the stirred-liquid bath, either of the concentric-tube type (Fig. 4-1) or the parallel-tube type (Fig. 4-2). The essential feature of both of these designs is the separation of the chamber within which the heat is supplied from that within which the thermometers are placed. As great a distance as possible is allowed between the point of generation of heat and the thermometers. Baths of the concentric tube type are best for the range down to $-150\,°C$ using isopentane as the stirred liquid; for the range 80 °C to 300 °C using various mineral oils, and for the range 200 °C to 600 °C using a mixture of molten salts. From 1 °C to 100 °C, very effective stirred water baths may easily be made using the parallel-tube design and having an electric immersion heater at the bottom of the heater tube. A uniformity of temperature within 1 mK can be obtained over 50 cm of

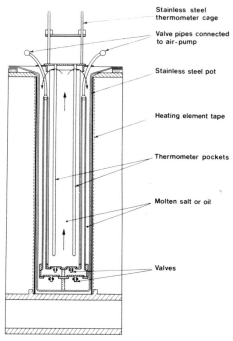

Fig. 4-1. Stirred-liquid bath, concentric-tube type, stirred by the action of an air pump.

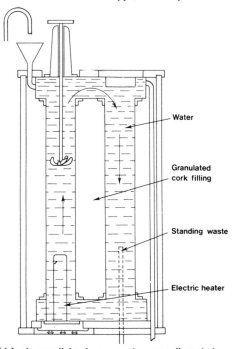

Fig. 4-2. Stirred-liquid bath, parallel-tube type using propeller stirring.

immersion at a temperature of 50 °C, and a uniformity of 3 mK over the same depth at 90 °C.

The use of isopentane requires a double-walled vessel immersed in liquid nitrogen. When air is present in the space between the walls, rapid cooling of the isopentane is obtained which can be arrested by evacuation of the air when the required temperature is reached. The residual cooling can subsequently be controlled by a bare-wire heater immersed directly in the liquid. A propeller is used to circulate the isopentane up the central tube and down the outer tube.

Above the range of the water bath, light mineral oils are used up to about 200 °C, and heavy oils for higher temperatures up to about 300 °C. The upper limit for ordinary oils is set by their flash point or by their tendency to carbonize, and for silicone oils by the emission of dangerous fumes above about 200 °C. The lower limit for the use of any oil is of course simply the temperature at which the viscosity has become too high for efficient heat transfer to be maintained. For those applications near room temperature in which the presence of water is excluded for one reason or another, there exists a range of light paraffin or silicone oils that are perfectly suitable. At the lower end of the temperature range, the uniformity of temperature that can be achieved in an oil bath is hardly inferior to that obtained in a water bath. Above 100 °C, the best uniformity that can be obtained is about 10 mK over a 50 cm immersion, and above 200 °C, it is about 50 mK over the same depth.

At temperatures above about 300 °C, there exist no suitable oils and, instead, a mixture of equal parts of potassium nitrate and sodium nitrite is employed. Such a mixture works well over the range 150 °C to 600 °C. Mixtures of these salts are very corrosive, and the baths, along with everything else that comes into contact with the hot salt, must be made of a suitable corrosion-resistant material such as austenitic stainless steel. It must also be emphasized that water or moisture must be excluded at all costs from contact with the molten salt since even small quantities can cause a serious explosion. It is also essential to avoid bringing any other easily oxidizable material, such as aluminium, into contact with the molten salt. Before building or operating a salt bath, reference should be made to an appropriate industrial safety code such as "Precautions in the use of nitrate salt baths" (HMSO Pamphlet No. 27, London 1964).

Above 600 °C it is necessary to turn to devices other than stirred liquid baths for the provision of uniform temperature enclosures for the comparison of thermometers. One such device is the fluidized bed, illustrated in Fig. 4-3. The great advantage of the fluidized-bed bath is the wide range of temperature accessible with a single material. For example, the system illustrated in Fig. 4-3, using alumina powder fluidized by the passage of air can be operated from room temperature to 900 °C. The disadvantages of the fluidized-bed bath are twofold: first, it is quite difficult to avoid depositing a fine layer of alumina powder over everything in the vicinity and second, the uniformity of temperature is inferior to that obtainable in stirred-liquid baths or resistance

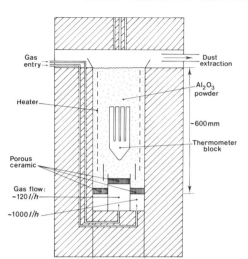

Fig. 4-3. Fluidized-bed calibration bath (schematic) (after Bousson, 1978).

wound furnaces. At 400 °C, a temperature difference of 0.4 °C over 25 cm might be expected in the fluidized-bed bath of Fig. 4-3, while at 900 °C, this would rise to at least 1 °C. Of course for the comparison of thermometers, a metal block would normally be placed in the fluidized bed, which would lead to a rather better uniformity of temperature being obtained.

More widely used than the fluidized bed is the resistance-wound furnace, of which many varieties exist from the simple nichrome-wound furnace for use up to about 1100°C, to the much more complicated molybdenum-wound high-temperature furnace which must be operated in an inert atmosphere. For temperatures up to 1100 °C, the design illustrated in Fig. 4-4 is quite suitable.

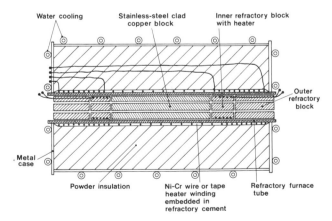

Fig. 4-4. Furnace for comparison of thermometers up to 1100 °C.

The heater winding is a strip of nichrome (80/20 Ni–Cr alloy), and the furnace tube is one of the proprietary refractory tubes suitable for operation in air at 1100 °C. The heater is normally a single winding, but to improve the distribution of temperature along the tube, it may be split into three sections, permitting a reduction in the current in the centre section by means of a shunt. Depending upon the ratio of length to diameter of the furnace tube, it may be necessary to provide additional means of heating inside the tube at each end of the comparison metal block, as illustrated in Fig. 4-4. Temperature control is best carried out using a commercial temperature controller operating on the current passing through the main heater only. To avoid undue complication, the shunt and the end-heaters are normally set manually at an appropriate fraction of the main heater current. In building a furnace of the type shown in Fig. 4-4 it is essential to ensure that the powder insulation is very firmly packed to avoid the formation of voids in the vicinity of the heater element. The presence of such voids is one of the most frequent causes of heater failure. The metal block, placed in the furnace tube to improve temperature uniformity around the thermometers, may be made of pure copper, stainless-steel-clad copper, nickel or even silver depending upon the particular requirements and temperature range. Stainless-steel clad copper is particularly advantageous for use at temperatures up to about 1050 °C (m.p. of copper 1084 °C). A cladding of stainless steel only about 1.5 mm thick will protect copper in air for very long periods at this temperature. It may be used both for simple blocks for the comparison of thermocouples or resistance thermometers (Fig. 4-5), or for blackbody cavities used for the calibration of radiation pyrometers (see Chapter 7). The use of a heat pipe for the provision of a uniform temperature enclosure in a furnace is discussed below.

The temperature range of furnaces of the type shown in Fig. 4-4 can be extended as far as about 1250 °C using Kanthal alloy heaters. For higher temperatures, however, it is necessary to employ heaters made from alloys of platinum and rhodium. Platinum-10% rhodium can be used up to about 1600 °C in the form of wires (0.5 mm diameter) or tapes (12 mm × 0.025 mm), the latter being less expensive but having a shorter life. Above 1600 °C, higher

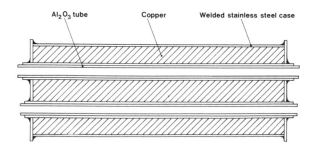

Fig. 4-5. Stainless-steel-clad copper block for use in the furnace shown in Fig. 4-4.

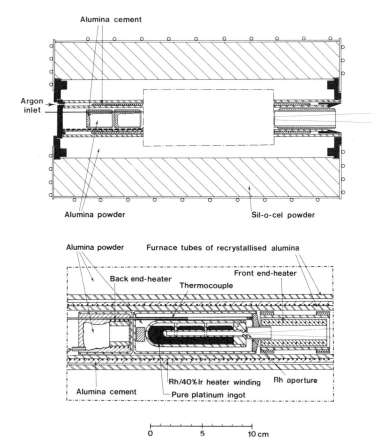

Fig. 4-6 (a and b). Furnace and blackbody assembly used in the determination of the freezing point of platinum (Quinn and Chandler, 1972).

alloys of platinum and rhodium or iridium can be used. Such a furnace is shown in Fig. 4-6 designed for a determination of the freezing point of platinum; it was also used for the calibration of thermocouples up to the melting point of platinum, 1769 °C (see Chapter 6), using an alumina blackbody cavity as reference. In furnaces operating at temperatures above about 1200 °C, the furnace tube and the refractory elements inside the furnace tube such as insulators, spacers or a blackbody must be made of high purity re-crystallized alumina.

Furnaces for still higher temperatures must be operated in an inert atmosphere, and therefore pose quite different problems. Heating elements may be made of tungsten, molybdenum or tantalum, and for use at temperatures above 2000 °C they are usually self-supporting to avoid reactions which would otherwise occur with refractory oxide supports. The whole question of compatibility of different materials becomes very difficult at

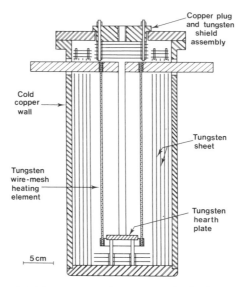

Fig. 4-7. The hot zone of a high-temperature furnace operating in ultra-high vacuum (after Burns and Hurst, 1971).

high temperatures, and the choice of refractory suitable for use with a given metal is often very limited. For example, above 2000 °C, only BeO, Y_2O_3, ThO_2 and HfO_2 can be considered reasonably stable (Droege *et al.*, 1972; Schich, 1966). An example of a high-temperature furnace designed for the calibration of thermocouples (Burns and Hurst, 1971) at temperatures up to 2700 °C is shown in Fig. 4-7. The heating element in this furnace is a tungsten mesh 6.4 cm in diameter and 30.5 cm long and is enclosed in tungsten radiation shields. The furnace can be operated either in a vacuum at a pressure of 10^{-6} Pa or in high-purity argon.

As an alternative to metallic heating elements, furnaces for use at high temperatures can employ heaters made from graphite or from doped mixtures of refractory oxides. It is usual in a heater of this type to adjust the cross-section (Groll and Neuer, 1972) or even the composition (McMahon and Rothwell, 1972) along its length to compensate for heat losses from the ends (Fig. 4-8). One of the more difficult problems to deal with, when using heaters of this sort at very high temperatures, is the need to allow for the thermal expansion of the heater. Expansions totalling some millimetres are common, and must be accommodated without putting undue stress on the heating element while at the same time maintaining good electrical contact.

4-3 Heat pipes

In what has gone before we have been concerned with various ways of

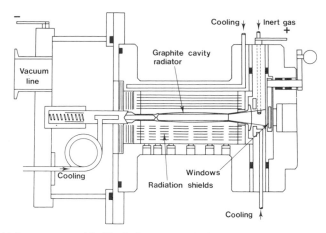

Fig. 4-8. A high-temperature blackbody furnace employing a graphite cavity (after Groll and Neuer, 1972).

producing a uniform temperature zone within which thermometers can be compared. In recent years, a new device has become available for this purpose which makes use of the large heat-of-vaporization of fluids. It is called a "heat pipe", and was originally developed as a way of transporting very large amounts of heat from narrow confined spaces. A heat pipe behaves as a tube exhibiting a very large thermal conductivity, as much as 10^4 times that of copper. It follows that if such a tube is placed in a region of nearly uniform temperature, its very large thermal conductivity will lead to a highly uniform temperature being established within it. Herein lies the interest for thermometrists.

The basic principles of a gas-controlled heat pipe for the production of a uniform temperature zone are illustrated in Fig. 4-9. In operation, heat is

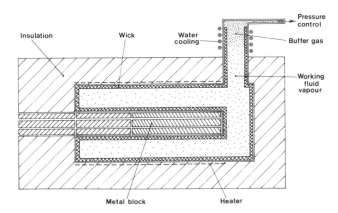

Fig. 4-9. A gas-controlled heat pipe for the comparison of thermometers (schematic).

supplied to the outside of a chamber containing a small quantity of a volatile fluid so that at the required temperature, the vapour pressure of the fluid lies within the range 25 kPa to 300 kPa. The whole of the outside of the chamber is well insulated except for a pipe which leads to a pressure control system containing an inert gas. The working fluid will condense in this tube at the point where the temperature is such as to make the vapour pressure of the fluid equal the pressure applied to the inert gas. The condensed vapour will be drawn back into the hot enclosure by a porous wick attached over the whole of the inside of the heat pipe. In this way, a continuous flow of fluid is maintained through the vapour and liquid phases. The condensing vapour carries with it the latent heat of vaporization, and will condense and deposit this heat wherever the temperature in the heat pipe falls below that of the vapour/gas interface. Provided that the insulation surrounding the heat pipe is good, a relatively modest vapour flow will be sufficient to maintain a uniform temperature throughout the whole of the inside of the heat pipe surface. The temperature is controlled by the pressure applied by the inert "buffer" gas. An increase in pressure will push the vapour/gas interface up the temperature gradient in the condenser tube and lead the whole of the heat pipe to take up the new, higher, temperature of the vapour/gas interface. An increase in heater power will increase the rate of evaporation of fluid and will advance the vapour/gas interface towards the lower, temperature end of the condenser. If the buffer-gas pressure is maintained constant, however, the temperature of the heat pipe must remain constant. Therefore the advance of the vapour/gas interface will proceed just as far as is necessary for the heat required to raise the temperature of the part of the condenser now at the vapour/gas interface to equal the extra heat brought out from the heat pipe by the increase in vapour condensation. In this way, variations in heater power or in heat losses from the heat pipe are compensated for by movements in the position of the vapour/gas interface to maintain a constant and uniform temperature over the whole of the condensing surface.

Under nearly ideal conditions, the principal heat loss from the heat pipe will be that which takes place at the vapour/gas interface, and this will be small if the internal surface area at the point of condensation is small. Even though the overall heat flow is small, there will be a small temperature gradient through the vapour from the point of evaporation to the point of condensation. This arises from the pressure gradient that must be present between the evaporation and condensation regions to provide the driving force for the movement of vapour. Nevertheless, the resulting temperature drop is extremely small and accounts for the usefulness of the heat pipe as a device for maintaining a uniform temperature zone. As an indication of the quantity of heat that can be carried by a vapour stream under only a minute temperature gradient (Fig. 4-10), a heat flux of 1 kW can occur in a water vapour heat pipe 1 m long and 5 cm in diameter for a total temperature drop of less than 0.1 mK!

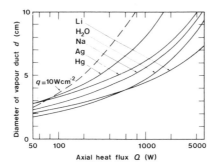

Fig. 4-10. The dynamic temperature-drop in a heat pipe of length 100 cm at a pressure of one atmosphere and with $\Delta T/T = 2 \times 10^{-7}$, for various fluids (after Busse *et al.*, 1975).

There are, of course, other factors to be considered that affect the overall performance of a heat pipe, and real temperature gradients can far exceed that resulting from the simple vapour flow. Among these are the thermal resistance of the wick and of the walls of the heat pipe, variations in the position within the wick of the liquid/vapour interface at the point of evaporation and hydrostatic pressure variations within the vapour column. In addition, capillarity failures resulting from the presence of impurities can lead to non-wetting of certain parts of the internal surface. Although the principles of the gas-operated heat pipe are straightforward, the details of its performance and its quantitative evaluation are extremely complex and reference should be made to original papers for further details (Sun and Tien, 1975; Grover *et al.*, 1964; Busse *et al.*, 1975).

It is, of course, possible to operate a heat pipe without the presence of a buffer gas. In this case, the temperature of the heat pipe will be controlled by the furnace within which it is placed (Busse *et al.*, 1975). The heat-pipe temperature will follow that of the coldest part of the furnace accessible to it. The internal vapour pressure and temperature will rise and fall according to variations in the outside temperature while maintaining a very-nearly uniform temperature over the whole of the internal surface. Various commercial heat pipes of this type are available, and are designed to be placed in a furnace to provide a uniform temperature zone.

From what has been said so far about the performance of heat pipes, one could draw the conclusion that the heat pipe is the ideal solution for a large number of thermometry problems. This may be so, but with some qualifications. One of the most important is the relatively narrow range of temperature over which a given working fluid has a useful vapour pressure. The exponential rise of vapour pressure with increase in temperature means that the temperature range within which the vapour pressure is sufficiently high to provide adequate heat transfer yet not so high as to pose severe mechanical problems of containment, is very limited. Account must also be taken of the need for compatibility of wall and wick material and working

Table 4-1. Working fluid/wall material combinations for heat pipes (after Neuer and Brost, 1975).

Working fluid	Wall material	Useful temperature range (K)
Liquid nitrogen	Stainless steel	70–110
Methane	copper, aluminium	100–150
Carbon tetrafluoride	copper, aluminium	100–200
Freons	copper, aluminium	120–300
Ammonia	stainless steel, nickel aluminium	230–330
Acetone	copper, stainless steel	230–420
Methanol	copper, nickel	240–420
Water	copper, nickel, titanium, stainless steel	300–550
Organic fluids	stainless steel, super alloys, carbon steel	400–600
Mercury (with additives)	stainless steel	450–800
Potassium	stainless steel, nickel	700–1000
Sodium	stainless steel, nickel	800–1350

fluid. In Table 4-1 are listed some of the possible combinations and their temperature range. The most useful for the purposes of thermometer calibration is probably the sodium/stainless steel combination. Over the range 500 °C to 1100 °C, it is becoming increasingly used in both metal fixed point and comparison furnaces.

4-4   Boiling points of water and sulphur

The previous discussion of gas-controlled heat pipes leads on naturally to some remarks concerning the realization of the boiling points of water and sulphur. The only difference between the gas-controlled heat pipe, as described above, and the classical apparatus for the realization of the steam and sulphur points is the absence in the latter of a wick covering the whole of the inside surface. Instead, the function of the wick, viz. to return the condensed vapour to the region of vaporization, is carried out simply by gravity. Although no longer a defining fixed point of IPTS-68, the sulphur boiling point (444 °C) remains useful in that it provides a convenient means of comparing thermometers at or near the zinc point. The apparatuses commonly used to realize the steam and sulphur points are shown in Figs 4-11 and 4-12. The modifications required to the designs shown in these two figures to permit a

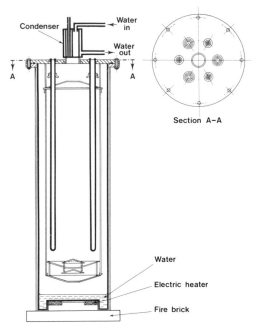

Fig. 4-11. Steam-point apparatus made from stainless steel.

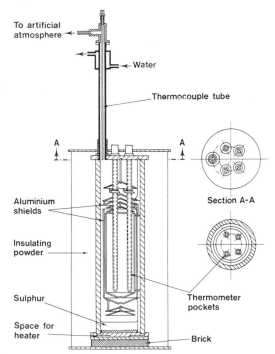

Fig. 4-12. Sulphur-point apparatus made from pure aluminium.

range of temperatures to be covered need be no more than a pressure-control system attached to the vapour outlet tube and operating on an inert gas.

Even though the freezing point of tin is present in the IPTS-68 as an alternative for the definition of the scale, a steam point apparatus remains essential for a metrological laboratory. This is because few low-temperature capsule-type platinum resistance thermometers can be operated at temperatures above the tin point, 231 °C, and thus a determination of both the α-coefficient and the δ-coefficient requires measurement at 100 °C, and at one other temperature above 100 °C (see Chapter 5). As was noted at the very beginning of this chapter, the demands of pressure measurement become increasingly difficult to meet as the accuracy required of a boiling point measurement increases. However, when a boiling-point apparatus is used simply as a comparison bath, an absolute knowledge of the pressure is not required and the necessary pressure stability is fairly easily ensured.

4-5   The low temperature boiling and triple points

For the purposes of precision thermometry, the low-temperature boiling or triple points of most interest are those of helium, hydrogen, neon, oxygen, argon and methane. The general principles to be followed in the realization of one of these boiling or triple points are common to them all. They will be illustrated in the course of a description of the apparatus and techniques used to realize the triple and boiling points of hydrogen followed by a few comments on the particular points applicable to the other gases in turn. The vapour pressures of ^{3}He and ^{4}He occupy a special place in that they provide internationally agreed temperature scales in their own right. The derivation and realization of these vapour pressure scales have already been discussed in Chapter 2, and therefore helium will not be mentioned further in this chapter.

4-5-1   The triple and boiling points of hydrogen

Two types of hydrogen molecule exist: ortho-hydrogen, in which the spins of the two protons are parallel; and para-hydrogen, in which they are opposite or anti-parallel. In the case of ortho-hydrogen, the nuclear spin moment takes a value of 1 and can therefore take any one of the three values, 1, 0 or −1, relative to the angular momentum vector of the whole molecule. In the case of para-hydrogen, the nuclear spin moment is zero, and therefore only this one value is possible for the spin of the whole molecule. The state of lowest energy is that of para-hydrogen for which a rotational quantum number of zero is accessible, i.e. the lowest of the even quantum numbers. Ortho-hydrogen has access only to the odd quantum numbers. At low temperatures, therefore, the existence of para-hydrogen is more favoured and, indeed, as the temperature falls, the

Table 4-2. Proportions of ortho- and para-hydrogen as a function of temperature.

T/K	% para-hydrogen	Specific heat/kJ Kg^{-1} K^{-1}		
		para-	ortho-	normal
0	100	10.32	10.32	10.32
20	99.8	10.40	10.32	10.34
50	76.9	13.41	10.47	11.20
100	38.5	16.08	12.69	13.54
200	26	14.85	14.14	14.32
300	25.1			
∞	25			

proportion of para-hydrogen increases. At high temperatures, the proportions of ortho- and para-hydrogen tend towards the values imposed by the relative probabilities of the spin states, namely 3:1 in favour of ortho-hydrogen. The approximate proportions of ortho- and para-hydrogen as a function of temperature are shown in Table 4-2 (Woolley et al., 1948).

Hydrogen having the ortho-para proportions of 3:1 is known as "normal" hydrogen and hydrogen having the equilibrium proportions of ortho- and para-hydrogen for a particular temperature is known as "equilibrium" hydrogen. The existence of ortho- and para-hydrogen would be of only minor interest to thermometrists were it not for the fact that the rate of conversion of normal hydrogen to equilibrium hydrogen is extremely slow unless special steps are taken to accelerate it. Complete spontaneous conversion at the triple point, 13.81 K can take many weeks, and since the boiling and triple points of ortho- and para-hydrogen differ (Table 4-3), the slow change in the proportions of ortho- and para-hydrogen lead to slow changes in the observed triple and boiling points, Fig. 4-13.

Non-equilibrium mixtures of ortho- and para-hydrogen have triple and boiling points between the values given in Table 4-3. As a temperature reference point, therefore, the state of the hydrogen must be specified. Since the

Table 4-3. Boiling and triple points of normal and equilibrium hydrogen.

	Normal hydrogen (75 % ortho)	Equilibrium hydrogen (0.21 % ortho)
triple point	13.956 K	13.81 K
triple-point vapour pressure	7210.3 Pa	7034.5 Pa
boiling point	20.397 K	20.28 K

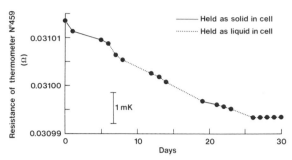

Fig. 4-13. Change in hydrogen triple-point temperature resulting from ortho→para conversion in the sample chamber (after Kemp and Kemp, 1979).

ortho→para conversion is in the direction of a lower energy state, the conversion to equilibrium from a state equilibrated at higher temperature is accompanied by a liberation of heat, about 1300 J mol^{-1} at 20 K. An unfortunate consequence of this is that if a vessel is filled with unconverted hydrogen straight from a liquefier, the heat subsequently released on conversion is sufficient to boil off more than half of the original amount of liquid present. It is therefore desirable to include a conversion catalyst between the liquefier and the storage vessel so that most of the conversion takes place before the liquid enters the storage vessel.

A particularly efficient conversion catalyst is ferric hydroxide in a finely divided form, although oxides of other magnetic elements such as chromic oxide or those of iron–nickel alloys are also effective. In fact, it is worth noting that the complete avoidance of a catalyst, for those experiments in which totally unconverted normal hydrogen is required, poses considerable problems. The most effective use of ferric hydroxide as a catalyst requires intimate contact between the surface of the ferric hydroxide and the liquid hydrogen. Since the rate of diffusion of one form of hydrogen through the other is very slow, it does not suffice simply to have a layer of ferric hydroxide at the bottom of the vessel containing the liquid. The various ways of distributing the ferric hydroxide in a typical cryostat have been studied in some detail by Ancsin (1977). He concluded that the ferric hydroxide must be distributed throughout the whole of the low temperature part of the cryostat. His results are illustrated in Fig. 4-14; the most rapid ortho→para conversion rates occurred after 1 g of catalyst had been poured down the chamber entry tube, and the same tube subsequently plugged with a gauze holding a further 0.1 g of catalyst. Under these conditions, ortho→para conversion appeared to be complete within minutes of cooling down. It is necessary to activate the ferric hydroxide by heating in vacuum to about 150 °C for some 20–40 h. This has the effect of driving off absorbed water and gas molecules, thereby increasing the effective surface area.

Having ensured proper ortho-para conversion, the realization of the hydrogen triple and boiling points is relatively straightforward. The observed

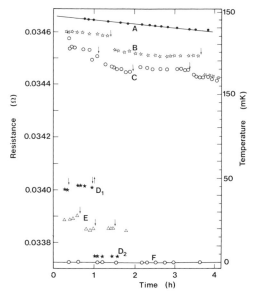

Fig. 4-14. The influence of hydrated ferric oxide catalyst, in various amounts and locations within the sample system, on the melting point of hydrogen. A, no catalyst; B, 0.78 g of catalyst in the catalyst chamber; C, 0.072 g of catalyst in the catalyst chamber; D1, D2, 0.147 g of catalyst in the vapour phase; E, 0.147 g of catalyst in the vapour phase touching the condensed sample; F, 1.0 g of catalyst poured down into the sample chamber. An arrow ↑ indicates re-freezing and ↓ indicates vaporizing and then re-freezing (after Ancsın, 1977).

effects of impurities on the melting point and on the vapour pressure of hydrogen are very small. This is partly because it is easy to ensure that few of the most important impurities ever reach the sample chamber. For instance, those gases whose boiling points are above that of nitrogen condense in regions remote from the liquid hydrogen. The most likely impurities are nitrogen, neon and helium. Helium is apparently insoluble in both liquid and solid hydrogen and has therefore no effect on the triple point. It was shown by Ancsin (1977) that neon is likely to be the most troublesome impurity, both for the melting point and the vapour pressure.

A typical apparatus used to realize the hydrogen triple point is illustrated in Fig. 4-15. For the cryostat shown in the figure, (1) is an outer vacuum case, (2) a temperature-controlled outer shield, and (3) the cell containing the sample. The cell consists of three parts: (a) a lower gas-cooled refrigerator (4), to which is soldered a copper tube (5), forming the effective thermal outer wall. This assembly is heated by a carbon heater (6), and its temperature monitored by a miniature platinum resistance thermometer (6a); (b) an upper gas-cooled refrigerator (7) soldered to a thick-walled copper thermometer pocket (8) in which the test thermometer is inserted with Apiezon N grease. The thermometer pocket may be heated by the carbon resistor (10); (c) the outer wall of the cell (11) consists of a thin-walled stainless steel tube 25 mm in

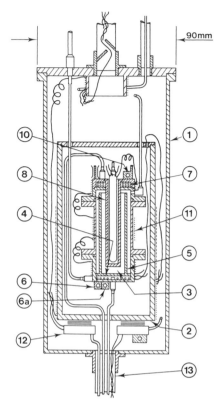

Fig. 4-15. A cryostat for the realization of triple points, see text for key (after Kemp *et al.*, 1976).

diameter which isolates the test thermometer from the heated outer wall (5) and the bottom of the cell. The three parts of the cell are sealed together with indium-wire seals. The radiation shield (2) is controlled adiabatically with respect to the cell using a gas-cooled refrigerator (12) as heat sink. The cold gas for the refrigerators is drawn up from the liquid helium through vacuum insulated tubes (13). The cryostat is suspended in a 100 cm deep helium dewar such that the base of the cryostat is some 50 cm above the bottom of the dewar. During operation, the level of liquid helium is about 20 cm below the base of the cryostat.

The principles of operation are very simple (Kemp *et al.*, 1976; Kemp and Kemp, 1979; Quinn and Compton, 1975). A nearly-adiabatic chamber is filled with liquid hydrogen (either already converted or with adequate ferric hydroxide distributed around the chamber) cooled below the freezing point, and the hydrogen allowed to solidify. The sample is then heated gently towards the triple point at a decreasing rate until the temperature of the solid hydrogen is only about 100 mK below its triple point. The heating is then stopped to allow thermal equilibrium to be established. The sample is then

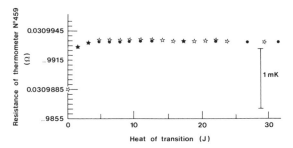

Fig. 4-16. A triple-point plateau of hydrogen in equilibrium at 13.81 K (after Kemp and Kemp, 1979).

melted by applying small heat pulses allowing thermal equilibrium to be re-established between each pulse. After each heat pulse, a measurement of thermometer resistance is made. A typical melting plateau at the triple point of hydrogen is shown in Fig. 4-16. The fact that melting takes place over a negligible temperature range, less than 0.05 mK in Fig. 4-16, allows us to identify unambiguously this melting point with the triple point.

For the measurement of the vapour pressure of liquid hydrogen and hence its boiling points, the cryostat has a different form from that shown in Fig. 4-15 for the triple point. A typical vapour pressure cryostat is shown in Fig. 4-17.

The sample chamber (1) and thermometer wells (2) are drilled in a block of OFHC copper (3) enclosed in a radiation shield (4) bolted to the base of the block. This assembly is connected to a gas-cooled refrigerator (5) enclosed in a further radiation shield (6), which is also bolted to the refrigerator. Stainless steel shims (7) reduce the thermal contact between the block assembly and the refrigerator. The whole assembly is contained in an enclosure (8) suspended from the top plate of a dewar by means of a thin-walled stainless steel tube (9) 12.5 cm in diameter. The sample is filled through the stainless-steel tube (10) which enters the sample chamber through the radiation trap (11) and a subsidiary chamber (12) in which catalyst may be placed. Hydrogen passes into the sample chamber through a sintered stainless-steel disc (13). The vapour-pressure-transmitting tube (14) enters the sample chamber through a radiation trap (15). Thermometer leads are anchored at (16) and (17). Temperature control is by means of a carbon resistance heater (19), platinum resistance thermometer (18) and an external electronic controller. The heat sink for the controller is (5) whose temperature is monitored by a carbon resistance thermometer (20). The temperature of the outer case is controlled with the aid of a platinum resistance thermometer (21) and carbon resistance heater (22).

The need to make absolute measurements of pressure increases the complexity of a boiling-point determination compared with that of a triple point. During a vapour-pressure measurement, the temperature of the sample has to be controlled by outside means so as to remain as stable as possible. This calls for a relatively massive OFHC copper block within which the

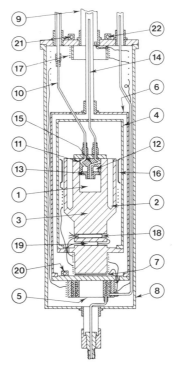

Fig. 4-17. A cryostat used to measure hydrogen vapour pressures, see text for key (after Kemp and Kemp, 1979).

sample chamber and thermometers are placed. A triple-point realization, on the other hand, relies upon the inherent temperature stability during melting and hence the relatively light-weight adiabatic calorimeter. The slope of the vapour pressure/temperature relation of hydrogen increases from 13 Pa mK^{-1} at 17 K to 30 Pa mK^{-1} at 20.28 K. For the precise determination of the 17 K point, therefore, rather more care must be taken with the pressure measurement. The cryostat must be designed so that the coldest spot is always at the sample chamber and never in the pressure-sensing tube through which the pressure is transmitted to the pressure-measuring device outside. A correction may be necessary for the pressure gradient existing in the pressure-sensing tube due to the aerostatic head of gas. The effect is proportional to the density of the gas and hence is inversely proportional to the temperature (see equations 3-30 and 3-31 Chapter 3; and Chapter 3, Section 2-4). This correction is very small for vapour pressure measurements of hydrogen, neon and for most gases having a higher boiling point. The relatively low vapour pressure/temperature slope for oxygen however, 11 Pa mK^{-1}, calls for a small correction if the highest accuracy is being sought. The effect is largest at low temperatures and, as we have seen in Chapter 2, must be measured carefully in precise helium-vapour-pressure work. The most reliable way of reducing errors from this source is to arrange that most of the pressure-sensing tube is made from copper maintained near room temperature, and that the temperature gradient is almost entirely confined to a short section of thin-walled stainless steel tube joining the copper tube to the low-temperature cryostat. In this stainless steel tube, the temperature gradient can be relied upon to be sufficiently linear for errors to be small.

Of course, it is possible, but technically rather more difficult, to reduce the aerostatic head correction strictly to zero. This can be done by arranging that the stainless-steel tube, along which the temperature gradient exists, is horizontal. Without adopting this latter method, the magnitude of the aerostatic head correction in a typical hydrogen-vapour-pressure cryostat is of the order of 3 Pa at 17 K and 1 Pa at 20 K. Care must always be taken in making vapour-pressure measurements of hydrogen to avoid errors due to the entry of unconverted or partially converted gas. If the temperature of the cryostat is reduced, for example, gas will be drawn into the measuring chamber, and it must be ensured that sufficient ferric hydroxide is present to effect rapid conversion.

For safe use of hydrogen in any laboratory, attention must be paid to the inherent dangers involved. These stem largely from the very wide range of explosive air/hydrogen mixtures that exist, and it must therefore always be ensured that hydrogen gas boiled off from storage dewars or experimental cryostats is vented to the outside using properly designed spark-free extraction equipment (Bailey, 1971; Scott et al., 1964).

4-5-2 Boiling points of neon (27.102 K) and oxygen (90.188 K)

A cryostat used for the precise realization of the boiling point of neon (Kemp and Kemp, 1978a) is shown in Fig. 4-18. In the figure, items (1) to (8) serve the same function as in Fig. 4-17, while (9) is a gas-cooled refrigerator whose temperature is monitored by the platinum resistance thermometer (19). The cryostat is suspended from the top plate of the dewar by means of a thin-walled stainless-steel tube (10). The filling tube (11) enters the sample chamber through a radiation trap (12); this tube also acts as the pressure-transmitting tube, and carries three miniature platinum resistance thermometers, one of which is shown (13). The leads from the four test platinum resistance thermometers are thermally anchored at (14) and (15). A platinum resistance thermometer (16) and carbon resistance heater (17) are used in the control of the block (3). The temperature of the sink (5) for this control is monitored by a platinum resistance thermometer (18), and is controlled by adjusting the flow of helium gas.

The main uncertainty associated with the realization of the neon boiling

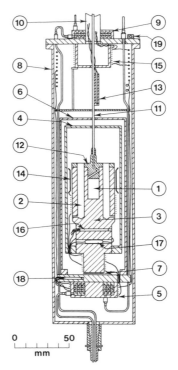

Fig. 4-18. A cryostat for the realization of the boiling point of neon, see text for key (after Kemp and Kemp, 1978a).

point stems from the slight uncertainty surrounding the isotopic composition of natural neon. In the original 1968 edition of IPTS-68, the composition of natural neon was given as 90.9 % ^{20}Ne, 0.26 % ^{21}Ne and 8.8 % ^{22}Ne, based upon measurements made in 1950 (Nier, 1950). More recent work (Walton and Cameron, 1966) suggests that the composition of natural neon is more likely to be 90.5 % ^{20}Ne, 0.26 % ^{21}Ne and 9.26 % ^{22}Ne, and the 1975 edition of IPTS-68 refers to these new values. The presence of heavier fractions in natural neon, i.e. ^{21}Ne and ^{22}Ne, leads to a weak dependence of the vapour pressure on the liquid/vapour ratio and on the evaporation or condensation of liquid in the sample. The temperature on vanishing vapour fraction is called the boiling point, and on vanishing liquid fraction, the dew point. On adding neon to the sample chamber, there is a difference of 0.4 mK between the boiling point (where the liquid has the natural composition) and the dew point (where the vapour has the natural composition). There is some evidence however that on removing neon from the chamber, that is to say, passing from a boiling point to a dew point, a difference of as much as 2 mK is observed (Kemp and Kemp, 1978a).

The effects of impurities on the boiling point of neon are not too serious. Helium is easily removed by freezing and pumping on the solid neon, although hydrogen does not respond to this treatment. The presence of 20 ppm of hydrogen will depress the neon boiling point by 0.1 mK. It is not easy to remove hydrogen from neon, but Ancsin (1978a) has demonstrated that with his design of cryostat, in which there is a large volume of vapour separated from the sample chamber by a narrow tube, the hydrogen appears to be pumped quickly away leaving an uncontaminated neon liquid/vapour surface. The presence of nitrogen and other non-volatile gases in the neon is relatively easy to avoid by ensuring that on initially condensing neon in the sample chamber, there are parts of the entry tube sufficiently cold to trap these other gases.

The boiling point of oxygen is rather more difficult to realize accurately. Its pressure sensitivity is only one third that of neon and, as we have already mentioned, this means that the aerostatic head correction must be taken into account. Impurities are also more likely and at the same time are more difficult to deal with. The very measurement of the purity of oxygen is made difficult by the fact that it reacts so readily with, for example, the filaments of mass spectrometers (Compton and Ward, 1974). Nevertheless extensive studies have been carried out of the effects of impurities on both the boiling and triple points of oxygen (Compton and Ward, 1976; Kemp et al., 1976; Ancsin, 1973a). It has been concluded that CO_2 and H_2O are unimportant because they condense too far away from the sample chamber, and that He and Ne are insoluble in liquid oxygen and are therefore easily pumped away. The most important impurities are nitrogen (as might be expected) and carbon monoxide. Their effects, together with those of argon and krypton, on the boiling point of oxygen are given in Table 4-4.

One of the difficulties in making measurements of the effects of impurities on the boiling point of oxygen, and of the other gases we are considering as well, is that of ensuring strict equilibrium of composition throughout the liquid/vapour system. It is almost always the case that the impurity apparently behaves in a different way upon condensation and vaporization. As was pointed out above in the case of neon, the presence of an impurity elsewhere in the system other than at the liquid–vapour interface does not affect the vapour pressure of the liquid.

Because impurity concentration differs between liquid and vapour phases, incremental condensation or evaporation causes a transient change in the

Table 4-4. Effect of impurity on the boiling point of oxygen.

Impurity	Change in the boiling point of oxygen per volume ppm
N_2	$-19\ \mu K$
CO	$-17\ \mu K$
Ar	$-5\ \mu K$
Kr	$+6\ \mu K$

vapour composition over the liquid surface. Gaseous diffusion eventually re-establishes uniformity of composition, but in the meantime, the boiling point is perturbed. The direction of perturbation depends upon the relative volatility of the impurity and upon whether condensation or evaporation has occurred. Thus for volatile impurities such as nitrogen, apparent boiling points during condensation are consistently lower than during vaporization. A cryostat suitable for the realization of the oxygen point differs little from that shown in Fig. 4-18. A detailed description may be found in Kemp *et al.* (1976) or Compton and Ward (1976).

4-5-3   The triple points of neon (24.561 K), oxygen (54.361 K), nitrogen (63.146 K), argon (83.798 K), methane (90.686 K), krypton (115.763 K) and xenon (161.391 K)

Much work has been carried out in recent years on the realization of the triple points of *neon* (Ancsin, 1978a; Kemp and Kemp, 1978b), *oxygen* (Kemp *et al.*, 1976; Compton and Ward, 1976; Ancsin, 1973a; Pavese, 1978), *nitrogen* (Kemp and Kemp, 1978b), *argon* (Kemp and Kemp, 1978b; Pavese, 1978; Ancsin, 1973b), *methane* (Bonhoure and Pello, 1978), *krypton* (Kemp and Kemp, 1978b) and *xenon* (Ancsin, 1978b). It has now become generally agreed that triple points provide better fixed points of temperature than do boiling points. There are two reasons for this, the first being the absence of the necessity of pressure measurement and the second, stemming from the first, being the recent development of sealed triple point cells. Before coming to the question of sealed cells, however, we will look briefly at the methods used in the realization of the triple points of these gases in the classic triple-point cryostat illustrated in Fig. 4-15.

The general principles to be followed in using a cryostat of the type shown in Fig. 4-15 are common to all of these gases, and differ very little from those already described for hydrogen. Heat leaks to the nearly-adiabatic sample chamber are kept as small as possible by careful control of the heat shield and vacuum case. As in the case of hydrogen, the sample chamber is filled, cooled just below the triple point and maintained there for some hours to allow equilibrium to be established. The melting curve is then obtained in the same way as for hydrogen, that is, by supplying successive heat pulses. The size of each heat pulse would normally be between 1 and 10 % of the total heat necessary to melt the sample completely. The optimum size of heat pulse along with the time necessary to establish thermal equilibrium after each pulse must be found by trial and error for each gas. The approximate latent heat of melting for some of these gases is given in Table 4-5.

In Fig. 4-19 are shown measurements made during the melting of nitrogen in the cryostat illustrated in Fig. 4-15. Three melting plateaux are illustrated (Kemp and Kemp, 1978b): those marked A and B were obtained by using

Table 4-5. Latent heat of melting of various gases.

Gas	Latent heat of melting/kJ mol^{-1}
Hydrogen	0.11
Neon	0.34
Oxygen	0.44
Nitrogen	0.72
Argon	1.18
Xenon	2.30
Krypton	1.64

heating pulses of 10% of the total melting energy. A heat pulse lasted 25 min and a recovery period of 138 min was allowed after each pulse. Early in the melting process, the application of the heat pulse led to an overheating of about 1 mK, but towards the end, the overheating had risen to about 10 mK. This is illustrated in the figure by the four solid traces taken at different times during the melt. It was found that the overall melting range could be reduced from the 0.4 mK, observed for plateaux A and B, to 0.1 mK by reducing the amount of power in the first few heat pulses. This is shown in plateau C, during which the initial heat pulses carried only about 2% of the total melting energy.

The reproducibility of the triple points of argon, nitrogen and methane realized in this way is at least as good as ± 0.1 mK. For neon and krypton, however, the reproducibility is less good, only about ± 0.2 mK. This is probably because isotopic effects in these two gases intervene. For those gases such as argon, nitrogen, oxygen and hydrogen exhibiting melting plateaux covering a very small temperature range, less than 0.5 mK, it is easy to identify and reproduce the flat portion of the plateau. It is not easy for those gases such

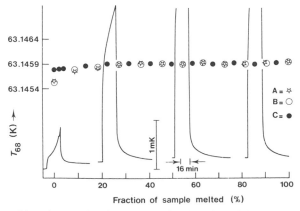

Fig. 4-19. Three melting plateaux for nitrogen, showing that the melting range is decreased by using heat pulses which add only 2.25% of the heat of transition in 25-min periods at the beginning of the transition plateau C. A = ☆, B = ○, C = ● (after Kemp and Kemp, 1978b).

as neon and krypton having melting ranges of 0.8 and 1.5 mK, respectively, and this accounts for their triple points having a somewhat poorer reproducibility as temperature-fixed points. The triple point of xenon is in a rather different category since it shows a melting range greater than 4 mK, and it is quite unsuitable as a fixed point of temperature. This is because of the large number of natural isotopes, none of which is predominant and the wide spread of mass number of the major components, 29 % having mass number $\leqslant 129$ and 19 % mass number $\geqslant 134$.

4-5-4   Sealed cells for the realization of triple points

In 1975, a novel method of realizing the triple point of argon was demonstrated by Bonnier, Malassis and Pavese (Bonnier and Malassis, 1975; Bonnier, 1975; Pavese, 1975). Instead of using a cryostat of the sort shown in Fig. 4-17, they proposed to seal permanently a given quantity of gas in a cell. The cell, shown in Fig. 4-20, is intended to be used as a highly reproducible, easily transportable, triple-point device. It is designed to be operated simply by being suspended, nearly-adiabatically, in a temperature-controlled cryostat and used in exactly the same way as an ordinary triple point apparatus. It is, of course, necessary to design the cell to withstand an internal pressure of some 80 atmospheres present when the cell is being stored at room temperature.

Fundamental to the performance of sealed triple-point cells is the question of purity of the gas and effects of long-term contamination. During construction and filling of the cell, it is necessary to treat it as an ultra-high-vacuum system. This implies careful degreasing and cleaning of the inner surface and degassing for extended periods at high temperatures before filling with high-purity gas. The final sealing off of the cell can be accomplished quite successfully by pinching the capillary, using standard pinching tools. It is then usual to embed the pinched end of the capillary in solder.

Experience since 1975 has confirmed the utility of the sealed cell as a method for realizing triple points, not only of argon, but also of nitrogen, neon, methane, hydrogen, krypton and oxygen (Bonnier and Hermier, 1982; Pavese and Ferri, 1982). The long-term stability of such sealed cells appears to be excellent, and the accuracy of realization of the triple point seems to depend solely upon the care with which the initial cleaning and filling are carried out. The fact that they are extremely robust and quite unaffected by mechanical shocks (in strong contrast to platinum resistance thermometers) has led to their being used for the comparison of fixed-point realizations among national laboratories (Ancsin, 1978c; Bonhoure and Pello, 1980). In view of the conclusions reached by Ward and Compton (Chapter 2), that fixed point differences were the main source of disagreement between various realizations of IPTS-68 at low temperatures, the increasing use of sealed triple-point cells is likely to lead to a significant improvement in this respect (Pavese, 1982; Furukawa, 1982).

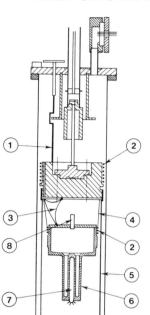

Fig. 4-20. A cryostat containing a sealed triple-point cell. The sealed cell (6) is suspended in the cryostat by nylon wires (4) inside a gold plated radiation shield (5). A silver wire (1) links this screen with an external liquid nitrogen bath. The temperature of (5) is adjusted with the aid of a differential thermocouple (3) and a heater (2). The sealed cell itself is made of stainless steel and carries a copper sealing-off tube (8). The thermometer is placed in a well (7) about which the solid/liquid gas mixture condenses during the realization of the triple point (after Bonnier, 1975).

The sealed cells so far discussed are designed to accommodate capsule-type thermometers. For the calibration of long-stem thermometers at the triple point of argon, now an alternative to the boiling point of oxygen, an equivalent sealed cell has been demonstrated (Bonnier, 1975). The cell itself is shown together with the arrangement for cooling and realizing the argon triple point in Fig. 4-21. The internal pressure in the cell while at room temperature is about 56 atmospheres. It is filled with just sufficient argon to fill the lower part of the cell with solid/liquid mixture at the triple point. In operation, the cell is first immersed in liquid nitrogen to a level just sufficient to condense solid argon in the lower part of the cell. When the argon is fully solidified, more liquid nitrogen is added to the vessel until the cell is completely immersed in the liquid. The vessel is then sealed and the internal pressure, due to the evaporation of nitrogen, is allowed to rise to the triple-point temperature of argon. The valve on the top of the vessel is adjusted to maintain an internal temperature just above that of the triple point of argon (83.798 K). The increase in pressure required to raise the temperature of the boiling point of nitrogen from 77.344 K at 101 325 Pa to 83.798 K is about 130 000 Pa. In

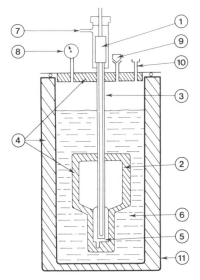

Fig. 4-21. Sealed triple-point cell and cryostat for argon designed for use with long-stemmed thermometers. (1) Thermometer, (2) stainless steel cell, (3) thermometer-support tube, (4) expanded polyurethane, (5) solid argon, (6) liquid nitrogen, (7) entry for helium gas, (8) manometer, (9) valve, (10) filling tube and (11) dewar vessel (after Bonnier, 1975).

this way, an argon triple point may be realized and observed using a long-stem platinum resistance thermometer. The sealed cell is covered with a layer of expanded polyethylene to reduce the effects on the cell of inhomogeneities in the temperature of the liquid nitrogen. The accuracy of realization of an argon triple point carried out in this way is not as good as that obtainable with the sealed cells designed for capsule thermometers because the uniformity of temperature is not as good. Nevertheless, to within about 1 mK, a sealed cell of the type shown in Fig. 4-21 provides an excellent alternative to an oxygen-boiling-point apparatus.

4-6 Superconducting transition points

The superconducting transition in a Type I superconductor is a second-order phase transition occurring at a temperature T_c which is characteristic of the metal, its purity, state of anneal and applied magnetic field. For a particular sample, therefore, in zero magnetic field the superconducting transition temperature should provide a fixed point of temperature. This assumes that the width of the transition is sufficiently small and that suitable means can be found for observing the transition. These questions have been examined in some detail at the NBS and the conclusions were sufficiently encouraging for a fixed-point device to be developed (Schooley *et al.*, 1972) containing five metals

Table 4-6. Superconducting transition temperatures on EPT-76 of SRM 767.

Element	$T_{c(76)}$ (K)	Transition width (mK)	Source of material
Lead	7.1999	1→2	Cominco 99.9999 % Pb HPM 9284
Indium	3.4145	1→2	Indium Corporation of America 99.999 % In
Aluminium	1.1796	1.5→3.5	Cominco 99.9999 % Al HPM 5831
Zinc	0.851	1.5→6	NBS SRM 682
Cadmium	0.519	1→3	NBS — Office of Standard Reference Materials

whose superconducting transition temperatures ranged from about 0.5 K to 7.2 K. The metals and their assigned values of T_c on EPT-76 are shown in Table 4-6 for the NBS Superconducting Fixed Point Device SRM 767 (Schooley and Soulen, 1982).

The superconducting temperature is defined as the mid-point of the transition and appears to be independent of the means of observing it, whether it be by mutual inductance, resistance or heat capacity (Soulen and Colwell, 1971), (Fig. 4-22). The method most commonly used for the observation of the transition is that of ac mutual inductance. In a Type I superconductor below the transition temperature, all magnetic flux is excluded from the metal: the Meissner effect. This can readily be observed using a mutual inductance bridge and external Helmholtz coils to cancel external magnetic fields. The current in

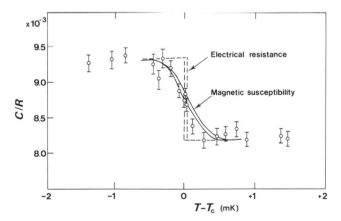

Fig. 4-22. The temperature dependence of the heat capacity (open circles with uncertainty bars), magnetic susceptibility (solid lines) and electrical resistivity (broken lines), near the superconductive transition of an indium sample, determined by two separate experiments. The equality of the experimental temperature axes was carefully established using a second indium sample (after Soulen and Colwell, 1971).

the Helmholtz coils may be adjusted to give a maximum value for T_c, the condition for zero magnetic field at the superconductor.

Using this method, it is possible to reproduce the centre of the transition to within about 0.1 mK without too much difficulty. The long-term reproducibility of the T_c values appears to be excellent, better than 0.3 mK for each of the metals in SRM 767.

The values of T_c and the widths of the superconducting transitions observed in different samples of SRM 767, however, remain a little disappointing. Differences in T_c of up to 1 mK have been observed between different samples, and this is undoubtedly a result of minor differences in annealing and purity. Improvements can be expected in this respect as more experience is gained in the construction and use of these devices. It should be noted that EPT-76 is already based in part upon the fixed point temperatures of SRM 767 (Chapter 2). For temperatures below 0 K, a further set of superconductive transition points is available in SRM-768 (Schooley and Soulen, 1982). These cover the range 0.015 K to 0.2 K. For further information on thermometry in this range see Loumasmaa (1974).

In contrast to a first-order phase transition, such as the melting or boiling point, there is no latent heat associated with a second-order phase transition. Such a transition therefore acts as an indicator of temperature, but in no way provides any control over the temperature. It should be compared with the freezing of a pure metal, which we discuss below, during which a substantial ingot will remain at the freezing temperature while the surroundings continue to cool. In the case of superconducting transitions, however, the absence of a latent heat of transition does not pose a problem. This is because at low temperatures, it is easy to provide an adequate independent temperature control, and thermal conductivities and specific heats are such that non-uniformity of temperature and response time do not lead to difficulties.

4-7   The melting and freezing of metals

4-7-1   Introduction

Very extensive studies have been carried out, notably by McLaren (1957a, b; 1958a, b) and by McLaren and Murdock (1960a, b; 1963; 1968a, b) of the methods and techniques required for the precise realization of metal melting and freezing points. The limit to the reproducibility with which metal freezing points can be realized is set by the performance of the thermometers used to observe them rather than by the metals themselves. An intractable problem, and one which we discuss further in Chapter 5, is the difficulty of ensuring a sufficient immersion of a platinum resistance thermometer in the medium whose temperature we are trying to measure. Depending upon the design of thermometer, an immersion in a uniform temperature zone of between 10 cm and 20 cm is required for the resistance element to approach within 0.5 mK of the temperature of the thermometer well. Since the difference, ΔT, between the temperature of the resistance element and the well decreases exponentially with immersion, it does not make a great deal of difference to the required immersion if the well is at the ice point, tin point or even gold point. Each increase of 1.5 cm or 3 cm (depending upon thermometer design) decreases ΔT by about a factor of ten. While it is possible to obtain sufficient immersion in a metal freezing-point cell, when it comes to using the platinum resistance thermometer to measure the temperature of something other than a metal freezing point, the limitation is always the uniformity of temperature of the body whose temperature is being measured. Thus above about 500 °C it becomes increasingly difficult to measure the temperature of a body to an accuracy better than about 50 mK by means of a platinum resistance thermometer. It is worth noting in this connection the potential advantages of a heat pipe for producing extended regions of very uniform temperature.

The melting and freezing of an ideal pure metal takes place at unique temperature and involves the absorption or liberation of the latent heat of melting. If a sufficiently large quantity of metal is used (150 cm^3 is a typical volume of a melting point ingot), the latent heat is sufficient to maintain the ingot and the thermometer immersed in it at a constant temperature for some hours while melting or freezing is taking place. The presence of small amounts of impurity, however, in the form of dissolved metals will both change the temperature at which the bulk metal appears to melt and freeze, and lead to melting and freezing taking place over a range of temperature. Although all of the metals of interest to the thermometrists as fixed points, which include gallium, indium, cadmium, lead, tin, zinc, antimony, aluminium, silver and gold, are available having purities sufficient for our purposes, it is not easy to preserve the purity during use. The effects of impurities therefore cannot be ignored since contamination of the metal by the material of the crucible, by extraneous matter present during filling or by gaseous impurities at high temperatures, must be guarded against. For this reason, we begin by a brief elementary outline of the effects of small amounts of impurity on the melting and freezing of an otherwise pure metal. In this discussion there is no need to enter into the various theories of the microscopic processes which occur on melting. For these and a discussion of the various precursor effects which are exhibited on either side of the solid/liquid transformation, reference should be made to Ubbelohde (1978) and Ziman (1979).

4-7-2 *Melting and freezing in terms of the phase equilibrium diagram*

Figure 4-23(a) shows a small portion of the A-rich end of the phase-equilibrium diagram of a binary alloy A–B. Hume-Rothery and Raynor (1956)

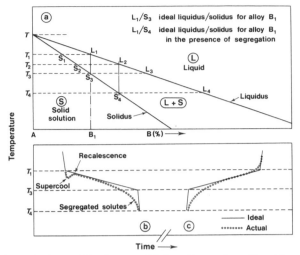

Fig. 4-23. (a) Schematic of part of the phase-equilibrium diagram for a dilute binary-alloy system. (b) Freezing and (c) melting curves for a simple binary alloy.

should be consulted for the background to phase-equilibrium diagrams. Instead of melting and freezing at a unique temperature T_A, an alloy of B in A having a composition B_1 melts and freezes over a range of temperature, ideally from T_1 to T_3. Figure 4-23(a) is drawn for the case of a solute, B, that lowers the melting point of A. Note that both T_1 and T_3 are lower than the melting point of the pure metal A. On cooling an alloy of composition B_1 from a temperature within the liquid range, and assuming for the moment that super-cooling does not take place, solid first appears at a temperature T_1. The solid which appears at this temperature has the composition S_1 and leaves a liquid of composition L_1. On further cooling, more solid precipitates having a composition that moves along the solidus and leaving a liquid whose composition moves along the liquidus. At a temperature T_2, solid of composition S_2 and liquid of composition L_2 are present and at a lower temperature T_3, solid of composition S_3 is in equilibrium with liquid of composition L_3. For the purposes of the argument so far we have assumed that the rate of cooling is infinitely slow so as to allow compositional equilibrium always to be maintained. That is to say the solid of composition S_1 which was first precipitated has time to transform, by diffusion, into solid of composition S_3 by the time the temperature has fallen to T_3. Since solid-state diffusion is always slow and cooling rates cannot be infinitely slow, compositional equilibrium in the solid is never achieved, with the result that at temperatures below T_3 there remains solid having a composition between S_1 and S_3 and thus liquid with excess B which does not finally freeze until the temperature has reached T_4.

Under equilibrium conditions, the solidification of the alloy of composition B_1 is complete at a temperature T_3 when the last liquid, of composition L_3, has solidified. The equilibrium freezing curve is shown in Fig. 4-23(b) along with the equilibrium melting curve of the same alloy. During melting, the first liquid (of composition L_3) appears at temperature T_3, and the last solid (of composition S_1) melts at a temperature T_1.

The slopes of the solidus and liquidus for a given solute, B, in a pure metal, A, depend upon the nature of A and B (Smithells, 1976), and in particular upon the relative solubilities of B in the solid and liquid A. A solute B that is more soluble in the liquid $A–B$ than in the solid depresses the freezing point as in Fig. 4-24(a), while one that is more soluble in the solid raises the freezing point as in Fig. 4-24(b). This is expressed analytically by defining a distribution coefficient $(k_0)_B$ for a solute B so that

$$(k_0)_B = B_S/\bar{B_O} = \Delta T_2/\Delta T_3,$$

for those solutes that depress the freezing point $k_0 < 1$, and for those solutes that raise the freezing point $k_0 > 1$.

If we now move on from equilibrium melting and freezing to near-equilibrium melting and freezing as observed in practice, we find that the ideal melting and freezing curves of Fig. 4-23(b) and (c) take on a slightly different form. The very slow rate of diffusion of B in solid $A–B$ leads to a composition gradient in the solid in the direction taken by the advancing solid/liquid interface. Those solutes having values of $k_0 < 1$ are concentrated in the last liquid to freeze while those for which $k_0 > 1$ are concentrated in the first solid to freeze. The greater the departure of k_0 from unity for a particular solute, the greater is its tendency to segregate on freezing. The dotted lines in Fig. 4-23(b) and (c) show the result of non-equilibrium melting and freezing. Although, on

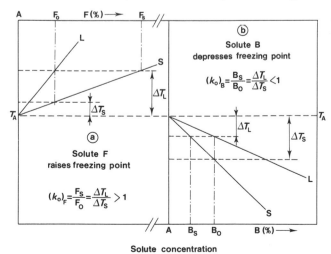

Fig. 4-24. Phase-equilibrium diagrams and equilibrium distribution coefficients for solutes that (a) lower and (b) raise alloy freezing points from that of the pure element A.

freezing, the liquidus point is not affected by solute segregation, the temperature at which the last liquid solidifies is very much lower than it would be without segregation. On subsequent melting, the temperature at which the first melting occurs is, in consequence, much lower than in an unsegregated alloy. In fact, the melting range is a very sensitive indicator of the presence of impurities. The narrower the melting range, the purer the sample. Since segregation is the result of non-equilibrium freezing, it is always instructive to examine the melting range following a series of freezes carried out at different rates. Segregation will clearly be more evident following a fast freeze than following a slow one.

A number of different methods are available for making detailed comparisons between melting ranges; one of the most useful makes use of the "inverse" melting curve in which a histogram is plotted which approximates the temperature derivative of the melting curve. The percentage of the total melting time for which the ingot remains in a given temperature interval is plotted against average temperature of that interval. For slow heating rates the furnace temperature remains nearly constant during the time it takes for the whole ingot to melt, so that the rate of heat input to the ingot is nearly constant. Under these conditions, the percentage of the total melting time spent in a given temperature interval is close to the percentage of the total ingot which melts during that interval. Another technique is to compare the fraction of the total melting time spent in a given melting range after both fast and slow freezes.

Information on the effects of various contaminants on the melting and freezing points of the metals mentioned earlier is to be found in compilations of binary-alloy phase diagrams (Smithells, 1976; Hanson and Anderko, 1958). Care should be exercised in interpreting these phase diagrams at very low alloy concentrations since experimental information is rarely available at the limit of dilute solid solution (Connolly and McAllen, 1980). The solidus and liquidus are usually simply extrapolated to meet at the melting point of the major constituent. Under certain circumstances the slopes may be in error if the nearest experimental information is at a minor constituent composition of, say, 5 %.

4-7-3 Practical realization of the melting and freezing points of metals

Having outlined what is to be expected in the melting and freezing of metals, we come now to a description of the equipment and methods which must be used to make precise measurements. The size of the ingot of metal depends ultimately upon the size of the platinum resistance thermometer to be used for the temperature measurement. The heat conducted out of the ingot by the thermometer via its leads and supports must always be negligible compared with the latent heat being released by the freezing ingot, i.e. the immersion must be sufficient. If this condition is not achieved, then temperature gradients will be set up, destroying any semblance of equilibrium in the ingot, quite apart from non-equilibrium resulting from finite freezing rates. The purity of the metal must be maintained and, again, a common solution is found in using a crucible made from high-purity graphite. During melting and freezing, it is important that the heat passing either to or from the ingot be closely controlled and nearly uniform in all directions. This requires a furnace having a sufficiently-extended zone of uniform temperature to include the whole of the graphite crucible. For metals melting in the temperature range extending up to about 450 °C, the furnace shown in Fig. 4-25 has been shown by Chattle (1972) to be quite suitable. For the higher-melting-point metals, a rather more complex arrangement is required (Chattle, 1972) such as that shown in Fig. 4-26 unless a heat pipe is used in which case the simple furnace of Fig. 4-25 may be used right up to 1100 °C. The sodium/stainless steel heat pipe is simply put in the place of the nickel block. Further information on the design and construction of furnaces and crucibles to contain very pure metals is to be

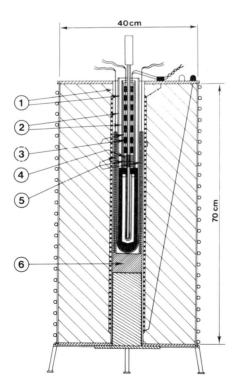

Fig. 4-25. Zinc-point furnace: 1, silica-wool; 2, glass tubes; 3, alternate graphite/ceramic discs; 4, platinum-wound heater on alumina former; 5, thermocouples; 6, aluminium block (after Chattle, 1972).

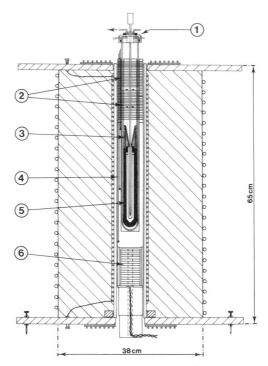

Fig. 4-26. Gold-point furnace: 1, gas-tight cap (water cooled) with connections to cylinder of argon; 2, platinum-wound heaters on silica-brick former; 3, platinum-wound heater on alumina former; 4, nickel block; 5, graphite crucible; 6, platinum-wound heater on silica-brick former (after Chattle, 1972).

found in Furakawa *et al.* (1972), McAllen and Ammar (1972) and McLachlan *et al.* (1972). Figure 4-27 shows a typical freeze obtained using an ingot of gold in the furnace illustrated in Fig. 4-26. The length of the freezing plateau is controlled by the rate at which the furnace is allowed to cool after the metal ingot has entered the freeze. A freeze lasting many hours can be obtained without difficulty, provided that the nucleation of the freeze is properly carried out so that excessive undercooling is avoided.

4-7-4   Methods of nucleating the freeze

Solidification can take place in a metal when the free energy of the solid falls below that of the liquid. The temperature at which this occurs is the freezing temperature or liquidus temperature in the case of an alloy. Freezing, however, requires the formation of small particles of solid surrounded by liquid, and the detailed mechanisms which are involved in producing these small solid particles and allowing them to grow are complex. At temperatures close to and

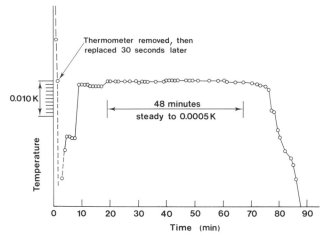

Fig. 4-27. A gold-point freeze observed with a high-temperature platinum resistance thermometer (after Chattle, 1972).

only just below the freezing temperature, the difference between the free energy of the solid and that of the liquid is small, and therefore the driving force to transform from liquid to solid is small. When a particle of solid forms, there is a decrease in free energy due to the transformation to the solid phase, but there is also an increase in free energy resulting from the presence of the solid–liquid interface. It is only when the former effect exceeds the latter that the small solid particle can grow. When this happens, the freeze is said to nucleate, and the solid spreads rapidly through the liquid, releasing sufficient latent heat to raise the temperature of the whole to the freezing temperature. The magnitude of the undercool that can occur before spontaneous nucleation takes place is a function of the thermodynamic properties of each metal.

The techniques used to nucleate the freeze and bring the ingot out of the undercool depend on the metal being used and the size of the natural undercool. For lead, cadmium, zinc, indium, silver and gold, the natural undercool is small and is generally less than 1 K. For these metals, a perfectly satisfactory freeze is obtained if nucleation is triggered, as the ingot enters the undercool, by removing the thermometer from the re-entrant well in the ingot, allowing it to cool outside the furnace for about half a minute, and then re-inserting the cold thermometer into the re-entrant well, as in Fig. 4-27.

This procedure cools the graphite well sufficiently for a mantle of frozen metal to form around it. Freezing then continues slowly, as heat is lost from the outside of the ingot, by the growth of solid from the outside walls of the crucible; typical rates for the advance of solid into the melt are 2–5 mm per hour (McLaren and Murdock, 1960a). At this rate of solidification, in metals having only a few ppm of impurity, the solute concentration in the newly frozen solid is such that the depression of the freezing point is less than about

0.1 mK, and thus a freezing plateau is observed with the platinum resistance thermometer.

For metals having a much larger natural undercool before spontaneous nucleation takes place, such as gallium, tin and antimony, it is not sufficient just to cool the re-entrant thermometer well as described above. The temperature drop thus obtained in the walls of the well is not great enough to trigger crystallization, since these metals can remain in the supercooled liquid state: in the case of antimony, some 40 K below the equilibrium freezing temperature. A method of overcoming this tendency to massive undercooling is described by Chattle (1972) in which provision is made for cooling the outside of the crucible by means of a flow of argon or nitrogen. By this means the crucible, but not to any great extent the furnace, can be cooled rapidly by some tens of degrees, sufficient to induce nucleation of the freeze over the whole of the inside of the crucible wall. The latent heat released is sufficient to raise the ingot and crucible to the temperature of the freezing plateau within a few minutes. On reaching the freezing plateau, the solidification by rapid growth of dendrites, which always occurs during freezing from the supercooled state, ceases and the remaining metal freezes with a smooth solid/liquid interface slowly advancing inwards towards the thermometer well. An alternative method (McLaren and Murdock, 1968a) of inducing nucleation for metals such as tin and antimony is to remove the crucible assembly completely from the furnace when the freezing temperature is first reached and place it in another furnace held at a temperature some ninety degrees lower. As soon as nucleation has taken place and sufficient latent heat has been released to stop the forced cooling of the ingot, the whole assembly is replaced in the original furnace whose temperature is still only a few degrees below the freezing temperature. The success of a procedure such as this provides a very striking demonstration of the energy released on passing from the liquid to the solid state.

Both these techniques provide sufficient cooling of the ingot for nucleation to take place without allowing the whole furnace to cool so far below the freezing temperature that the latent heat released by the freezing metal is unable to raise the temperature of the whole assembly to the freezing point. If natural nucleation is allowed to take place spontaneously, then the freezing plateau is drastically shortened, sometimes to such an extent that it completely disappears.

In studies of the freezing point of antimony, McLaren (1968a, b) found that if, in addition to triggering nucleation by removing the ingot to another furnace, he followed this by cooling the thermometer well while the ingot was still coming out of the undercool, then the temperature of the subsequent freezing plateau was lower by 0.6 mK. He suggests that the explanation for this increase in apparent freezing temperature is to be found in the effect on the freezing temperature of a very small radius of curvature of the advancing solid interface. Table 4-7 shows the magnitude of this effect for antimony and tin

Table 4-7. Depression of the freezing point with radius of curvature of solid–liquid interface.

	$\Delta T/\text{mK}$		
Metal	$r = 5$ mm	$r = 0.5$ mm	$r = 0.1$ mm
Antimony	0.04	0.35	1.8
Tin	0.04	0.26	1.3

calculated, for a solid/liquid surface of radius r, from the expression (Chalmers, 1964; Winegard, 1964)

$$\Delta T = T_E - T_r = \frac{2\sigma T_E}{\rho L r}$$

where σ is the solid–liquid interfacial energy, ρ the density of liquid at the freezing temperature, L the latent heat of fusion, T_E the equilibrium freezing temperature for plane solid–liquid interface, and T_r that for a solid–liquid interface of radius r. The important point to note is that if the procedure adopted by McLaren results in a shell of solid being formed at the thermometer well while the ingot is in the supercooled state, then this shell of solid will be made of a mass of sharp dendritic crystals. The radius of curvature of the solid/liquid interface will thus be very small and will remain so since the rest of the ingot will freeze from the outside inwards provided that the heat lost down the thermometer stem is small.

While the freezing plateau is used to provide the most reproducible fixed point temperature, the melting plateau, as we have seen, provides much information on the content and distribution of impurities in the ingot.

So far we have not mentioned the effect of oxygen contamination. It is now common practice to seal the metal ingot under an inert atmosphere in order to exclude oxygen during melting and freezing. It is essential to do this in the case of silver since oxygen is soluble in molten silver and leads to a depression of the freezing point that can amount to as much as 5 mK (Bongiovanni *et al.*, 1975). For gold, aluminium, zinc and platinum (1769 °C), the effects of dissolved oxygen are not significant, while for copper (Coates and Andrews, 1978) (1084.88 °C) and palladium (Jones and Hall, 1979) (1555 °C), care must be taken to exclude oxygen. In silver, the presence of oxygen at the level of 0.5 ppm by weight, is shown by the melting curve exhibiting a small negative slope. This probably arises because oxygen, insoluble in the solid but trapped at grain boundaries, is released on melting and dissolves in the liquid, progressively lowering its melting point. To remove oxygen from molten silver held in a graphite crucible, it is necessary to maintain it at about 1000 °C for some 100 h, either under vacuum or under an inert gas.

4-7-5   The melting and freezing of eutectic alloys

In an attempt to establish a fixed point between the freezing point of antimony and silver and between zinc and antimony, the melting and freezing behaviour of some eutectic alloys has been studied. The ones that have been examined in most detail are the Cu–Ag eutectic (Bongiovanni *et al.*, 1972) which melts at about 780 °C and the Al–CuAl$_2$ eutectic (McAllen, 1972) which melts at about 548 °C. The processes which take place in a freezing eutectic alloy are much more complicated than those in a freezing dilute alloy. In addition, no eutectic composition is known to better than about 0.1 %. In the case of the Al–CuAl$_2$ eutectic, of nominal composition 66.85 % Al, alloys having compositions lying between 66.80 % Al and 66.96 % Al gave melting plateaux differing by 10 mK. During freezing, two phases of quite different composition separate out from the liquid. Material rejected by one phase on freezing must diffuse to the second phase and be absorbed so that the average composition of the solid remains near to the eutectic value. As a result of the time required for this process to proceed, the structure of the solid is very sensitive to the rate at which solidification takes place. The dimensions of the individual particles of the two phases which separate out on melting are very small, and thus the effects of curvature of the solid–liquid interface referred to earlier are much larger. For reasons of this sort, there seems little prospect of using the freezing plateau of a eutectic as a reproducible fixed point. The melting plateau, however, shows rather more promise. Melting plateaus which follow freezing at a very slow rate (freezing lasting 12 h or more) can be reproduced to within about 5 mK. Whether this precision is going to be thought worth the considerable trouble needed to achieve it remains to be seen. I am doubtful about the advantages of using these less-accurate fixed points, albeit at useful temperatures, in view of the considerable difficulties involved in setting them up. For a review of this question see McAllan (1982).

4-8   The triple point of water, 273.16 K

It is not difficult to prepare a triple-point-of-water cell that will realize the defining fixed point of the IPTS-68 to within one or two tenths of a millikelvin. Details of how such cells may be made have been given by Barber *et al.* (1954) and more recently by Ambrose *et al.* (1973). The latter authors give a somewhat simplified procedure, but claim that the accuracy of realization of the triple point in cells manufactured according to their recipe is at least as good as ±0.1 mK. Although triple-point cells are available commercially, a brief outline of Ambrose's method of manufacture will be given since the apparatus and techniques required are very simple. The question of the purity of the water and the effect on the triple point of variations in the proportion of deuterium present are discussed in some detail by Barber *et al.* They conclude

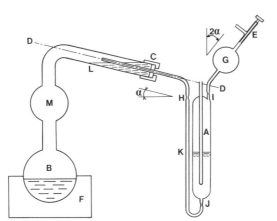

Fig. 4-28. Apparatus used for filling triple-point-of-water cells. (A) the triple-point cell, (B) boiling flask, (C) screwed compression joint (Quickfit and Quartz SQ 24), (DD) axis of rotation of cell assembly between cleaning and filling positions, (E) valve (Fisher and Porter Inc.), (F) heater, (H), (J) and (I) sealing-off constrictions, (G) reservoir, (L) water seal, (M) anti-splash chamber (after Ambrose *et al.*, 1973).

that natural variations in deuterium content among different samples of water are not likely to lead to variations in the triple-point temperature exceeding 0.1 mK. They point out, however, that care must be taken in filling the cells to avoid undue fractionation of the deuterium. For a discussion on the effects of variations in pressure on the water triple point see McAllan (1982).

Commercially-made triple-point cells range in length from 38 cm to 43 cm, in width from 4 to 6.5 cm and have internal wells of diameter 1 cm to 1.3 cm. There is no reason why much larger or smaller cells should not be made except that with very large cells, difficulties are encountered in forming the ice sheath around the central well. Conversely with very small cells, the thermal mass of the water and ice may be insufficient compared with that of the glass, and thus the time for which the triple point can be maintained may be very short. Attempts to make triple-point of water cells from metal, similar to those used for the low-temperature triple points, have not yet been successful (Bonnier and Hermier, 1982). The apparatus used by Ambrose is shown in Fig. 4.28; the cell itself together with the glassware used in its filling and cleaning are made of pyrex.

During the manufacture of triple-point cells, it is essential to avoid the use of greased joints, and in the equipment shown in Fig. 4-28, only PTFE joints and valves are used. Before connecting the cell to the flask at C, all of the glassware is cleaned by filling with a saturated solution of chromic acid in sulphuric acid and leaving for a few minutes. It is then rinsed out with distilled water and connected together with about one litre of distilled water in the flask B and the valve E open. The joint C is rotated so that the triple point cell is inverted and the water in B is boiled gently for about 2 h. The cell is then rotated to the vertical position so that steam from B begins to condense in the cell. The rate of boiling is kept at such a level that steam bubbles through the condensed water and maintains the cell free of air. When the water level has risen to within one or two centimetres of the top of the cell, the heater at F is switched off and the valve E closed. After the internal pressure has fallen below one atmosphere, the constriction H is sealed off followed by that at I. The cell is then inverted and finally sealed off at J.

Completed triple-point-of-water cells should be handled carefully since the absence of air in the bulb allows free movement of water which can slap against the end of the cell with a shock sufficient to fracture the glass if the cell is suddenly up-ended.

In order to realize the triple point of water, it is first necessary to freeze a mantle of ice about the central well. This may be done by first cooling the whole cell to near 0 °C in a bath of wet crushed ice and then placing a cold metal rod in the central well. The metal rod should be a close fit in the well and it should be sufficiently cold to form a sheath of ice at least 10 mm thick around the well. Having formed the sheath of ice, it is necessary to establish equilibrium between the three phases of water by melting a thin layer of ice around the well. This is easily done by introducing warm water into the well for a few seconds until the ice mantle is free to rotate about the well. That the ice sheath is detached can be verified by giving the cell a sudden rotatory movement and observing that the ice sheath spins. The whole cell is then immersed in wet crushed ice, and after a few hours, the triple point is established in the thermometer well to within a ten-thousandth of a degree. Provided that the cell is kept immersed in the ice bath, the triple point may be maintained for up to about a week without further attention.

4-9   The triple point of gallium (29.774 °C)

Gallium melts at a temperature close to 29.772 °C and has a triple point 2 m °C higher at 29.774 °C. Studies of the melting behaviour of gallium (Mangum and Thornton, 1979) showed that it provides an extremely reproducible and convenient fixed point. Differences in the triple point from sample to sample rarely exceed 0.05 mK, and the reproducibility of an individual sample is probably better than that of the best platinum resistance thermometer used to observe it, one of the limiting factors being the stability of the heating effect in the thermometer.

Gallium expands on melting by about 3 % and thus a fixed-point cell to contain it should have flexible walls to avoid the danger of breaking. It was shown by Thornton and Mangum that a cell made from PTFE and having a re-entrant thermometer well of nylon, is perfectly satisfactory. Such a cell, containing about 1 kg of gallium, is shown in Fig. 4-29. The thermometer well is made of nylon as it was found by Mangum and Thornton that supercooled

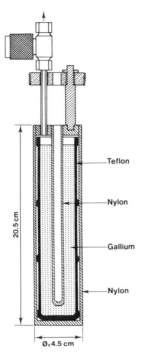

Fig. 4-29. A triple-point-of-gallium cell.

liquid gallium readily nucleates at a machined surface of nylon and so large undercoals are avoided.

The results of a comparison of gallium triple-point temperatures among samples of different origin show a spread of values not exceeding 0.2 mK, a result which is very satisfactory (Chattle *et al.*, 1982; Mangum, 1982).

References

Ambrose, D., Collerson, R. R. and Ellender, J. H. (1973). A simple method of filling water triple point cells. *J. Phys. E.*, **6**, 975–977.

Ancsin, J. (1973a) Dew points, boiling points and triple points of pure and impure oxygen. *Metrologia*, **9**, 26–39.

Ancsin, J. (1973b) Studies of phase changes in argon. *Metrologia*, **9**, 147–154.

Ancsin, J. (1977). The thermometric fixed points of hydrogen. *Metrologia*, **12**, 79–86.

Ancsin, J. (1978a). Vapour pressures and triple point of neon and the influence of impurities on these properties. *Metrologia*, **14**, 1–7.

Ancsin, J. (1978b). Note concerning the suitability of xenon as a temperature fixed point. *Metrologia*, **14**, 45–46.

Ancsin, J. (1978c). Intercomparison of triple points of argon and oxygen of INM, IMGC and NRC. *Metrologia*, **14**, 79–81.

Bailey, C. A. (1971). "Advanced Cryogenics", Plenum Press, New York.

Barber, C. R., Handley, R. and Herrington, E. F. G. (1954). The preparation and use of cells for the realization of the triple point of water. *Brit. Jn. App. Phys.*, **5**, 41–44.

Bongiovanni, G., Crovini, L. and Marcarino, P. (1972). Freezing and melting of silver–copper eutectic alloys at very slow rates. *High Temperatures — High Pressures*, **4**, 573–587.

Bongiovanni, G., Crovini, L. and Marcarino, P. (1975). Effects of dissolved oxygen and freezing techniques on the silver point. *Metrologia*, **11**, 125–132.

Bonhoure, J. and Pello, R. (1978). The temperature of the triple point of methane. *Metrologia*, **14**, 175–178.

Bonhoure, J. and Pello, R. (1980). Points triples de l'argon et du méthane: Utilisation de cellules scellées. *Metrologia*, **16**, 95–103.

Bonnier, G. (1975). Point triple de l'argon (83.798 K) référence de transfert. *Bulletin Bureau National de Métrologie (Paris)*, 14–18.

Bonnier, G. and Hermier, Y. (1982). The thermal behaviour of thermometric sealed cells and of a multi-component cell. *TMSCI*, **5**, 231–238.

Bonnier, G. and Malassis, R. (1975). Realisation d'un nouveau type de cellule scellée destinée aux étalonnages cryogéniques. *Bulletin Bureau National de Métrologie, Paris*, Oct., 19–20.

Bousson, G. (1978) Bain à lit fluidisé utilisé pour l'étalonnage des capteurs de température. *Bulletin BNM Paris*, **1**, 56–60.

Burns, G. W. and Hurst, W. S. (1971). A program in refractory metal thermocouple research. *J. Res. Nat. Bur. Stds.* **75C**, 99–106.

Busse, C. A., Labrande, J. P. and Bassani, C. (1975). The gas-controlled heat pipe: a temperature–pressure transducer, *Temperature*, **75**, 428–438.

Chalmers, B. (1964). "Principles of Solidification", J. Wiley.

Chattle, M. V. (1972). Platinum Resistance Thermometry up to the gold point. *TMCSI*, **4**, Part 2, 907–918.

Chattle, M. V., Rusby, R. L., Bonnier, G., Moser, A., Renaot, E., Marcarino, P., Bongiovanni, G. and Frassineti, G. (1982). An intercomparison of gallium fixed point cells. *TMCSI*, **5**, 311–316.

Coates, P. B. and Andrews, J. W. (1978). A precise determination of the freezing point of copper. *J. Phys. E.* **8**, 277–285.

Compton, J. P. and Ward, S. D. (1974). Determination of impurities in oxygen by mass spectrometry. *Analyst*, **99**, 214–217.

Compton, J. P. and Ward, S. D. (1976). Realization of the boiling and triple points of oxygen. *Metrologia*, **12**, 101–113.

Connolly, J. J. and McAllen, J. V. (1980), Limitations on metal fixed points caused by trace impurities. *Metrologia*, **16**, 127–132.

Droege, J. W., Schimek, M. and Ward, J. J. (1972). Chemical compatibility in the use of the refractory metal thermocouples. *TMCSI*, **4**, Part 3, 1767–1779.

Furukawa, G. T. (1982). Reproducibility of the triple point of argon in a transportable sealed cell. *TMCSI*, **5**, 239–248.

Furukawa, G. T., Riddle, J. L. and Bigge, R. (1972). Investigations of freezing temperatures of National Bureau of Standards Tin Standards. *TMCSI*, **4**, Part 1, 247–263.

Groll, M. and Neuer, G. (1972). A new graphite cavity radiator as a blackbody for high temperatures. *TMSCI*, **4**, Part 1, 449–456.

Grover, G. M., Cotter, T. P. and Erickson, G. F. (1964). Structures of very high thermal conductance. *J. Appl. Phys.*, **35**, 196–212.

Hanson, M. and Anderko, K. (1958). "Constitution of Binary Alloys", McGraw Hill, (supplements 1965 and 1969).

Hume-Rothery, W. and Raynor, G. V. (1956). "The Structure of Metals and Alloys",

Institute of Metals, London.

Jones, T. P. and Hall, K. G. (1979). The melting point of palladium and its dependence on oxygen. *Metrologia*, **15**, 161–163.

Kemp, R. C. and Kemp, W. R. G. (1978a). The realization of the normal boiling point of neon. *Metrologia*, **14**, 9–13.

Kemp, R. C. and Kemp, W. R. G. (1978b). The triple points of krypton, argon, nitrogen and neon. *Metrologia*, **14**, 83–88.

Kemp, R. C. and Kemp, W. R. G. (1979). The triple point, boiling point and 17 K point of equilibrium hydrogen *Metrologia*, **15**, 155–159.

Kemp, R. C., Kemp, W. R. G. and Cowan, J. A. (1976). The boiling points and triple points of oxygen and argon. *Metrologia*, **12**, 93–100.

Lounasmaa, O. V. (1974). "Experimental Principles and Methods below 1 K". Academic Press, London and New York.

Mangum, B. W. and Thornton, D. D. (1979). Determination of the triple-point temperature of gallium. *Metrologia*, **15**, 201–215.

Mangum, B. W. (1982). Triple point of gallium as a temperature fixed point. *T MCSI*, **5**, 299–310.

McAllan, J. V. (1972). Metal binary eutectics as fixed temperature points. *T MCSI*, **4**, 265–274.

McAllan, J. V. (1982). Reference temperatures near 800 °C. *T MCSI*, **5**, 371–376.

McAllan, J. V. (1982). The effect of pressure on the water triple point temperature. *T MCSI*, **5**, 285–290.

McAllan, J. V. and Ammar, M. M. (1972). Comparison of the freezing points of aluminium and antimony. *T MCSI*, **4**, Part 1, 275–285.

McLachlan, A. D., Uchiyama, H., Saino, T. and Nakaya, S. (1972). The stability of the freezing point of copper as a temperature standard. *T MCSI*, **4**, Part 1, 287–293.

McLaren, E. H. The freezing points of high purity metals as precision temperature standards: I, *Can. Jn. Phys.* **35**, 37–80, (1957a), (Resistance thermometry); II, *Can. Jn. Phys.* **35**, 1086–1106, (1957b), (Zinc, cadmium and tin); III, *Can. Jn. Phys.* **36**, 585–598, (1958a), (Zinc); IV, *Can. Jn. Phys.* **36**, 1131–1147, (1958b), (Indium and tin).

McLaren, E. H. (1962). The freezing points of high-purity metals as precision temperature standards. *T MCSI*, **3**, Part 1, 185–198.

McLaren, E. H. and Murdock, E. G. The freezing points of high purity metals as precision temperature standards: V, *Can. Jn. Phys.* **38**, 100–118, (1960a), (Tin); VI, *Can. Jn. Phys.* **38**, 577–587 (1960b), (Lead); VII, *Can. Jn. Phys.* **41**, 95–112, (1963), (Bismuth); VIIIa, *Can. Jn. Phys.* **46**, 369–400, (1968a), (Antimony); VIIIb, *Can. Jn. Phys.* **46**, 401–444, (1968b), (Antimony, temperature scale realization from 0 °C to 631 °C).

McMahon, B. J. and Rothwell, E. (1972). The electrical resistivity of castable zirconias. *High Temperatures–High Pressures*, **4**, 513–521.

Mitsui, K. and Inaba, A. (1978). A sealed portable cell made of copper for realizing triple points of condensed gases Doc. CCT/78–28, CCT 1978 (available from BIPM).

Neuer, G. and Brost, O. (1975). Heat pipes for the realization of isothermal conditions at temperature reference sources. Temperature-75, 446–452.

Nier, A. O. C. (1950). A redetermination of the relative abundances of the isotopes of neon, krypton, rubidium, xenon and mercury. *Phys. Rev.*, **79**, 450–454.

Pavese, F. (1975). Realization of the IPTS-68 between 54.361 K and 273.15 K. Temperature-75, 70–79.

Pavese, F. (1978). The triple points of argon and oxygen. *Metrologia*, **14**, 93–103.

Pavese, F. (1982). On the use of first generation sealed cells in an international comparison of triple-point temperature of gases. *T MCSI*, **5**, 209–216.

Pavese, F. and Ferri, D. (1982). Ten years of research on sealed cells for phase-transition studies of gases at IMGC, *TMCSI*, **5**, 217–228.

Quinn, T. J. and Chandler, T. R. (1972). The freezing point of platinum determined by the NPL photoelectric pyrometer. *TMCSI*, **4**, Part 1, 295–309.

Quinn, T. J. and Compton, J. P. (1975). The foundations of thermometry. *Rev. Prog. Phys.*, **38**, 151–239.

Schich, H. L. (1966). "Thermodynamics of Certain Refractory Compounds". Academic Press, London and New York.

Schooley, J. F. and Soulen, R. J. (1982). Superconductive thermometric fixed points. *TMCSI*, **5**, 251–260.

Schooley, J. F., Soulen, R. J. and Evans, G. A. (1972). Preparation and use of superconductive fixed point devices SRM 767, NBS Special Publication 250-44.

Scott, R. B., Denton, W. N. and Nicholls, C. M. (1964). "Technology and Uses of Liquid Hydrogen", Pergamon Press, London.

Smithells, C. J. (ed.) (1976). "Metals Reference Book", 5th Edition, Butterworths, London.

Soulen, R. J. and Colwell, J. H. (1971). The equivalence of the T_c of pure indium as determined by electrical resistance, magnetic susceptibility and heat capacity measurements. *J. Low Temp. Phys.* **5**, 325.

Sun, K. H. and Tien, C. L. (1975). Thermal performance characteristics of heat pipes. *Int. J. Heat Mass Transfer*, **18**, 363–380.

Ubbelohde, A. R. (1978). "The Molten State of Matter", J. Wiley, London.

Walton, J. R. and Cameron, A. E. (1966). The isotopic composition of atmospheric neon. *Zeit. Naturforsch.*, **21a**, 115–119.

Winegard, W. C. (1964). "An Introduction to the Solidification of Metals", Institute of Metals, London.

Woolley, H. W., Scott, R. B. and Brickwedde, F. G. (1948). Compilation of thermal properties of hydrogen in its various isotopic and ortho-para modifications. *J. Res. NBS* **41**, 379–475.

Ziman, J. M. (1979). "Models of Disorder", Cambridge University Press.

5

Resistance Thermometry

▬▬▬▬▬▬▬▬▬▬▬▬▬▬

5-1   The electrical resistance of metals, alloys and semi-
        conductors

Since the early days of resistance thermometry and the work of Callendar on
the platinum resistance thermometer, the subject of resistance thermometry
has undergone considerable changes. In addition to the classical platinum
resistance thermometer used for work of the highest accuracy over an
increasingly wide range of temperature, we now find extensive industrial use of
resistance thermometers employing wire elements of either platinum, copper
or nickel, or screen-printed thick-film elements of platinum. Thermistors can
now provide an excellent low-cost means of precise thermometry in the room
temperature range. For scientific applications at low temperatures we have
resistance thermometers with elements of rhodium–iron, germanium, carbon
or carbon–glass. In many industrial applications the resistance thermometer is
replacing the thermocouple as the main process-control instrument. At
temperatures below about 700 °C, the best industrial resistance thermometers
are now more accurate and more reliable than any available thermocouple. In
addition, the increasing use of microprocessors in instrumentation is allowing
a much more rapid and sophisticated use to be made of the information
contained in the signal from the thermometer, than was previously possible.

   A detailed understanding of the electrical conduction process is clearly not a
prerequisite for the proper use of a resistance thermometer to measure
temperature. Nevertheless, investigations aimed at improving its
reproducibility, extending its range or using it to the very highest accuracy are
unlikely to be very productive without at least a passing acquaintance with the
underlying theory of what is observed. Before embarking upon a description of
the behaviour and applications of the more important resistance
thermometers, we shall therefore review briefly the theory of electrical
conduction in pure metals, alloys and semiconductors.

The theory of electrical conduction in mixed-oxide semi-conductors (thermistors) is altogether too complex and is insufficiently well understood for it to be worthwhile entering into in this introductory section. Instead, an outline of the main contributing factors to the conduction process will be given in Section 5-8, which deals with their properties and applications in thermometry.

It must be made clear at the outset that the processes involved in the conduction of electricity are extremely complex. While most of the processes taking part in electrical conduction are fairly well understood, and the general shape of the resistance/temperature curve for metals, alloys and semiconductors can be predicted, quantitative theoretical calculations are not good enough for them to be adopted, unmodified, as the basis for resistance thermometry. One of the difficulties in making quantitative predictions is that of calculating, sufficiently accurately, the relative magnitudes of the various competing processes involved.

Although the simple electron-gas model of Drude in 1900 successfully predicted Ohm's law and the Wiedermann-Franz law, it did not explain why electrical conductivity depends upon temperature, how magnetic properties arise nor why the electrons' heat capacity is much lower than the classical value of 3R. As we shall see, it is now quite clear why the electrical resistivity of the purest available metals falls from a typical value of about 10 $\mu\Omega$ cm at room temperature to less than 10^{-3} $\mu\Omega$ cm at liquid helium temperatures, and why that of a concentrated alloy falls only by about a factor of two over the same temperature range. The behaviour of a semi-conductor is also well understood; the resistivity rises exponentially on cooling, and at very low temperatures, a pure semiconductor becomes a good insulator. The addition of traces of impurity into a semiconductor leads, in general, to a substantial decrease in the resistivity, in contrast to the case of a pure metal in which a trace of impurity always leads to an increase in resistivity.

We shall begin by taking the very simple view that an electric current is the result of a small drift velocity being imparted to the free electrons by the applied electric field. In a metal, the number of electrons available for the conduction process depends upon the crystal structure, but for a monovalent metal, it is one per atom. In dealing with the behaviour of electrons in solids, it has been found convenient to develop the theory in terms of a three-dimensional co-ordinate system for which the three cartesian co-ordinates are the components of wavenumber, k, of the electron, namely k_x, k_y and k_z. An electron of energy E and momentum p has a wavenumber k such that, from de Broglie's relation, $p = \hbar k$, where $\hbar$ is the Planck constant upon 2π, and $E = p^2/2m$. The position of an electron in so-called "k-space" is thus represented by a vector of length k, proportional to the momentum of the electron. In a metal of N free electrons at the absolute zero of temperature, the $N/2$ lowest energy states will be occupied ($N/2$ because the Pauli exclusion principle allows electrons of opposite spin to occupy the same state). Thus all the states up to a certain energy will be occupied. In k-space this leads to the presence of a constant energy surface centred on the origin. All states inside this surface are occupied and all those outside it are empty; such a surface is known as a Fermi surface, and the energy of the state at the Fermi surface is the Fermi energy, or Fermi level, E_F. The Fermi surface will be spherical only for those metals, such as the alkali metals, having the simplest electron configuration. The number of k-states per unit energy range is called the density of states $D(E)$.

In a metal at a temperature above absolute zero, the sharp break between occupied states and unoccupied states is blurred and the electrons are found over a range ($\approx kT$) of energy states straddling the Fermi level.

In the absence of an electric field the motion of the electrons is random, so that on average for every electron having momentum $+k$, there is another of momentum $-k$. The presence of an electric field modifies the distribution of electron momentum in k-space in a manner indicated in Fig. 5-1. Electrons on the right-hand side move into a region previously unoccupied, while those on the left leave vacant regions previously occupied. The extent of the movement of the Fermi surface is limited by the various scattering mechanisms which tend to destroy systematic motion in one direction, i.e. in terms of k-space, to return the region of occupied space to one symmetrical about the origin. Calculations show that the size of the displacement, $\delta k/k$, is extremely small of the order of 10^{-10}. The number of electrons that effectively take part in electric

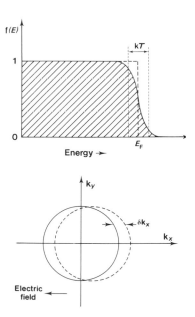

Fig. 5-1. The effect of an electric field on the distribution of occupied states in k-space; — before field is applied, --- after field has acted for a time long compared with the relaxation time.

conduction, namely those in the region of k-space δk of Fig. 5-1, is thus a minute fraction of the total number of electrons in the metal. These "conduction electrons" are only those very close to the Fermi surface. For all the rest, an electron of momentum $+k$ continues to be cancelled by one of momentum $-k$. Theoretical calculations of the conduction process rely, therefore, upon a detailed knowledge of the behaviour of electrons having energy close to E_F. In other words, we need to know the shape of the Fermi surface, which for most metals is complex and difficult to work out. For a proper introduction to the theory of solids and the conduction process, the interested reader is referred to the texts by Ziman (1972), Kittel (1976) or Dugdale (1977) upon which the following brief discussion is based.

If we consider each electron to have a mass m and charge e, it is straightforward (see for example Dugdale, 1977, p. 13 *et seq.*) to show that the conductivity of the metal σ (the reciprocal of the resistivity ρ) is given by

$$\sigma = \frac{ne^2\tau}{m} = \frac{1}{\rho} \qquad (5\text{-}1)$$

where n is the number of electrons per unit volume, e the electronic charge, m the electronic mass and τ is a characteristic time, known as the relaxation time. The relaxation time is related to the rate at which the drift velocity reaches a constant value following the application of an electric potential. Using σ for a typical metal at room temperature, and the known values of e and m for an electron and taking n to be equivalent to one electron per atom, we find that τ is extremely short, only of the order of 10^{-13} sec. The relaxation time is of crucial importance in the understanding of electrical conductivity. In pure metals, for example, it is the only parameter in equation (5-1) that changes with temperature. In semiconductors, the number of charge carriers is very sensitive to temperature, and variations in n have a much bigger effect on the conductivity than do changes in τ.

The basis of our present understanding of the conductivity of metals (Allen and Butler, 1978) is the idea, due to Bloch (1928), that the free electrons travel through the metal as plane waves modified by a function having the periodicity of the lattice. This immediately overcomes the objection to the simple electron-gas theory that the lattice atoms themselves must be the principal scattering centres for the conduction electrons. It thus allows mean-free paths of the order of millimeters, which are observed at low temperatures in pure metals, to become reasonable. The resistivity of the metal, according to the Bloch theory, arises only from imperfections in the lattice. Thus foreign atoms, point defects and grain boundaries all lead to additional scattering and hence to an increase in the resistivity. The electrons are also scattered by lattice vibrations, by emission and by absorption of a quantum of lattice vibrational energy, a phonon. Each of these scattering mechanisms has its own relaxation time τ_1, τ_2, etc., which to a first order, is independent of the others so that, since we are combining rates, we can write

$$\frac{1}{\tau} = \frac{1}{\tau_1} + \frac{1}{\tau_2} + \ldots \tag{5-2}$$

where τ is an overall relaxation time. Thus the total resistivity ρ (proportional to $1/\tau$) can be considered as simply the sum of the resitivities due to separate processes. This is an expression of Matthiesen's rule and for small concentrations of foreign atoms and defects provides an adequate description of what is observed. It is worth remarking that one of the reasons for the growth of low temperature studies of the solid state is that the electron/phonon scattering component becomes negligible at low temperatures leaving the other scattering processes, which do not vary very much with temperature, available for study.

The band structure of solids is a further consequence of the interaction of the electron wave function with the lattice. It is a way of representing the frequencies and directions for which the electron wave function can and cannot pass through the lattice. The changes in direction of the electron wave on reflection of the Bragg angles from the lattice planes are perfectly elastic and do not contribute to the electrical resistance. For a particular crystal and electronic configuration, the Bragg condition imposes certain restrictions on the direction and energy accessible to an electron wave. These forbidden directions and energies lead to gaps in the otherwise almost continuous range of energies and directions available. It is these gaps (of the order of 1 eV in a semiconductor and 5 eV or more in a good insulator) that lead to the striking difference in conductivity between metals, semi-conductors and insulators (Fig. 5-2). The characteristic of a metal is that the Fermi energy lies within a band having vacant energy levels. A semi-conductor has a completely filled band with a gap sufficiently small for a small number of electrons to be

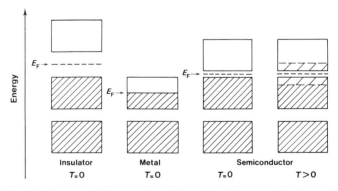

Fig. 5-2. Schematic diagrams of the energy bands for an insulator, a metal and a semi-conductor all at $T = 0$, and (at right) a semi-conductor at a finite temperature.

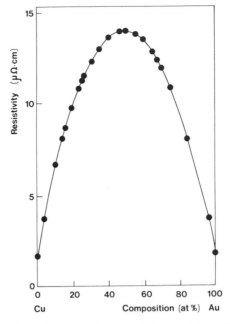

Fig. 5-3. The resistivity of disordered Cu–Au alloys (after Dugdale, 1977).

thermally excited into the empty band above. An insulator differs from a semi-conductor only in having a gap sufficiently wide for practically no thermally excited electron to cross it.

Before discussing in more detail the resistance/temperature characteristics of pure metals and semiconductors, we should note in passing the behaviour of concentrated alloys. The introduction of a significant proportion of foreign atoms into solid solution in an otherwise pure metal leads to a disordered lattice structure. The result is a nearly temperature-independent scattering term, the magnitude of which can be many times that of the original electron/phonon contribution of the pure metal. The variation with composition of residual resistivity of disordered Cu–Au alloys, illustrated in Fig. 5-3, can be described approximately by the simple relation

$$\rho = Ax(1 - x) \tag{5-3}$$

where x is the concentration of one of the components, and A is a constant which depends upon the amount of disorder introduced into the lattice by the addition of the second component. Equation (5-3) is based upon very simple assumptions about the behaviour of the electrons in the disordered lattice, since very little is known about the detailed conduction processes under such conditions. The constant A can be relatively small if the second component is a

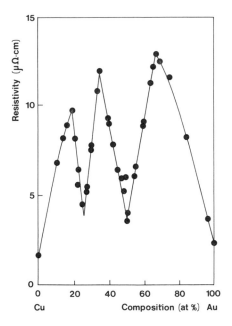

Fig. 5-4. The resistivity of ordered Cu–Au alloys (after Dugdale, 1977).

metal from the same column of the periodic table as the first. In this case, it would have the same number of valence electrons and may be very similar in size. If, at certain compositions and after appropriate heat treatment, ordering takes place then, of course, the resistivity drops considerably. This is illustrated in Fig. 5-4, which shows the same range of Cu–Au alloys as did Fig. 5-3, but after a suitable heat treatment to produce ordering at about 25 atomic % and 50 atomic % Au. One of the most useful techniques in the study of phase equilibrium diagrams (Hume-Rothery and Raynor, 1956) is that of resistivity measurement because it is such a sensitive indicator of the onset of ordering. Since, for obvious reasons, concentrated alloys are not of particular interest in resistance thermometry, we shall not dwell on their behaviour. It is worth noting however that studies of their resistance/temperature behaviour are of great importance in another field of metrology, that of electrical resistance standards, where the problem is to produce alloys having practically no change in resistance with temperature. The best resistance standards now have temperature coefficients of only one or two parts in 10^6 per kelvin at room temperature. The advent of the Josephson-junction voltage standard has led to a search for alloys having very low-temperature coefficients of resistance at liquid helium temperatures while at the same time having resistivities sufficiently large to permit resistances of 10 Ω or 100 Ω to be made (March and Thurley, 1979; Macfarlane and Collins, 1978).

5-2   The resistivity of a pure metal as a function of temperature

In discussing the resistivity of a pure metal as a function of temperature, it is convenient to distinguish three ranges of temperature in relation to a characteristic temperature θ_R, related to its Debye temperature, θ_D. The first of these is the high-temperature range, $T > \theta_R$, in which the phonons are practically all at the maximum cut-off frequency ω_m, having energy $\hbar\omega_m = k\theta_R$. The second is that which includes θ_R and extends to moderately low temperatures. In this range, the phonons have energies extending up to $k\theta_R$. Finally there is the very low-temperature range, $T \ll \theta_R$, in which the phonon energy does not, of course, exceed kT.

At high temperatures the vibrating lattice atoms can be considered as independent random scattering centres, and hence the scattering probability depends upon the mean square amplitude of the lattice vibrations $\overline{X^2}$. Provided that these are harmonic, their mean square amplitude is simply proportional to T. Thus, ignoring thermal expansion, the resistivity of a pure metal at high temperatures should be proportional to T. In fact for a simple harmonic oscillator of mass M, the equipartition of energy theorem leads to

$$\tfrac{1}{2}M\omega^2 \overline{X^2} = \tfrac{1}{2}kT$$

but we know that $\omega_m = k\theta_R/\hbar$ so that

$$\overline{X^2} = \frac{\hbar^2 T}{k\theta_R^2 M}$$

therefore

$$\rho \propto \frac{T}{k\theta_R^2 M}. \tag{5-4}$$

Although a simple T dependence is indeed approximately what is observed for most simple metals for $T \geqslant \theta_R$, the remaining differences cannot be completely accounted for, even taking into account known anharmonic effects, thermal expansion and a more detailed phonon frequency spectrum. We will come back to high temperature behaviour later on when we discuss the effects observed in transition metals.

At very low temperatures, $T \ll \theta_R$, the simple equipartition theorem, of course, no longer applies and the quantization of lattice vibrations must be taken into account. It is approximately correct to say that at a given low temperature, only those modes for which $\hbar\omega \leqslant kT$ are excited and that the number of modes excited at a given frequency varies as ω^2. This leads to the conclusion $\hbar\omega \approx kT$. The factors which contribute to the temperature dependence of the resistivity at low temperatures are the number of phonons

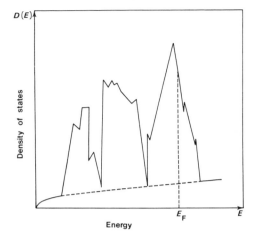

Fig. 5-5. Density of states *vs* energy (schematic) for a transition metal such as Pt or Pd (after Dugdale, 1977).

available for scattering, the scattering probability and the scattering angle. The combination of all of these factors leads to a T^5 dependence of resistivity upon temperature, although the details of the argument needed to reach this conclusion are too involved to go into here (see Ziman, 1972, p. 220 *et seq.*; Dugdale, 1977, p. 164 *et seq.*).

If we now consider metals of interest in resistance thermometry, we find ourselves dealing with transition metals in which there is a partly filled *d*-band as well as an *s*-band ("*s*" and "*d*" refer to the second quantum number of the electron having values of 0 and 2 respectively; see Hume-Rothery and Raynor, 1956). Because the *d*-electrons are more localized than the *s*-electrons, the conductivity is mainly due to the latter. However, the probability that the *s*-electrons will be scattered into the *d*-band is high because the density of *d*-states at the Fermi level is high (Fig. 5-5) and the resistivity of transition metals is therefore higher than that of non-transition metals. The *d*-band may also modify the temperature dependence of the resistivity. At high temperatures, kT may not be negligible compared with the energy difference between the Fermi level and the top or bottom of the *d*-band. The assumption that the Fermi surface provides a sharp distinction between occupied and unoccupied states is no longer valid and, for a paraboloid *d*-band, the resistivity is modified by a factor $(1 - BT^2)$, where B is a constant. However, the density of states in the *d*-band is far from being a smooth function of energy (Fig. 5-5), and the effect will be complicated by variations in the density of states within kT of the Fermi level. The deviation of the temperature dependence from the simple T-dependence may be positive or negative.

A further departure from a simple T dependence of the resistivity at high temperatures results from thermal expansion. The characteristic temperature falls and hence the amplitude of the lattice vibrations increases. Equation (5-4) must then be modified by the addition of a term in T^2. Thus for platinum, which has a θ_R of about 240 K, the resistance/temperature relation at room temperature and above should be quadratic for the reason of thermal expansion alone. In addition, the presence of higher order terms would not be unexpected in view of the complex nature of the density of states curve. In fact, a quadratic resistance/temperature relation provides quite a good fit to experimental data up to about 900 K. At still higher temperatures, a further term must be added resulting from the effects of lattice vacancies. The equilibrium concentration of vacancies in a metal leads to an increase in resistivity $\Delta\rho$ given by an equation of the form

$$\Delta\rho \propto \exp[-E_r/kT] \tag{5-5}$$

where E_r is energy of formation of a lattice vacancy, which in platinum is about 1.5 eV (Jackson, 1965; Berry, 1972). This is discussed further in Section 5.4.4.

At low temperatures, transition metals show an effect resulting from electron/electron scattering, that leads to a T^2 term in the resistivity/temperature relation. This type of large angle electron/electron scattering, sometimes called Baber scattering (see Dugdale, 1977, p. 250 *et seq.*), can arise when the Fermi surface is non-spherical or has contributions from more than one band. For most transition metals, the T^2 term becomes predominant below 10 K. For the ferromagnetic metals, a further T^2 term, indistinguishable from the Baber term, is also to be expected arising from the scattering of conduction electrons by magnetic spin waves. In addition, all of the ferromagnetic metals show anomalous resistivity behaviour near the Curie point.

We have already mentioned that at low temperatures in nearly pure metals the resistivity is strongly dependent upon the concentration of impurities and lattice defects. Particular effects are observed if a very small amount of a magnetic metal is dissolved in certain non-magnetic metals. These effects arise when the dissolved magnetic impurities form what are called "localized magnetic moments". Whether or not a localized moment is formed in a particular dilute alloy system is a question too complex to be entered into here (see Dugdale, 1977, p. 189 *et seq.*), but the effects of a localized moment are interesting. Under certain circumstances the presence of a localized moment can lead to a resistance minimum occurring at low temperatures: the Kondo effect. The reasons for the presence of a resistance minimum are very subtle, but broadly it occurs because the spin-dependent contribution to the resistivity ρ_{spin} contains a term proportional to $J \ln T$, where J is the $s \rightarrow d$ exchange-coupling constant. Thus when J is negative, ρ_{spin} increases with decreasing temperature and the total resistivity exhibits a minimum at a temperature near 10 K where the temperature coefficient of the lattice (or other) resistivity term is small.

The temperature of the resistance minimum will be only a weak function of

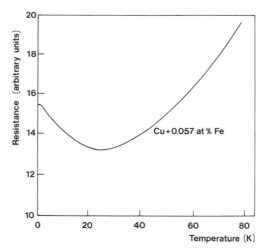

Fig. 5-6. Temperature dependence of resistance for an alloy of Cu + 0.057 % Fe (after Dugdale, 1977).

the concentration of the magnetic ions (Fig. 5-6). The depth of the resistance minimum will be roughly proportional to the concentration.

Of particular interest in thermometry, and the subject of Section 5-6 below, is the anomalous behaviour of dilute alloys of iron in rhodium. Instead of a resistance minimum, we find a monatonic decrease in resistivity as shown in Fig. 5.30. In this alloy, the host contains a d-band and the d-states of the iron impurity atom cannot be considered to be localized. However, fluctuations in spin density are enhanced in the neighbourhood of the impurity atoms and the magnetic properties of the alloy are similar to those of Kondo alloys. The impurity resistivity also has a form similar to the Kondo effect, the positive temperature coefficient being essentially a consequence of the similarity between iron and rhodium atoms and the d-band in particular (Rivier and Zlatic, 1972).

5-3   The resistivity of a semi-conductor as a function of temperature

In a semi-conductor with a band gap $E_g \approx 1$ eV, the number of electrons, n, having sufficient energy at room temperature, where $kT \approx 2 \times 10^{-3}$ eV, to bring them into the conduction band is very small. The Fermi level for a semi-conductor is, by convention, measured from the top of the valence band rather than from the bottom as for a metal, and is given the symbol μ. Since $(E_g - \mu)$ is large compared with kT the expression for the number of electrons reaching the conduction band is given by

$$n = 2\left(\frac{m_e k T}{2\pi\hbar^2}\right)^{3/2} e^{(\mu - E_g)/kT} \tag{5-6}$$

where m_e is the effective mass of the electron (see Ziman, 1972 or Dugdale, 1977) for a fuller discussion and derivation of these equations). Similarly the number of holes in the valence band is given by

$$p = 2\left(\frac{m_h k T}{2\pi\hbar^2}\right)^{3/2} e^{-\mu/kT} \tag{5-7}$$

where m_h is the effective mass of the hole. Since, in an intrinsic semiconductor, $n = p$ we can write

$$n = p = (np)^{1/2} = 2\left(\frac{kT}{2\pi\hbar^2}\right)^{3/2} (m_e m_h)^{3/4} e^{-E_g/2kT}. \tag{5-8}$$

The electrical conductivity σ is proportional to the mobilities of the electrons and holes which are in turn proportional to the relaxation times τ_e and τ_h for electron/phonon and hole/phonon interactions respectively. We can therefore write

$$\sigma = \frac{ne^2\tau_e}{m_e} + \frac{pe^2\tau_h}{m_h}. \tag{5-9}$$

In contrast to the behaviour of a pure metal, we find that it is n and p that are the dominating temperature-dependent factors in equation (5-9) and not τ_e and τ_h.

An intrinsic semiconductor, therefore, exhibits a variation of resistivity with temperature which is given by

$$\rho = A T^{-3/2} e^{E_g/2kT}. \tag{5-10}$$

The exponential function of equation (5-10) is sufficiently large, in the temperature range usually of interest, for the $T^{-3/2}$ part to be effectively constant and hence we normally write

$$\rho \propto e^{E_g/2kT}. \tag{5-11}$$

The controlled addition of impurities to a semiconductor can make a very large change in its resistivity, both in absolute value and in its rate of change with temperature. If a very small amount of impurity is added to a semiconductor such that each impurity (donor) atom, has a valency one more than that of the host atom, then for each impurity or donor atom an extra

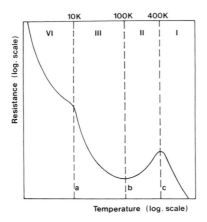

Fig. 5-7. Schematic temperature dependence of the resistance of a germanium thermometer (after Blakemore, 1962).

electron becomes available for excitation into the conduction band. The energy required for excitation is much smaller than, say, kT at room temperature. Similarly, if the impurity atom has a valency one less than that of the host atom, it acts as a sink for an electron and is called an acceptor atom. Its presence leads to the possibility of an extra hole to aid conduction in the valence band. Depending upon whether conduction is mainly through the addition of electrons or holes, the semiconduction is called n-type or p-type respectively.

The variation of resistivity with temperature (Blakemore, 1962) of a semiconductor to which a small amount of impurity has been added is illustrated in Fig. 5-7. In practice, there are always both donor and acceptor atoms present and the choice open to the designer of semi-conducting resistance thermometers is that of deciding the relative proportions of each. For the purposes of describing the processes of conduction we shall consider germanium to which has been added a concentration N_d of arsenic donor atoms and N_a of unspecified acceptor atoms. Four ranges of temperature can be distinguished in Fig. 5-7, one conduction process being predominant in each. Taking first the high temperature range (I), we find that conduction is predominantly by electrons thermally excited from the valence band into the conduction band according to equation (5-8), all the impurity atoms having long since been ionized. This is the region of intrinsic conduction and for germanium begins just above 400 K. This range is of little interest for germanium resistance thermometry.

As the temperature falls, we find that intrinsic conduction ceases to be of importance and in ranges II and III, conduction is by electrons released by the $(N_d - N_a)$ donor atoms. The number n of such electrons ionized into the conduction band is a complicated function of the relative number and nature of the donor and acceptor atoms, but is of the form (Blakemore, 1962):

$$n \approx \frac{N_d - N_a}{1 + BN_a e^{\beta/kT}} \tag{5-12}$$

where B and β are constants depending upon the donor impurities used; the conductivity, therefore, is given by

$$\sigma = \frac{ne^2\tau_e}{m_e} \approx \frac{(N_d - N_a)e^2\tau_e}{m_e(1 + BN_a e^{\beta/kT})}. \tag{5-13}$$

Above about 100 K, all of the $(N_d - N_a)$ electrons are ionized and available in the conduction band, $e^{\beta/kT}$ is small and equation (5-13) reduces to

$$\sigma \approx \frac{(N_d - N_a)e^2\tau_e}{m_e}. \tag{5-14}$$

The conduction thus depends upon the variation of τ_e with temperature. As would be expected, on increasing the temperature τ_e falls, and so in Fig. 5-7 the resistivity rises above 100 K until the intrinsic conductivity of range I takes over. Below 100 K, the ionization of the $(N_d - N_a)$ donor atoms ceases to be complete and so n falls according to equation (5-12). The resistivity thus begins to rise and continues to do so as the temperature falls until at a temperature near 10 K, ionization has practically ceased so that none of the carriers are excited. For the lower part of this range, we can write

$$\sigma \propto n \approx \frac{(N_d - N_a)}{N_a} e^{-\beta/kT}. \tag{5-15}$$

Below about 10 K we find the final range in which conduction takes place by electrons jumping directly from one impurity atom to the next. This is known as impurity conduction and is proportional to the excess number of donor atoms $(N_d - N_a)$ and temperature so that

$$\sigma \approx D(N_d - N_a)e^{-\gamma/kT} \tag{5-16}$$

where D and γ depend upon the species of the donor and acceptor atoms.

Of particular interest in thermometry is the junction between ranges III and IV which occurs in the region where germanium resistance thermometers are most commonly employed. According to Fig. 5-7, the junction between these two ranges, described by equations (5-15) and (5-16), appears quite abrupt. It is possible, however, by adjusting the concentration and species of the donor impurity to soften this abrupt change in slope. This is illustrated in Figs. 5-8 and 5-9.

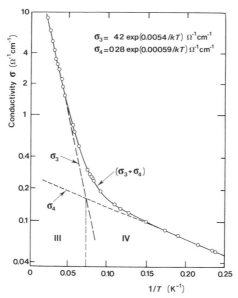

Fig. 5-8. A "hard" or abrupt transition between ranges III and IV for germanium doped with 3.75×10^{16} atoms of gallium per cm^3, of which 8 % are compensated with antimony. σ_3 and σ_4 represent equations (5-15) and (5-16) (after Blakemore, 1962).

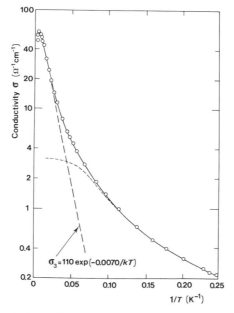

Fig. 5-9. A "soft" or gradual transition between ranges III and IV for germanium doped with 1.8×10^{17} arsenic atoms per cm^3 of which 8% are compensated with gallium (after Blakemore, 1962) σ_3 represents equation (5-15).

5-4   The high precision platinum resistance thermometer

The platinum resistance thermometer is the preferred instrument for the most accurate measurement of temperature over a range extending from the triple point of hydrogen, at 13.81 K, up to the freezing point of antimony, 903.89 K. The attributes of pure platinum as a resistance-thermometer element are that it is chemically inert, even at high temperature, has a high melting point, is readily available as fine wire in a highly pure condition and has a high resistivity ($\sim 10 \ \mu\Omega$ cm at room temperature). The property of malleability, although useful in drawing down fine wires, leads to difficulties since in the fully annealed state, pure platinum is very soft and thus poses certain problems as regards mounting. The very strong resistance to oxidation of platinum at high temperatures makes it an attractive material for high temperature thermometry both as a resistance thermometer and, as we shall see, as a thermocouple element.

5-4-1   The resistance/temperature relation for the platinum resistance thermometer

As is suggested by theory, the resistance of a platinum resistance thermometer follows quite closely a quadratic function of temperature over a wide range extending upwards from room temperature. We can therefore write

$$R_t = R_0(1 + At + Bt^2) \tag{5-17}$$

where R_t is the resistance of the thermometer at a temperature t and R_0 that at a reference temperature t_0, usually taken as 0 °C. In order to avoid the requirement for precise absolute measurements of resistance, the calibration of thermometers is always in terms of the ratio R_t/R_0, known as $W(t)$. Thus, in effect, resistivities are measured rather than resistances. There are also other advantages in using $W(t)$ as the variable. These stem from the relative insensitivity of $W(t)$ to small changes in resistance due to strain or contamination of the wire, a consequence of Matthiessen's rule being nearly obeyed. In the early days of platinum thermometry, the notion of "platinum temperature" t' was introduced. It was a linear function of resistance between 0 °C and 100 °C so that $R_{t'} = R_0(1 + \alpha t')$ where $\alpha = (R_{100} - R_0)/100 \ °C \ R_0$ hence

$$t' = 100 \ °C \left\{ \frac{R_{t'} - R_0}{R_{100} - R_0} \right\}. \tag{5-18}$$

Since $t' \approx t$, we can substitute for $R_{t'}$, from (5-17) and we find

$$t' = 100 \ ^\circ C \left\{ \frac{At + Bt^2}{100 \ ^\circ CA + 10^4 \ ^\circ C^2 \ B} \right\} = \left[\frac{At + Bt^2}{A + 100 \ ^\circ C \ B} \right] ^\circ C \qquad (5\text{-}19)$$

from which

$$t - t' = \frac{10^4 \ ^\circ C^2 \ B}{A + 100 \ ^\circ C \ B} \left\{ \frac{t}{100 \ ^\circ C} - \left(\frac{t}{100 \ ^\circ C} \right)^2 \right\} ^\circ C \qquad (5\text{-}20)$$

or

$$t - t' = -\delta \left\{ \frac{t}{100 \ ^\circ C} - \left(\frac{t}{100 \ ^\circ C} \right)^2 \right\} \qquad (5\text{-}21)$$

$$= \delta \left(\frac{t}{100 \ ^\circ C} \right) \left(\frac{t}{100 \ ^\circ C} - 1 \right). \qquad (5\text{-}22)$$

The relationships between the quantities A, B, α and δ are

$$A = \alpha \left(1 + \frac{\delta}{100 \ ^\circ C} \right) \qquad \alpha = A + 100 \ ^\circ C \ B$$

$$B = -10^{-4} \alpha \delta \ ^\circ C^{-2} \qquad \delta = \frac{-10^4 \ ^\circ C^2 \ B}{A + 100 \ ^\circ C \ B}.$$

These are the formulae originally derived by Callendar, and over the years, they have become very familiar to all those who use platinum resistance thermometers (Barber and Hall, 1962). They continue to be useful, in that the quantities α and δ are a characteristic of each thermometer showing, respectively, the mean slope of the resistance/temperature curve between 0 °C and 100 °C and the departure from linearity in the same range.

In the IPTS-68 a correction term was added to take into account departures from thermodynamic temperature of the original quadratic equation:

$$t_{68} = t' + 0.045 \ ^\circ C \left(\frac{t'}{100 \ ^\circ C} \right) \left(\frac{t}{100 \ ^\circ C} - 1 \right) \left(\frac{t'}{419.58 \ ^\circ C} - 1 \right) \left(\frac{t'}{630.74 \ ^\circ C} - 1 \right).$$

$$(5\text{-}23)$$

Since then, new work has shown that the departure from thermodynamic temperature of the simple quadratic is not of the form given by (5-23). This is discussed in more detail in Chapter 2.

That α and δ are characteristics of the thermometer is just what would be expected from our earlier discussion of the theory. The slope of the resistance/temperature relation is, from equation (5-1) inversely proportional to the overall relaxation time, τ. The major contribution to τ is that from electron/phonon interactions which is inversely proportional to temperature,

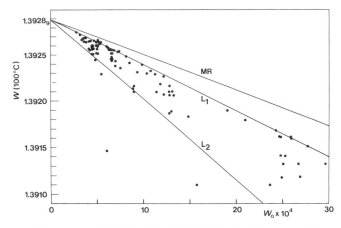

Fig. 5-10. Correlation between W(373.15 K) and W(0 K). The lines L_1 and L_2 define an envelope within which most experimental points lie, and MR is the line below which all experimental points should lie if Matthiessen's rule is obeyed. The value of α (ideal) is given by 10^{-2} [W(373.15 K) -1] (after Berry, 1963).

but it also contains contributions from electron/impurity, electron/vacancy and electron/grain boundary relaxation times. These are temperature dependent and therefore α should also be a good indication of the purity and state of anneal of the thermometer, as indeed it is. The departure from linearity, δ, will be a function of the coefficients of the T^2 and higher order terms which, as we have seen, depend upon thermal expansion and the density of states curve near the Fermi energy. Both of these quantities depend upon the purity of the wire, and indicate that δ should be related to α.

The resistance at absolute zero, when the electron/phonon resistivity has vanished, depends upon impurities and state of anneal and, as would be expected, there is a clear relation between α and W(0 K), Fig. 5-10. W(0 K) is, of course, obtained from measurements made at liquid helium temperatures since W(0 K) $\approx$ W(4.2 K). Careful measurements and calculations (Berry, 1963) indicate that very pure, well annealed platinum should have an α-coefficient of 0.003 928 9 °C^{-1} and a W(0 K) of 3×10^{-4}.

For temperatures below 0 °C, the departure from linearity of $W(t)$ becomes too great for the simple δ correction term to be sufficient. The original Callendar equation was, therefore, modified by Van Dusen in 1925 as follows:

$$t - t' = \delta \left(\frac{t}{100\ °C} - 1 \right) \frac{t}{100\ °C} + \beta \left(\frac{t}{100\ °C} - 1 \right) \left(\frac{t}{100\ °C} \right)^3 \quad (5\text{-}24)$$

which is equivalent to

$$R_t = R_0 (1 + At + Bt^2 + C(t - 100\ °C)t^3) \quad (5\text{-}25)$$

where

$$C = \frac{-\alpha\beta}{10^8 \ {}^\circ C} \quad \text{and} \quad \beta = -\frac{10^8 C \ {}^\circ C^4}{A + 100 \ {}^\circ C \ B}$$

The magnitudes of δ and β are shown in Fig. 5-11, the β coefficient being determined by a calibration at the boiling point of oxygen.

The introduction of the IPTS-68 led to changes in the way the resistance/temperature relation for platinum was expressed, particularly in the low temperature range. The Callendar–van Dusen equation was found to be inadequate to describe the behaviour of platinum resistance thermometers down to 13.81 K, the triple point of hydrogen. Instead the temperature is now given in terms of the difference between a reference function, covering the range 13.81 K to 273.15 K, and a series of deviation functions covering individual parts of this range (Bedford and Kirby, 1969). We now have

$$T_{68} = \sum_{j=0}^{20} a_j \left(\frac{\ln W_{\text{CCT-68}}(T_{68}) + 3.28}{3.28} \right)^j \text{ kelvins} \tag{5-26}$$

where $W_{\text{CCT-68}}(T_{68})$ is a reference resistance ratio, where the coefficients a_j are given in Appendix II (the text of the IPTS-68). The reference function which appears in the 1975 edition of IPTS-68 differs from that in the original 1968 version of the Scale. As first formulated, the reference function of T_{68} was given by

$$T_{68} = \left\{ A_0 + \sum_{j=1}^{20} A_j (\ln W_{\text{CCT-68}}(T_{68}))^j \right\} \text{kelvins} \tag{5-27}$$

in which the twenty coefficients A_j each contained sixteen digits all of which were necessary for the calculation of T_{68} to an accuracy of 0.1 mK. It was clear that this must be a very inefficient way of calculating a number having no more

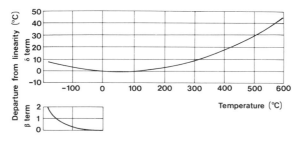

Fig. 5-11. The magnitude of the δ and β terms in the Callendar (5-22) and Callendar–Van Dusen (5-24) equations for $\delta = 1.491$ and $\beta = 0.11$.

than seven significant figures. In the 1975 edition the reference function (5-26) is a simple linear transformation of (5-27) in which the transformed variable

$$\frac{\ln W_{CCT-68}(T_{68}) + 3.28}{3.28}$$

is chosen so that it varies only within the range -1 to $+1$. The new coefficients of a_j of (5-26) thus have the property that they are in units of kelvins and may therefore be rounded depending upon the accuracy required in the calculation of T_{68}. All twenty-one terms of the transformed polynomial are still required however.

An even more efficient formulation of the reference function has been given (Quinn, 1975) in which

$$T_{68} = \tfrac{1}{2}C_0 + \sum_{n=1}^{20} c_n T_n(X) \qquad (5\text{-}28)$$

where $T_n(X)$ is the Chebyshev polynomial of degree n, and X is given by

$$X = \frac{2 \ln W_{CCT-68}(T_{68}) + \ln W_{CCT-68}(13.81)}{\ln W_{CCT-68}(13.81)}. \qquad (5\text{-}29)$$

In (5-28), the coefficients C_n (see Appendix II) are also in units of kelvins, but in addition the series itself may be truncated since the coefficients decrease with increasing n.

The resistance ratio $W(T_{68})$ for a real thermometer at a temperature T_{68} is defined by:

$$W(T_{68}) = W_{CCT-68}(T_{68}) + \Delta W_i(T_{68}) \qquad (5\text{-}30)$$

where $\Delta W_i(T_{68})$ is a deviation function expressed as a polynomial in T_{68}.

The range between 13.81 K and 273.15 K is divided into four parts, in each of which $\Delta W(T_{68})$ is defined by a different polynomial in T_{68}. The constants of the polynomial for a particular range are determined from the values of $\Delta W(T_{68})$ at the fixed points within this range and the condition that there be no discontinuity in the first derivative of $\Delta W(T_{68})$ with respect to temperature at the junctions of the ranges. The details of the deviation functions are given in Appendix II.

An inconvenience of the present form of IPTS-68, is that T_{68} appears in both the reference and deviations functions. Therefore, in order to calculate T_{68} from a measured value of $W(T_{68})$, an iterative procedure has to be used. This is not too serious a problem since experience shows that the procedure rapidly converges and not more than two iterations are necessary. The first step in the iteration is to substitute $W(T_{68})$ into the reference function. The value of T_{68} so obtained is then used to calculate a first approximation to $\Delta W(T_{68})$ from the deviation function. This leads to a value of $W_{CCT\text{-}68}(T_{68})$ from which a new calculation of T_{68} can be made.

Although the need for iteration is a minor inconvenience, a much more severe one stems from the requirement that the deviation functions join smoothly with one another at the junctions of the four parts of the low temperature range. It is specified that each deviation function must match the slope at the junction with the next higher range. Thus a thermometer must be calibrated over the whole range to the lowest temperature at which it is intended to be used. For those thermometers that are to be used over the range previously covered by the Callendar–Van Dusen equation, the deviation function is

$$\Delta W(T_{68}) = b_4(T_{68} - 273.15 \text{ K}) + e_4(T_{68} - 273.15 \text{ K})^3(T_{68} - 373.15 \text{ K})$$

$$= b_4 t_{68} - e_4 t_{68}^3 (t_{68} - 100 \text{ °C}) \qquad (5\text{-}31)$$

in which the constants are determined by the measured deviations at the boiling point of oxygen and the boiling point of water (or the freezing points of tin and zinc). The 1975 edition of IPTS-68 offers the alternative of the triple point of argon instead of the boiling point of oxygen, although the slopes of the deviation functions must continue to be matched at the oxygen point.

Although we have been concerned here only with IPTS-68, it is worth noting that various suggestions have been made for devising, in a much simpler way, a temperature scale below 0 °C based upon a platinum resistance thermometer (see for example Kemp et al., 1981). It is likely that in any replacement of IPTS-68, one of these much simpler schemes will be used.

5-4-2   Platinum resistance thermometers for use at low temperatures

More than one design of platinum resistance thermometer is required to cover the range 13 K to 903 K because the requirements for low-temperature thermometry are very different from those at high temperatures. In most low-temperature apparatus, for example, one needs to be able to enclose the thermometer completely within radiation shields and to arrange for thermal anchoring of the leads to the same temperature as the thermometer. These and other requirements have led to the adoption of a design typified by the so-called "capsule" thermometer shown in Fig. 5-12. It is common practice to use platinum wire of about 0.07 mm diameter coiled to give a total resistance of 25 ohms at 0 °C. This requires some 60 cm of wire which, in the most commonly used design, is coiled and supported in a pair of twisted glass tubes and sealed

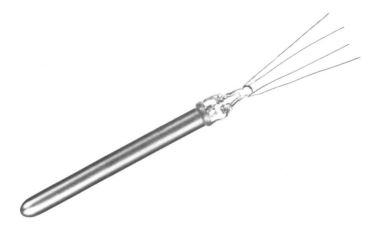

Fig. 5-12. A "capsule" platinum resistance thermometer (courtesy of H. Tinsley Ltd).

by a lead–glass seal into a platinum sheath under a pressure of helium of about 30 kPa at room temperature.

The platinum sheath is made from hard-drawn tube of outside diameter about 5 mm, wall thickness 0.1 mm and overall length, including glass seal, of about 60 mm. All high-precision platinum resistance thermometers are of the four-lead type having a pair of current leads and a pair of potential leads. The four platinum leads, some 50 mm in length, are taken out through the lead–glass seal and then bound to the outside of the seal. By far the most frequent cause of failure of such a thermometer is the breaking of a lead at the seal. This can sometimes be repaired by soldering but not always and so it is well worth taking great care to avoid flexure of the platinum at the point of exit from the glass. Other causes of failure are shorting of the coils and leakage of the gas.

Although the thermal conductivity of platinum is relatively low and the wire diameter is small, care is still needed in the thermal anchoring of the thermometer and leads when the thermometer is being used at low temperatures. The thermometer itself is usually placed in a close fitting well, often with a layer of grease (such as Apiezon N) to aid thermal contact. Common practice is to bind and varnish the copper leads, over a length of about 30 cm, to a thermal anchor at nominally the same temperature as the thermometer.

The heating effect of the measuring current must always be taken into account when accurate measurements are being made. For a helium-filled 25 ohm capsule thermometer, a measuring current of 1 mA is usual at about room temperature. This can lead to a self-heating of between 1 mK and 3 mK depending upon the design of the thermometer. Only a small pressure of helium is required in the thermometer to ensure good thermal contact since, of course, the thermal conductivity of a gas is nearly independent of pressure provided that the mean-free path is much smaller than the dimensions of the container, hence the modest filling pressure of only about 30 kPa at room temperature. For a given measuring current, the heating effect falls with temperature because the resistance of the thermometer falls considerably faster than the thermal conductivity. At about 14 K; for example, the heating effect of a 1 mA current is only about 0.01 mK, and at these temperatures larger currents are used. Although the heating effect is a small correction, it should always be measured. An unusually large heating effect in a new thermometer or an increase in the heating effect in an old thermometer is always an indication of the absence or loss of helium. If this happens, the thermometer is of little further use. The calculation of the resistance of the thermometer at zero current, or the temperature of the thermometer at zero current, is straightforward. The usual way is to increase the measuring current from I to $\sqrt{2}\,I$. On the basis of the heating effect being proportional to I^2, the observed change in temperature, or resistance, is the heating effect for a current I and should be deducted from subsequent measurements made using a current I. In addition to the proper thermal anchoring of the leads, attention must be paid to their inductive and capacitive effects and their d.c. resistance. This is discussed in Section 5-11.

Studies of the difference between the a.c. and d.c. resistivity of capsule type platinum thermometer have shown that it is only at very low temperatures that the effect is likely to be troublesome (Compton, 1972). Using the Automatic Systems A7 bridge operating at 375 Hz (see Section 5-11), the maximum difference between a.c. and d.c. measurements amounted to 3 mK at the triple point of hydrogen. The effect is probably the result of eddy currents in the pure-platinum sheath.

5-4-3   Platinum resistance thermometers for use up to 630 °C

High-precision platinum resistance thermometers for use above 100 °C are usually of the design shown in Fig. 5-13 and are generally referred to as long-stem thermometers. Although the capsule type of thermometer has many advantages, it ceases to be a practical proposition at high temperatures because the leakage resistance between the leads at the glass seal becomes too low. For this reason, the high temperature thermometer has leads that are insulated from one another by mica, silica or sapphire discs, or silica or sapphire tubes. The resistance element itself is usually made from 0.07 mm diameter platinum wire, as for the capsule thermometer, and has a resistance at 0 °C of 25 ohms. Common designs (Barber, 1950; Curtis, 1972; Riddle et al., 1973) are in the form of either a single wire in a bifilar winding on a mica cross, a coil wound in twisted silica tubes, or wires supported in alumina tubes, see Fig. 5-14. All of these designs are aimed at producing a strain-free thermometer element that is free to expand and contract on heating and

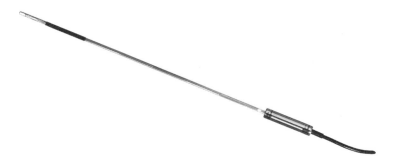

Fig. 5-13. A long-stem platinum resistance thermometer for use in the range −189 °C to 630 °C; o.d., 7 mm; length, 600 mm (courtesy of H. Tinsley Ltd).

cooling without the wire rubbing or being scratched by the support. Those designs having the wire close to the sheath, Fig. 5-14a and c, give improved thermal contact with the outside medium and show smaller heating effects than those in which the wire is further enclosed or is mounted close to the axis.

The thermometers are usually filled with dry air and sealed after preliminary annealing during manufacture. The temperature at which annealing is carried out varies from manufacturer to manufacturer, but is usually about 500 °C. At this temperature, a period of a few hundred hours is found necessary to stabilize new thermometers. The question of annealing procedures during the subsequent use of the thermometer is discussed later. Mica-insulated thermometers should only be used above 500 °C if phlogopite mica, $H_2KMg_3Al(SiO_4)_3$ sometimes known as India or ruby mica, is used. The other common mica, muscovite $H_2KAl_3(SiO_4)_3$, contains about 5 % water which is liberated at about 540 °C, and in so doing alters its shape. The disadvantage of phlogopite mica is that its insulation resistance at room temperature is slightly lower than that of muscovite, but it does not begin to dehydrate until a much higher temperature, ∼ 750 °C. A change in insulation resistance of 3×10^7 Ω at 500 °C is equivalent to an error in temperature measurement of about 1 mK. It was because of these problems with mica that alternative designs using alumina or sapphire or silica insulation were developed. Unfortunately the manufacture of the light and delicate supports from these materials to the fine dimensional tolerances required continues to be difficult and expensive.

One of the principal difficulties encountered in making high precision measurements with long-stem thermometers is that resulting from poor immersion characteristics of the thermometer. The relatively poor thermal contact between the thermometer element and the surroundings, together with the conduction and radiation down the stem, call for a very deep immersion. Figure 5-15 shows how the measured temperature, for this particular design of thermometer, depends upon immersion in a triple-point-of-water cell. The immersion characteristics are noticeably different for the

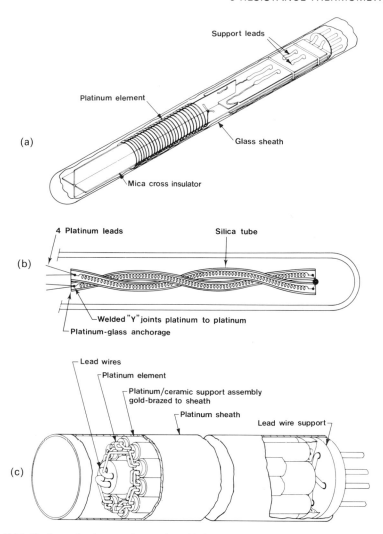

Fig. 5-14. Designs of resistance element found in long-stem platinum resistance thermometers (a) bifilar winding on a mica former (b) coil inside silica tubes (c) wires in alumina tubes; this type of element is usually mounted in a stainless steel tube (Curtis, 1972).

different types of thermometer and, as we would expect, depend upon how close the platinum wire is to the walls of the stem. The heating effects found in the three types of thermometer shown in Fig. 5-14a, b and c in a triple-point-of-water cell are 1 mK, 3 mK and 1 mK respectively for a 1 mA measuring current.

It is worth remarking here that if the ultimate in reproducibility is being sought, better than 20 μK for example, it is the stability of the heating effect that is likely to be the limiting factor. This arises because it becomes

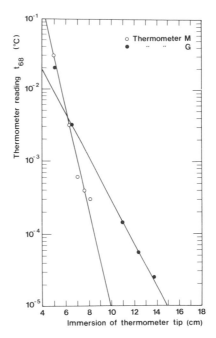

Fig. 5-15. Immersion characteristics in an ice bath of two long-stem platinum resistance thermometers; thermometer M is of the type shown in Fig. 5-14c, while thermometer G is of a type not shown in Fig. 5-14 but is made of a coiled coil would on a mica former (after Riddle *et al.*, 1973).

increasingly difficult to maintain the constancy of thermal contact between the resistance wire and the surrounding medium as the temperature resolution increases.

An immersion of 15 cm in a triple-point-of-water cell (which is demanded for measurements of the highest accuracy) is not difficult to achieve, but at higher temperatures, it becomes difficult to produce regions of uniform temperature which are sufficiently long. The immersion required for a given accuracy does not depend strongly upon temperature because of the logarithmic nature of the dependence. For example in Fig. 5-15 the temperature difference between thermometer G and the triple point decreases by a factor of ten for an increase in immersion of just over 3 cm. Thus, if the outside temperature differs by 250 °C, rather than 25 °C, from that of the medium being measured an additional 3 cm immersion is all that is required. Conversely, if the outside temperature differs by only 2.5 °C, only 3 cm less immersion is required. Thus if the temperature of a water or oil bath at 20 °C is being measured in a room at 22.5 °C, an immersion of 9 cm would be required for an accuracy of 0.1 mK, compared with an immersion of 12 cm in the triple-point-of-water cell. This also explains why it is so difficult to make accurate measurements of high temperatures. The uniformity of temperature demanded is much greater than might be expected. On the basis of a 3 cm immersion being required for each factor of ten in temperature difference, it is obvious that a temperature difference of 0.01 °C at a distance of 6 cm from the element will produce the same error as a difference of 1 °C at a distance of 12 cm.

In resistance thermometry at high temperatures, there is another factor affecting immersion characteristics which results from thermal radiation losses down the stem of the thermometer. For those thermometers having a silica sheath, a light-piping effect within the walls of the sheath has been shown (McLaren and Murdock, 1966) to lead to errors of as much as 80 mK at 600 °C. Fortunately, it is easy to suppress the heat loss by internal reflection, by lightly sand-blasting the outside surface of the sheath, or coating it with an absorbing medium such as aqua-dag, over a length of a few centimetres just behind the thermometer element (see Fig. 5-13). It is now standard practice to do this to all long-stem thermometers, including glass-sheathed ones, that are to be used above the tin point (~ 230 °C).

The effects of thermal radiation can, of course, act in the opposite direction and increase the temperature of the element if an external source of thermal radiation is visible from the position of the element. This can arise most easily if the temperature of a transparent fluid is being measured in a room illuminated by tungsten lamps. It should be remembered that the heating effect of a 1 mA measuring current is equivalent to 25 μW of power being developed in the element. A high-temperature source of thermal radiation, such as a single 150 W tungsten lamp at a distance of 3 m from the thermometer, could easily emit 20 W per steradian in the direction of the thermometer. In the absence of an intervening absorbing medium, a typical thermometer element could absorb 9 μW of thermal radiation: equivalent to an additional heating effect of nearly 1 mK in some designs of thermometer. The solution in this case is to enclose the thermometer in an opaque tube containing a light oil to maintain thermal contact. Care should be taken that no exothermic or endothermic chemical reactions can occur between any of the materials present. Some mineral oils begin to decompose at temperatures of 120 °C, and of course the oxidation of steel is well under way at 500 °C.

A test of adequate immersion of a long-stem thermometer in a metal-freezing-point cell is provided by a measurement of the variation of freezing temperature with depth. Calculations using the Clausius–Clapeyron equation lead to linear gradients which should be observed in antimony, zinc and tin of (5.4; 27 and 22) $\times 10^{-6}$ K cm^{-1} respectively. In a vertical freezing point apparatus, of the type shown in Fig. 4-25, the overall difference in temperature from the top to the bottom of the ingot while freezing is greatest for zinc, and amounts to 0.3 mK. Since the measurement of this hydrostatic head effect requires the withdrawal of the thermometer from the freezing ingot, only a thermometer having more than adequate immersion can be used to observe it. From Fig. 5-15, one can deduce that an ingot length of some 20 cm would be required for the hydrostatic head effect to be measurable over a length of about 8 cm. In fact, the thermal contact between thermometer element and the freezing ingot would not be good enough for this to occur, and for zinc, an ingot length of about 23 cm would be required.

5-4-4 Platinum resistance thermometers for use above 630 °C

As was noted in Chapter 2, it has long been the intention of the CCT to replace the Pt 10% Rh/Pt thermocouple as an interpolation instrument of the IPTS-68 by an extension of the range of the platinum resistance thermometer towards the gold point. There is little doubt that platinum itself is a perfectly suitable material for use in a resistance thermometer up to at least 1100 °C. The difficulty in designing a practical thermometer is simply that of finding a way of mounting and supporting the wire that avoids undue strain on heating and cooling and which has sufficiently high insulation resistance. Although electrical resistivity is an intrinsic property of a metal, as is thermoelectric power, the electrical resistance of a thermometer is an extrinsic property of the wire (unlike thermoelectric emf), and is therefore influenced by dimensional changes and even by scratches. Platinum at high temperatures is so soft that scratching, due to the relative movement of wire and support on heating and cooling, appears to be almost impossible to avoid. A gradual increase in resistance is thus observed that is not wholly compensated for by working with resistance ratios.

A number of different designs of high temperature thermometer have been tried, some of which are illustrated in Fig. 5-16 (Evans and Burns, 1962; Long Guang and Tao Hongtu,.1982; Sawada and Mochizuki, 1972). At the time of writing, none has shown overwhelming advantages in respect either of performance or ease of manufacture. The question of how best to make a high temperature platinum resistance thermometer thus remains open. Regardless of the details of the design eventually adopted, platinum resistance thermometry above 600 °C will be subject, as we shall see below, to the effects arising from the quenching-in of lattice defects on cooling and variations in thickness of the oxide layer on the platinum surface.

We saw in Section 5-2 that the equilibrium concentration of lattice defects in platinum requires that an extra term be added to the equation describing the variation of resistance with temperature. For a real thermometer, equation (5-5) may be written (Berry, 19'2)

$$\frac{\Delta R(0\ ^\circ C)}{R(0\ ^\circ C)} = 1.2 \times 10^3 \exp\left[-1.51\ eV/kT\right] \tag{5-32}$$

where $\Delta R(0\ ^\circ C)$ is the increase in resistance at 0 °C due to quenched-in vacancies after rapid cooling from a temperature T. If this equation represents the vacancy resistivity at a temperature T, i.e. if Matthiessen's rule holds, it predicts a value of the vacancy resistivity at the gold point of some 3 %. Because of the exponential nature of the effect it is negligible below about 950 °C. The existence of a vacancy resistivity term in the interpolation equation would not, of itself, be a problem and could easily be taken into account. The problem arises when the thermometer is cooled. Depending upon the rate of cooling and purity of the platinum, a more or less significant proportion of the lattice vacancies, in equilibrium at a high temperature, remain quenched in the

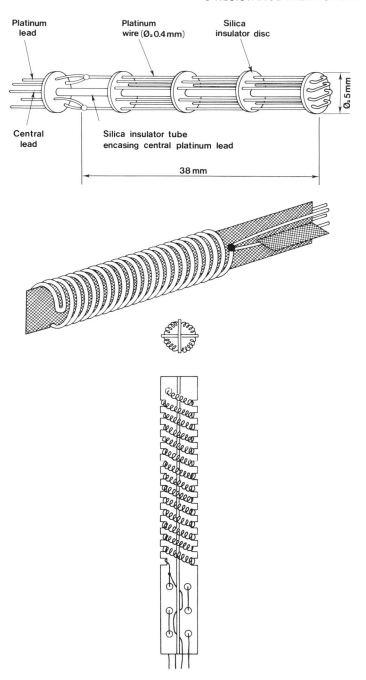

Fig. 5-16. Three examples of elements of platinum resistance thermometers designed for use up to the gold point: (a) the bird-cage design (Evans and Burns, 1962); (b) single coil design (Li Xumo *et al.*, 1982); (c) coiled design (Sawada and Mochizuki, 1972).

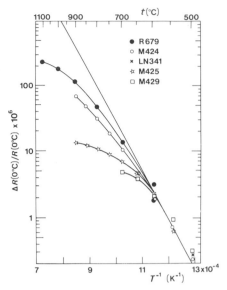

Fig. 5-17. The magnitude of the quenched-in resistance $\Delta R(°C)$ for various thermometers as a function of quenching temperature. Line J represents equation (5-32) (after Berry, 1972).

lattice at temperatures below 500 °C. In view of the fact that the variation of resistivity with temperature for platinum is about 4 parts in 10^6 per millikelvin, only a very small fraction of the 3 % equilibrium concentration of lattice defects at the gold point need be quenched in for serious problems to arise (Berry, 1966, 1972). A high rate of cooling is required to quench in a high proportion of the vacancies; probably a rate $> 10^4$ K sec^{-1} would be required if most of them were to remain. The cooling of a thermometer element from 1000 °C to below 500 °C is unlikely to take place in under about 20 sec and so a very small fraction of the vacancies initially present would be quenched in. Figure 5-17 illustrates the fractional increase in R_0 resulting from quenched-in vacancies as a function of initial temperature. What is remarkable about Fig. 5-17 is the wide variation in the magnitude of the quenched-in resistivity between different thermometers, despite a similar quenching rate being employed. These differences are related to the number and nature of the vacancy sinks present in the wire of each thermometer.

For example, thermometer R679 (Fig. 5-17) shows the largest quenched-in resistance. This thermometer is of the bird-cage construction designed for high temperature use, and is made of 0.5 mm diameter wire having a resistance ratio at 100 °C of 1.392 75. This is a high value for $W(100 °C)$ and indicates very pure platinum, and the relatively large diameter wire allows large grains to grow at high temperatures, with the result that relatively few vacancy sinks are present. Thermometer M425, on the other hand, shows a smaller quenched in resistance and is a thermometer having fine platinum wire of $W(100 °C)$, only 1.329 29. The other thermometers showing a large quenched-in resistance

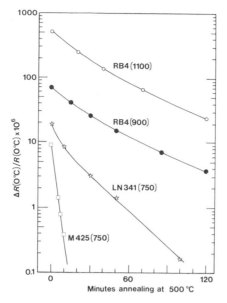

Fig. 5-18. The magnitude of the quenched-in resistance $\Delta R(^\circ C)$ remaining in various thermometers as a function of the annealing time at 500 °C. The quenching temperature is shown alongside each line (after Berry, 1972).

are of fine wire having a high value of $W(100\ ^\circ C)$ namely 1.392 66. The only thermometer in Fig. 5-17 not following the pattern is M429 which, despite a high value of $W(100\ ^\circ C)$, shows a small quenched-in resistance.

Evidence of the effect of impurity sinks on the rate of annealing of quenched-in resistance is shown in Fig. 5-18. Again the thermometer having the purest platinum behaves quite differently from those having less-pure platinum, the latter thermometers showing a smaller quenched-in resistance and a higher rate of annealing. It is worth remarking here that this is one of the rare instances when the very best material does not lead to the very best thermometer. There is a strong case, from the point of view of quenched-in resistance, for specifying that the platinum to be used for high temperature thermometers should have a lower $W(100\ ^\circ C)$ than that used for the best thermometers which are used up to 630 °C. It should be remembered that the amount of impurity needed to reduce $W(100\ ^\circ C)$ from 1.392 76 to 1.392 29 is very small, the amount depending on the particular impurity present. If the impurity were iron, then a concentration of only about 13 atomic parts per million would suffice. Tables showing how various impurities affect both $W(100\ ^\circ C)$ values and the thermoelectric power for platinum have been given by Cochrane (1972). The $W(100\ ^\circ C)$ specified in the IPTS-68 is 1.392 50 compared with that in the IPTS-48 of only 1.391 0.

The problem of choosing an adequate annealing procedure for thermometers that have been used above 630 °C is complicated by the likelihood that irreversible changes have taken place due to scratching of the wire. These irreversible changes lead to a slow increase in R_0 that can easily be confused with an inadequate annealing procedure. Various authors (Chattle, 1972; Evans and Wood, 1971; Berry, 1975; Marcarino and Crovini, 1975) have suggested procedures that differ in detail from one another, but all are based upon the principle that at temperatures below about 600 °C, the equilibrium concentration of vacancies is negligible, and that at a temperature of about 650 °C, the mobility of vacancies is such that any remaining from high temperatures disappear relatively quickly. A reasonable procedure appears to be the following: the thermometer should be cooled to 650 °C at whatever rate is convenient; annealed for one or two hours at this temperature; cooled at any convenient rate to 450 °C, and annealed at this temperature for about one hour to remove vacancies remaining from the 650 °C anneal; it is then cooled to room temperature by simply removing it from the furnace at 450 °C.

With the present definition of IPTS-68, a platinum resistance thermometer for use above 630 °C can only be calibrated on IPTS-68 by comparison with a Pt-10 % Rh/Pt thermocouple. Since the reproducibility of a resistance thermometer is considerably better than that of a thermocouple, even taking into account the problems of lattice defects and scratching of the wire; this is not a very satisfactory situation. The absence of an agreed interpolation equation is one of the factors that has impeded the more widespread use of the high-temperature thermometer. Pending definitive measurements of the differences between IPTS-68 and thermodynamic temperature in the range 630 °C to 1064 °C, any resistance thermometer interpolation equation can do no more than attempt to match the quadratic of emf/temperature employed for the thermocouple. Such an equation already exists and leads to a resistance thermometer scale that matches IPTS-68 to within the accuracy expected of a Pt-10 % Rh/Pt thermocouple, namely about ± 0.2 °C.

The equation, proposed by Evans and Wood (1971), is the following:

$$t = \theta - 54.84 \ ^\circ\text{C} \left(\frac{\theta}{1064.43 \ ^\circ\text{C}} - 1 \right) \left(\frac{\theta}{963.93 \ ^\circ\text{C}} - 1 \right) \left(\frac{\theta}{630.74 \ ^\circ\text{C}} - 1 \right)$$

$$\left(\frac{\theta}{480.081 \ ^\circ\text{C}} - 1 \right) \tag{5-33}$$

where a temperature θ is defined by $W(\theta) = A + B\theta + C\theta^2$, and A, B and C are constants.

The use of this equation with a particular thermometer requires calibration at 0 °C, the steam point (or tin point) and the freezing points of zinc, silver and gold. The values of $W(480.081 \ ^\circ\text{C})$ and $W(630.74 \ ^\circ\text{C})$ are obtained by calculation using the IPTS-68 interpolation equation (5-23).

5 4 5 The effects of oxidation on the resistance of platinum

Although platinum is usually considered to be highly resistant to oxidation,

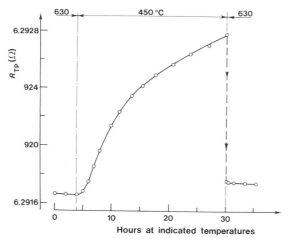

Fig. 5-19. Changes in $R(273.16 \text{ K})$ of a platinum resistance thermometer produced by heating at the indicated temperatures under pressure of 83 kPa of oxygen (after Berry, 1978).

the effects of oxidation can, in certain circumstances, be observed in the behaviour of a platinum resistance thermometer. At temperatures between 450 °C and 560 °C, a film of orthorhombic β-phase PtO_2 will form on the surface of platinum heated in air (Berry, 1978). It will dissociate rapidly in the temperature range 600 °C to 650 °C. The effect on the resistance of a platinum thermometer is illustrated in Fig. 5-19, in which an oxide film of thickness 11 nm was built up over a period of 30 h at 450 °C in air. These resistance changes can be explained on the basis of a model in which the current is carried by the wire and oxide in parallel, the resistance of the wire increasing in inverse proportion to its cross sectional area remaining after that of the outer layer of oxide has been subtracted. Since the changes in resistance can be ascribed simply to dimensional changes, the resistance ratio, $W(t)$, for a particular state of oxidation remains unchanged. Thus for resistance thermometry, oxidation need not interfere with accurate measurements, provided that both R_t and R_0 are measured with the wire in the same state of oxidation. This means that the value of R_0 to be associated with W_t, for 450 °C to 650 °C, must be the value of R_0 measured *after* R_t. From the point of view of oxidation, therefore, the difficult range lies between 450 °C and 650 °C. In the previous section we discussed the effects of quenched-in vacancies, and concluded that an anneal at 650 °C followed by a further anneal at 450 °C is desirable. In view of the build-up of oxide at 450 °C (albeit rather slow) the anneal at 450 °C should not exceed 1 h. There is also evidence that very much smaller changes in resistance occur as a result of oxidation that takes place in air at temperatures between 100 °C and 250 °C. These are probably the result of a two-dimensional oxide layer having an average thickness of only about one quarter of a monolayer which forms in this temperature range, but which dissociates at about 300 °C. The consequent changes in R_0 do not exceed 1 mK.

5-5 Industrial platinum resistance thermometry

The high-precision platinum resistance thermometer, which we have discussed in earlier Sections, is clearly a fragile and delicate instrument. Mechanical shocks, even if not sufficiently strong to break the sheath, introduce strain into the resistance element and increase its resistance. In some designs, repeated small shocks in the axial direction can lead to coils being compressed and eventually to adjacent coils actually touching one another. To distinguish between platinum resistance thermometers of this sort, intended for high precision work, and those manufactured to withstand normal industrial use, the latter have come to be called "industrial platinum resistance thermometers" (or "RTD's" in the USA). There is a very wide range of such thermometers on the market in many shapes and sizes. What is common to them all, however, is that the platinum resistance element is firmly supported, often by being completely embedded in glass or ceramic. This renders the thermometer extremely robust, but at the same time reduces the stability of its resistance. There are two reasons for the resistance being less stable than that of the standards laboratory thermometer. The first is that, on thermal cycling, differences between the thermal expansion coefficient of the platinum and that of the matrix in which it is embedded lead to extra strain resistance in the wire as well as to changes in resistance resulting from the consequent dimensional changes. While the extra strain resistance can be annealed out at sufficiently high temperatures, the effects of dimensional changes, of course, cannot. The second reason is that at high temperatures, changes in resistance occur because of contamination of the platinum due to diffusion from the surrounding material. Although the reproducibility of the industrial platinum resistance thermometer does not approach that of the high-precision platinum resistance thermometer, it is nevertheless significantly better than that of a thermocouple in an industrial environment. For this reason, many millions are used in a wide range of engineering, aeronautical and manufacturing industries.

Considerable effort has been devoted to producing an international standard which adequately describes the resistance/temperature relation of currently manufactured elements over the range $-200\,°C$ to $+850\,°C$. This is the temperature range within which most industrial applications occur. The lower limit of $-200\,°C$ is such as to include the temperatures encountered in the now-important industry concerned with the bulk transport and storage of liquefied gases. The upper limit is that above which few industrial platinum resistance thermometers are stable, mainly because at temperatures above about 850 °C, contamination of the platinum by the supporting matrix is too severe (McAllan, 1982). In a new IEC Draft International Standard (IEC, 1980; Chattle, 1975, 1977; Actis and Crovini, 1982), the resistance/temperature relation is based upon measurements made on a range of industrial resistance thermometers from different manufacturers throughout the world. This resistance/temperature relation is given in Appendix III.

There are two main types of industrial platinum resistance thermometer (Johnston, 1975); in the first, the element is a fine platinum wire, which may have a diameter as small as 0.01 mm, and in the second the element is a film of platinum. Examples of different constructions of the wire-type of element are shown in Figs 5-20 and 5-21. The design chosen for a particular application depends very much upon whether the requirements are for extreme ruggedness, to withstand vibration of the sort encountered for example in the measurement of temperature in power station steam lines, or perhaps for rapidity of response. In industrial thermometry, the sheath which encloses and supports the thermometer element is an integral part of the thermometer, along with the supporting pocket in which it is mounted for use. We shall discuss below some special applicatiɔns which show the importance of a proper matching of element, sheath and pocket. Thermometer elements of the film type are often made by the screen-printing process in which a platinum ink is printed on to an alumina substrate and then fired. Subsequently it may be coated with a glaze which, after further firing, provides a robust coating to the platinum element. Before glazing, the resistance of the element must be trimmed to the required value, a process which can conveniently be done using a laser. Examples of screen-printed platinum resistance elements are shown in

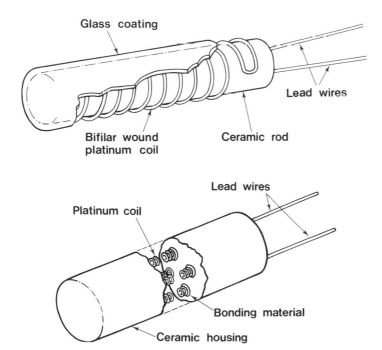

Fig. 5-20. Two designs of rugged industrial platinum resistance thermometers for general purpose use (after Johnston, 1975).

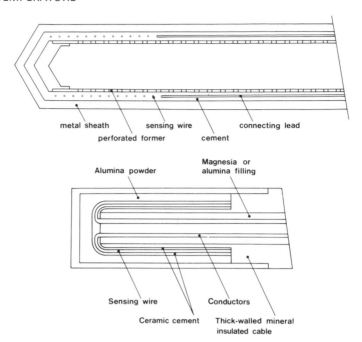

Fig. 5-21. Industrial platinum resistance thermometers having: (a) a hollow thermometer element for rapid response, (b) an element designed for very high vibration conditions (after Johnston, 1975).

Figs 5-22 and 5-23. Screen-printed elements cannot be used to quite such high temperatures as can wire wound elements, and a maximum operating temperature of about 500 °C is usually recommended. An alternative to the screen-printed element is the sputtered-film element which is now being used in increasing numbers.

The ultimate reproducibility (meaning here: short-term stability) of an industrial platinum resistance thermometer depends upon the care with which it is constructed, the purity of the original materials and, perhaps most of all, upon the temperature range over which it is made to operate (Curtis, 1982; Mangum and Evans, 1982). For thermometers covering the widest possible range and used up to temperatures above 600 °C, the tolerances laid down in the IEC specification give an estimate of the reproducibility which can be obtained. For thermometers used over a much more restricted range of temperature, the reproducibility can be very much better. Provided that the temperature does not exceed about 250 °C, a reproducibility of about 5 mK can be expected from selected thermometers of the type shown in Fig. 5-20. If the maximum operating temperature is within the range 25 °C to 150 °C, a reproducibility as good as 2 mK can be expected on some thermometers of this type. An initial ageing period of up to one week at 150 °C is usually necessary before such reproducibilities can be achieved.

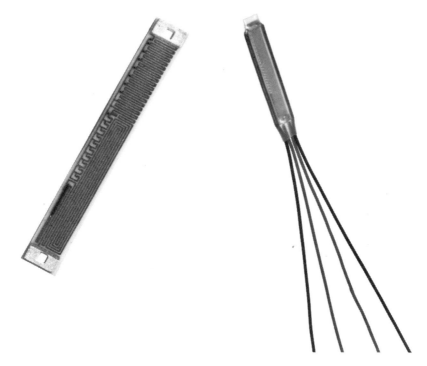

Fig. 5-22. A platinum resistance thermometer element made by the screen printing process (courtesy of Rosemount Engineering, Ltd).

Fig. 5-23. A platinum resistance thermometer using the screen printed element of Fig. 5-22 (courtesy of Rosemount Engineering Ltd).

For industrial applications, it is rarely the ultimate in reproducibility that is demanded but, rather, a modest reproducibility combined with good long-term stability under adverse conditions of perhaps vibration, pressure, thermal cycling or corrosive atmosphere together with interchangeability between thermometers made to the same specification. It is because of these requirements that the sheath and the manner in which the element is mounted within the sheath are so important. The great majority of thermometer failures encountered in industrial applications occur due to failures of the leads. Failure occurs as a result of strain induced by dimensional changes which are repeated on thermal cycling.

The most common construction of a general purpose sheathed industrial platinum resistance thermometer is that shown in Fig. 5-24d. The thermometer element, either a wire wound type or platinum film type, is firmly held at the bottom of the stainless steel or special-alloy protection sheath by means of a cement. The insulated leads passing up the sheath to the connection block may be held in place by insulating powder or cement or plastics filler

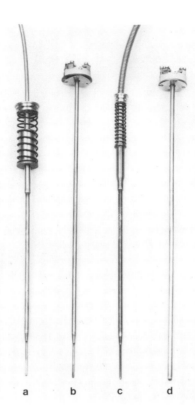

a   b   c   d

Fig. 5-24. Metal-sheathed industrial platinum resistance thermometers designed for the measurement of the temperature of steam temperatures in power stations. The details of each are given in the text (courtesy of Rosemount Engineering Ltd).

depending upon the level of vibration for which the thermometer is designed and its maximum operating temperature.

It is quite common to find this type of thermometer having a tip of reduced diameter to improve the response time, as is the case for the other thermometers of Fig. 5-24 for which we mention applications below.

5-5-1   The measurement of steam and turbine temperatures

The importance of proper matching of the thermometer sheath and pocket to the application can be well illustrated by looking at some of the requirements for temperature measurement in a modern coal-fired power station. For the measurement of steam temperatures, sheathed thermometer elements of the design show in Figs 5-24(a), (b), (c) and (d) are used. Type (a) is made for the measurement of the temperature of steam in superheater outlet pipes to the

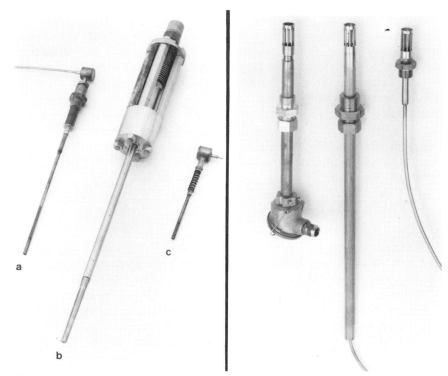

Fig. 5-25. Platinum resistance thermometers used for the measurement of the temperature in steam turbines; for description see text (courtesy of Rosemount Engineering Ltd).

Fig. 5-26. Platinum resistance thermometers designed for the measurement of high gas temperatures (courtesy of Rosemount Engineering Ltd).

high-pressure turbines where temperatures up to 600 °C both at the tip and at the head are encountered. Note that all of the thermometers are equipped with spring-loaded heads so that they can be securely mounted in their pockets. Type (b) is for measurements in areas where the ambient temperature does not exceed 100 °C and where no special precautions have to be taken with the wiring. Type (c) is for use in regions where the whole thermometer might reach temperatures of 500 °C. Type (d) is a general purpose thermometer most commonly used for test purposes.

The thermometer pockets in which these thermometers are mounted on steam pipes are designed to withstand extreme conditions of vibration, pressure and temperature. The vibration results chiefly from vortex shedding from the thermometer pockets themselves in the high-speed steam flow. Temperature measurements on inner turbine casings require thermometers that will withstand temperatures up to 550 °C over the whole of their length when inserted into pockets passing through the outer turbine casing. Such a thermometer, in its pocket and having mineral-insulated leads, is illustrated in Fig. 5-25 (a). For measurement of temperature gradients in the turbine casings, thermometers of the type shown in Fig. 5-25b are used. This thermometer contains three elements at different positions in the sheath. Again, the whole of the thermometer, including mineral-insulated leads, must withstand temperatures of 550 °C. The overall length of this particular thermometer is 1.3 m. Special spring-loaded thermometers of the type shown in Fig. 5-25c are used to measure turbine-bearing temperatures. The sensing element is in the chamfered tip which is maintained in press contact with the surface, the temperature of which is being measured. Finally, for the measurement of gas temperature (in the ducts from the boiler, for example) thermometer pockets are used having slotted openings to improve the thermal contact between the sheathed thermometer and the gas flow, Fig. 5-26.

5-5-2 The measurement of air temperature in aeronautics

The problem of the measurement of temperature in high-speed gas flows is of course a very important one in aeronautical engineering, but one into which we cannot delve very deeply in this book. The reader is referred to more specialized texts (Trenkle and Reinhardt, 1973; Stickney and Dutt, 1970) for a full discussion of the subject. As a particular application of industrial resistance thermometry, however, the measurement of the outside air temperature of an aircraft in flight provides an interesting contrast to that of power-station thermometry.

Before coming to the practical details of the construction and mode of operation of the air temperature sensors used in modern aircraft, we must be clear about what it is we are trying to measure. Except at low altitudes and low speeds in clean air, the term "outside air temperature" has no well-defined meaning. Instead we must define four temperatures:

(1)   Static air temperature T_s. The temperature of the undisturbed air through which the aircraft is about to fly.

(2)   Total air temperature T_t. The temperature which the air would have if all the kinetic energy of relative motion between the air and aircraft were to be converted to heat, by the acceleration of the air to the speed of the aircraft.

(3)   Recovery temperature T_r. The temperature of the air on a particular part of the aircraft skin due to incomplete recovery of the kinetic energy.

(4)   The measured temperature T_m. The temperature actually measured, which differs from T_r due to extraneous heat transfer.

The relationship between these four temperatures is illustrated in Fig. 5-27. Provided that a good measurement of the total air temperature can be made, thermodynamic relations are available which allow the static air temperature and other quantities to be calculated. Devices which aim to measure T_t are usually called "total temperature" sensors, and in modern designs T_m differs

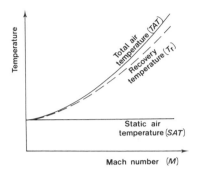

Fig. 5-27. The relationship between the three quantities related to temperature that must be distinguished when measuring the temperature of a high-speed air stream.

from T_t by less than 0.5 %. The principles of operation of total temperature sensors are straightforward, but the details of their design are, as we shall see, quite complex.

To measure T_t, it is necessary to place a thermometer element in air which has been accelerated to the speed of the aircraft. For T_m to be close to T_t, the thermometer element must be isolated from all heat sources or sinks other than the air surrounding it. It must therefore be shielded from radiation or conduction from the aircraft skin or from any deicing heater that may be present, and the heating effect of the measuring current must be low. In addition, the air passing over the thermometer element must be taken from outside the aircraft-skin boundary layer and not in the wake of any other device projecting from the aircraft. For supersonic aircraft, the detached bow shock wave must always be upstream of the total temperature sensor. In addition, there must be protection against damage from hail, insects and birds: a non-trivial problem!

The thermometer element itself is required to have a relatively short time constant; depending upon flight conditions, it may be between one and five sec. The construction of the element is illustrated in Fig. 5-28. The pure platinum wire of diameter 0.05 mm is wound in a helix and supported between two thin-walled platinum tubes, insulated from these by mica and embedded in cement. The total temperature sensor showing the position of the thermometer element is illustrated in Fig. 5-29. The airflow is made to turn through a sharp angle before it reaches the thermometer element so that any solid entrained particles escape through the rear exit port. The internal boundary layer is sucked away through the holes indicated to prevent flow separation at the sharp turn.

Having deduced T_t from a measurement of T_m, the value of T_s can be calculated from

$$\frac{T_t}{T_s} = 1 + \frac{\gamma - 1}{2} M^2 \qquad (5\text{-}34)$$

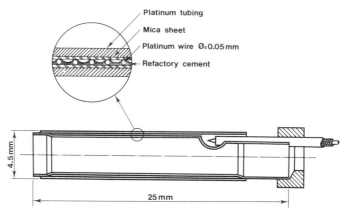

Platinum tubing

Mica sheet

Platinum wire Ø=0.05mm

Refactory cement

4.5 mm

25 mm

Fig. 5-28. Platinum resistance-sensing element for air-temperature measurement. Elements of this type are used in the total air sensor of Fig. 5-29 (courtesy of Rosemount Engineering Ltd).

where γ is the ratio of the specific heats of air and M is the Mach number. The true air speed, S_t, can then be found from

$$S_t = M \left[\frac{\gamma R T_s}{M_A} \right]^{1/2}$$

(5-35)

where M_A is the molecular weight of air.

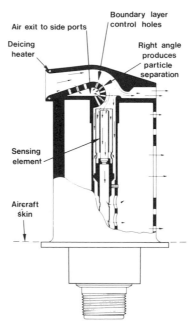

Boundary layer
control holes

Air exit to side ports

Deicing
heater

Right angle
produces
particle
separation

Sensing
element

Aircraft
skin

Fig. 5-29. A total-temperature sensor for the measurement of air-temperature on high speed aircraft (courtesy of Rosemount Engineering Ltd).

An additional important application of total temperature sensors is for the measurement of the inlet temperature of the compressor of a jet engine. An accurate knowledge of this is required for the proper selection of the engine pressure ratio, one of the critical control parameters used in preparing for take-off.

5-5-3 Industrial resistance thermometry — conclusions

In this short section on industrial resistance thermometry, a brief outline has been given of the main instruments and methods. No attempt has been made to give an exhaustive treatment of the subject, since that would occupy a volume in itself, nor has any mention been made of systems used to measure the resistance. The latter is a rapidly developing field due, as was mentioned at the beginning of this chapter, to the increasing application of microprocessor-controlled instruments.

The important point to be made in all of this, is that accuracy in industrial resistance thermometry calls for the application of exactly the same principles as in the most accurate standards thermometry. The additional requirements of industrial resistance thermometry (those of ruggedness, low cost and sometimes size) must be accommodated without degrading unduly the basic requirements of accurate thermometry, namely, good thermal contact with the object being measured, strain-free mounting of the resistance element, freedom from corrosion and the provision of means for checking that the thermometer is operating correctly.

5-6 The rhodium/0.5 % iron resistance thermometer

We remarked in Section 5-1 that the addition of a small proportion of magnetic impurity to certain metals leads to the formation of a local magnetic moment and consequently to a resistance minimum at low temperatures. It was suggested by Coles (1964), after studying the behaviour of dilute alloys of rhodium/iron, that the particular behaviour exhibited by these alloys might be useful in thermometry. Instead of giving rise to a resistance minimum, the addition of about 0.5 atomic percent of iron to pure rhodium leads to a low-temperature resistance anomaly which has a positive temperature coefficient. This suggestion was taken up by Rusby (1975) and has resulted in the development of the rhodium/iron resistance thermometer. This new thermometer is now showing promise of displacing the germanium resistance thermometer as the principal means of maintaining temperature scales in the range 1 K to 30 K. The resistance/temperature relation for the rhodium/iron alloy chosen for the thermometer, is shown in Fig. 5-30.

The construction of the rhodium/iron thermometer developed by Rusby

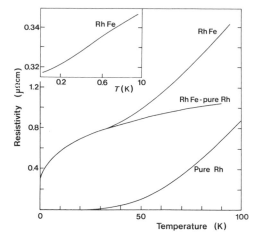

Fig. 5-30. Resistivity of Rh-0.5 % Fe after subtraction of the resistivity of Rh (after White and Woods, 1959 and Rusby, 1975).

and now manufactured commercially by the Tinsley Company (London) is illustrated in Fig. 5-31. It is very similar to the pure-platinum capsule-type thermometer. The wire, produced by powder metallurgical methods, is about 0.05 mm in diameter. The process for making the wire is roughly as follows: the iron is deposited chemically in fine rhodium powder which is then dried,

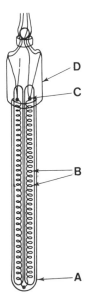

Fig. 5-31. Schematic diagram of a rhodium–iron thermometer: A, platinum sheath 5 mm in diameter and 50 mm long; B, two of the four glass tubes containing the coiled Rh–Fe wire; C, flame welds to platinum leads; D, glass-bulb seal (after Rusby, 1975).

sintered, hot swaged and hot drawn. Finally it is strain-annealed in hydrogen at 1100 °C. All hot processing is carried out in hydrogen. The aim is to produce an annealed recrystallized wire without excessive grain growth.

The principal difficulty encountered in the manufacture of the rhodium/iron wire is that of ensuring complete solution of the 0.5 % iron in the rhodium. If oxidation of the iron takes place during the mixing process, subsequent annealing can fail to lead to the solution in the rhodium of some of the iron particles encapsulated in oxide. Recrystallization, which normally takes place during an anneal at 1200 °C, is inhibited by the presence of undissolved iron particles and this, together with the different resistance/temperature behaviour resulting from the smaller proportion of dissolved iron, can lead to a wire unsuitable for thermometry.

In making the thermometer, the wire is wound in the form of a spiral and supported in four sections inside glass tubes. The current and potential leads of pure platinum are flame welded to the ends of the rhodium/iron wire and the whole assembly is annealed at 700 °C before being inserted into a platinum sheath and sealed, under a pressure of 30 kPa of helium, by means of a lead-glass seal. The dimensions of thermometers being currently manufactured are either 40 mm or 60 mm long by 5 mm in diameter. The larger size is usually preferred on the grounds that its higher resistance (100 Ω at 300 K compared with 50 Ω) leads to higher resolution at low temperatures.

In evaluating the suitability of rhodium–iron for resistance thermometry, we are interested in the fractional sensitivity, $(1/\rho)d\rho/dT$, which is an intrinsic property of the material, and the voltage sensitivity, dV/dT, which is a property of the thermometer and depends on the measuring current. The fractional sensitivity is plotted in Fig. 5-32 together with that of pure platinum. The voltage sensitivity of a 100 Ω thermometer is compared with that obtained with a capsule-type platinum thermometer, under real measuring conditions,

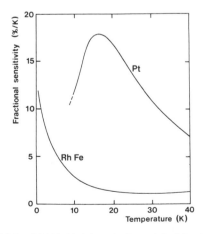

Fig. 5-32. Fractional sensitivity $(1/\rho)(d\rho/dT)$ for Rh–Fe and Pt (after Rusby, 1975).

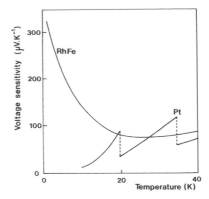

Fig. 5-33. Voltage sensitivity dV/dT of a Rh/Fe thermometer (Fig. 5-31) at 0.5 mA measuring current and of a 25 Ω platinum thermometer using a measuring current of 5 mA below 20 K, 2 mA from 20 K to 35 K, and 1 mA above 35 K (after Rusby, 1975).

in Fig. 5-33. We see from the voltage sensitivity that the higher resistivity of rhodium–iron more than compensates for the relatively low values of fractional sensitivity, especially below 20 K. Note that the fractional sensitivity of platinum is always higher than that of rhodium–iron above about 5 K, although they are not very different at 300 K. In the case of both platinum and rhodium–iron, the measuring current, and hence the voltage sensitivity, is limited by the self-heating. In rhodium–iron at 0.5 mA, this increases from a value of 0.2 mK at 20 K to reach a maximum of nearly 0.8 mK at the temperature at which the helium filling condenses to form superfluid liquid. Below 1 K, thermal contact deteriorates and the heating effect rises once again, leading to an effective lower limit to the useful range of the capsule-type thermometer of about 0.5 mK. A rhodium–iron wire directly immersed in ^{3}He provides a useful thermometer at temperatures down to 20 mK, below which thermal contact again becomes a problem.

In spite of the improved voltage sensitivity compared with that of platinum below 20 K, one might wonder why the rhodium–iron thermometer should be preferred to the germanium resistance thermometer in this temperature range. For, as we shall see, it is possible to obtain germanium thermometers having a fractional sensitivity well in excess of 100 % per kelvin at 2 K while having at the same time a resistance of some thousands of ohms. The answer lies in the superior long-term stability of the rhodium–iron thermometer. Unforeseen jumps in the resistance of germanium thermometers have always been a problem, the origins of which are discussed further in the section which follows. Experience so far with rhodium–iron thermometers shows that sudden jumps in resistance do not occur (nor would they be expected to), and that slow drifts do not amount to more than 2 ppm per year (~0.15 mK at 20 K). An additional advantage of the rhodium–iron thermometer over the germanium thermometer for the purposes of scale maintenance, stems from its much slower rate of change of resistance with temperature. This allows the same thermometer to be used, say, from 0.5 K to 273 K.

It is possible to fit polynomials of moderate order to the rhodium–iron characteristic over quite wide range of temperature (Rusby, 1982). For the range 0.5 K to 20 K, an eighth order polynomial may be fitted with a standard deviation of 0.2 mK, while for 0.5 K to 27 K, an eleventh order polynomial is required. The polynomials express temperature as a function of resistance. Less accurate data between 27 K and 273 K can be fitted with a standard deviation of about 1 mK provided that the independent variable is ln Z, where Z is the ratio $(R_T - R_{4.2})/(R_{273.16} - R_{4.2})$. Difficulties arise in fitting ranges spanning a temperature of 28 K because here, the low-temperature impurity-dominated resistivity is overtaken by the high-temperature phonon-scattering resistivity, and the total resistivity passes through a point of inflection.

5-7   The germanium resistance thermometer

We have already seen in 5-1 how the behaviour of a semi-conductor is modified by the introduction of a small quantity of impurity. The resistance/temperature relation is critically dependent upon the species and the quantity of the added impurity, and both can be varied to give approximately the desired properties. Referring to Fig. 5-7, it is evident that the ranges which can usefully be employed for resistance thermometry are III and IV. Although the slope of the resistance/temperature relation and the absolute value of the resistivity can be controlled approximately, the sensitivity of both of these parameters to small changes in amount of impurity do not allow good reproducibility to be obtained from batch to batch. There is no possibility, therefore, of establishing a reference function, or even a series of reference functions for different temperature ranges, equivalent to that which is available for platinum resistance thermometers. Purely empirical polynomials must be used, and each thermometer calibrated at a sufficient number of points within its range for all of the coefficients of the polynomial to be determined. Since a calibration from 1.6 K to 20 K would require twelfth order polynomial to achieve an accuracy of 0.3 mK, it is clear that the calibration of a germanium thermometer poses problems quite different from those of a platinum thermometer. Before looking in more detail, however, at the question of calibration of germanium thermometers we must first deal with their construction and performance.

The design of germanium thermometers as high precision instruments has changed little since they were pioneered by Kunzler and others about 1960 (Kunzler et al., 1962; Lindenfield, 1962). The doped germanium is cut in the form of a bridge, Fig. 5-34, to the arms of which are attached gold wires to provide current and potential leads. Germanium is strongly piezoelectric and so a strain-free mounting is essential, usually provided by the leads themselves.

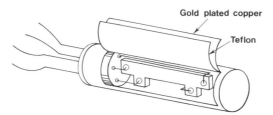

Fig. 5-34. Schematic view of a germanium resistance thermometer (after Blakemore *et al.*, 1962).

The germanium is hermetically sealed in a gold-plated capsule which is filled with helium to aid thermal contact. Despite the presence of helium, more than two thirds of the heat conduction to the germanium is via the leads. This implies that the temperature indicated by the thermometer is more a function of the thermal anchoring of the leads than of the temperature of the capsule itself. It is most important to take note of this in the design of low-temperature equipment (Hust, 1970). For platinum and rhodium/iron thermometers this is also true, but to a much lesser extent since the ratio of the surface area of the wire resistance element to its cross sectional area is much larger than for a germanium thermometer. As with other thermometers, the heating effect of the measuring current for a germanium thermometer is a function of the thermal contact with the surrounding medium. If the whole thermometer is immersed in liquid helium at 4.2 K, a typical heating effect for a measurement power of 0.2 μW is about 1 mK, increasing to 10 mK if the thermometer is simply in helium vapour at the same temperature. Similarly, the time constant of the thermometer increases from about 50 ms when immersed in liquid helium at 4.2 mK, to 400 ms when in helium vapour. For germanium thermometers attached to a copper block in the usual way, that is to say the thermometer itself is placed in a well using a grease (Apiezon N for example) to improve thermal contact and having its leads thermally anchored to the block, the heating effect is illustrated in Figs 5-35 and 5-36. The values given in these

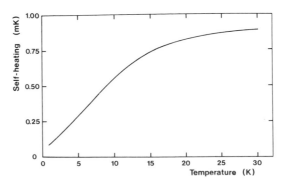

Fig. 5-35. Self-heating of a typical germanium thermometer for a potential drop across it of 4 mV (after Besley and Kemp, 1977).

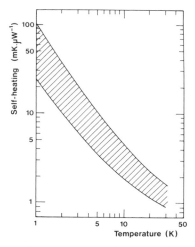

Fig. 5-36. Self-heating of germanium thermometers in $mK\mu W^{-1}$ (after Besley and Kemp, 1977).

figures (for a cryocal CR 1000) indicate how the heating effect of a typical well-anchored thermometer varies with power, current and temperature. These curves should not be taken to apply to a particular thermometer without experimental verification at at least two points within the range (Besley and Kemp, 1977; Anderson and Swenson, 1978).

An effect which is peculiar to germanium resistance thermometers arises because of the relatively large Peltier coefficient of doped germanium. This shows itself in a significant difference being observed between the resistance measured using a direct current and that measured using an alternating current (Swenson and Wolfendale, 1973; Kirby and Laubitz, 1973). The passage of a direct current through a germanium resistance thermometer leads to a temperature gradient being established from one end to the other due to the Peltier heat liberated and absorbed at the junctions with the leads. The temperature gradient gives rise to a small thermal emf between the potential leads, thus leading to an error in the measurement of resistance. If, instead of a direct current, an alternating current is used of frequency f, damped thermal waves propagate from each end of the resistor. The attenuation is exponential, having an exponent proportional to $\sqrt{f}$, so that as f increases, the thermal waves become more and more confined to the ends of the resistor. For four-terminal resistors of the bridge type, once the frequency is above the value necessary for the thermal waves to be attenuated to negligible proportions before they reach the potential leads, the effect vanishes and true resistance is measured. This frequency is a function of the diffusivity of the material and the geometry of the thermometer. The magnitude of the difference between a d.c. measurement and a true resistance measurement can be as high as 1.5 % at room temperature, 0.2% at 80 K, 0.02% at 50 K and less than 1 part in 10^6 below 10 K. The corresponding temperature errors are less than 0.1 mK below 20 K, but rise to 200 mK at 100 K.

It is important to screen germanium, and also carbon, thermometers from stray rf fields. At helium temperatures the effect of such fields, due, for example, to a nearby television transmitter, can amount to 0.3% of the resistance, and the effect becomes worse at lower temperatures (Ambler and Plumb, 1960).

5-7-1   The long-term stability of germanium thermometers

There is no reason to think that the stability of the resistance of either p- or n-type germanium is one of the limiting factors in the reproducibility of germanium thermometers. The sudden small jumps in resistance which are occasionally observed in thermal cycling are most likely to have their origin at the junction between the gold leads and the germanium. The stresses introduced by the difference in thermal expansion coefficient between gold and germanium are concentrated at these junctions. The high resistivity and piezoelectric coefficient of germanium, together with the difficulty of making electrical contact at a well-defined position, all combine to make the resistance of the thermometer particularly sensitive to mechanical strain.

A systematic investigation has been carried out of the stability on thermal cycling of sixty germanium thermometers (Besley and Plumb, 1978; Besley, 1978, 1980). The aim was to test, and perhaps quantify in some way, the generally held opinion that germanium thermometers occasionally show large instabilities even after periods of good reproducibility. The results show that about one out of every two new germanium thermometers will repeat a calibration at 20 K with a standard deviation of less than 1 mK during a series of one hundred thermal cycles from 20 K to room temperature. There was some evidence that a slight improvement in stability occurs during the first ten or so thermal cycles. It would therefore be a useful practice, on receipt of a new germanium thermometer, to carry out ten or so such cycles before any measurements are made. No significant difference appeared between the stability of p-type or n-type germanium thermometers.

For those interested in obtaining a thermometer having a stability rather better than 1 mK at 20 K, the task of choosing one is more difficult. Among the sixty thermometers tested, only 18 showed a standard deviation of less than 0.25 mK. Very few thermometers, however, showed a change in their pattern of behaviour during the series of measurements. Excluding the first ten thermal cycles, those thermometers which initially appeared very stable continued to be so and those thermometers which showed either a slow drift or any other sort of instability continued to do so also. Nevertheless, it was found that, from time to time, a thermometer which had shown a satisfactory reproducibility over a considerable number of thermal cycles would suddenly show a step change in calibration, Fig. 5-37. The effect is worse at higher temperatures where the resistance of the thermometer is lower. It is this behaviour, which is

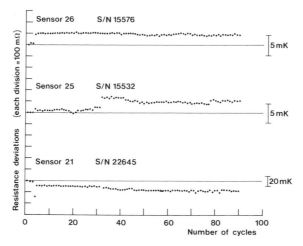

Fig. 5-37. Variation in the resistance at 20.28 K of three germanium resistance thermometers over 90 thermal cycles to room temperature. Thermometer No. 15532 shows jumps in resistance characteristics of what is found from time to time with such thermometers (after Besley, 1978).

absent in rhodium/iron thermometers, that has made the maintenance of low-temperature scales on germanium thermometers so difficult.

5-7-2   Calibration and interpolation equations for germanium thermometers

In discussing the theory of the conduction processes operating in doped germanium, we deduced a number of analytical expressions for the conductivity or resistivity in terms of atomic constants, the number and species of impurity atoms and the temperature. It was made clear that, although these expressions were in reasonable qualitative agreement with what is observed, they were not to be taken to represent in a quantitative way the behaviour of real materials. The details of the conduction processes are too complex for this to be possible. The fitting of experimental data of resistance *vs* temperature, therefore, must be carried out empirically with little help from physical theory, as, indeed, is the case for platinum. The fitting of such data for germanium resistance thermometers is rather more difficult than it is for platinum resistance thermometers for two reasons. The first is that the resistance/temperature relation is much more variable between samples made to the same specification than is the case for platinum. This is because the resistivity of doped germanium is, as we have seen, very sensitive to the number and species of the impurities. The second is that the resistivity is an exponential function of the temperature, and therefore varies much more rapidly with temperature than does that of platinum. Even though we cannot write down an explicit relation between resistance and temperature which is based upon physical theory, the fact that we are confident of the closely-exponential dependence of resistance upon temperature leads us to try first polynomials of the form

$$\ln R = \sum_{t=0}^{m} a_i (\ln T)^i \tag{5-36}$$

or

$$\ln T = \sum_{i=0}^{n} b_i (\ln R)^i. \tag{5-37}$$

The general problem of fitting analytic functions to experimental data is a far-reaching one which we cannot enter into in any depth here. A few general principles, however, need to be taken into account in the fitting of calibration data for germanium thermometers (Powell *et al.*, 1972).

If transformed variables are used, as is desirable in order to linearize the relation between R and T (equations (5-36) and (5-37) for example), it is important that the data be properly spaced in the transformed variable to avoid oscillations occurring in parts of the range. Broadly this means that for a germanium thermometer being calibrated between 1 K and 20 K, there should be as many data points between about 1 K and 2 K as there are between 10 K and 20 K if a polynomial is to be fitted to the data. If possible, data points should also be present just outside the eventual range of fitting if the standard error of the fit is not to deteriorate near the ends of the range. If this is not possible, the data should be rather more closely spaced at either end of the range than in the middle. The proper fitting of a polynomial using the method of least squares requires that the variance of the dependent transformed variable be constant over the range of fitting. This is rarely the case, unless the range of fitting is extremely short, and therefore weighting factors must be applied to the data to make it so. Since, in the case of germanium thermometry, both R and T have a variance which varies over the range, the weighting factor of the dependent variable should be inversely proportional to the total variance σ_t^2, given by

$$\sigma_t^2 = \sigma_{(\ln R)}^2 + \left[\frac{\delta(\ln R)}{\delta(\ln T)} \right]^2 \sigma_{(\ln T)}^2 \tag{5-38}$$

for equation (5-36) in which $(\ln R)$ is the dependent variable and $(\ln T)$ the independent variable, and $\sigma_{(\ln R)}^2$ and $\sigma_{(\ln T)}^2$ are the variances of the dependent and independent variables respectively.

The choice of the order of fit can also pose problems, since too high an order can lead to unwanted oscillations, while too low an order obviously degrades the representation of the experimental data. There do not appear to be any rigorous rules for determining the optimum order of fit for experimental data of the type obtained during thermometer calibrations. A useful guide is given, however, for the fitting of orthogonal polynomials by observing the rate of decrease in the magnitude of the coefficients of the polynomial obtained by the least-squares fitting procedure. Assuming that the experimental data comprise the real function plus a noise function, it can be assumed that the coefficients of the polynomial describing the real function will converge quite rapidly while those of the noise function will not converge at all. A plot of the coefficients *vs* order shows that for a germanium thermometer calibration fitted to an orthogonal polynomial of the form of equation (5-36), the coefficients decrease roughly as $1/n$. The level at which the noise coefficients dominate is easily distinguished, Fig. 5-38.

Experience has shown that germanium thermometer calibration data can equally well be fitted by an equation of the form $T = f(\ln R)$ using for example a Chebychev polynomial. It is not necessary, of course, that the whole range be fitted with a single polynomial; indeed the procedure of "spline fitting" can, in the limit, employ as many polynominals as there are data points (see for example in Chapter 7 Section 6-2 where we discuss the fitting of polynomials to the transmission curves of interference filters). It is not unusual to divide the range of the germanium thermometer into two sections to facilitate fitting, particularly if the range includes the point of inflection between zones III and IV at about 10 K (Fig. 5-38).

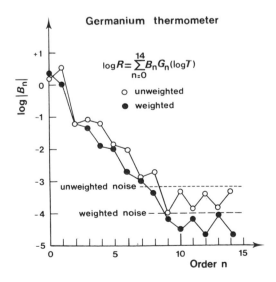

Fig. 5-38. Convergence of orthogonal coefficients for germanium thermometer calibration data. The noise coefficients are random and of low magnitude (after Powell *et al.*, 1972).

5-8 Thermistors

There exists a class of semiconductors, made up of mixed oxides of the transition metals, which appear under the generic name "thermistors": the word thermistor stemming from their original name "thermally sensitive resistors". Much of the stimulus for the development of thermistors arose from the need to provide a means of compensating for resistance changes in electrical circuits resulting from ambient temperature variations. Early thermistors were made from UO_2, but this was superseded in the early 1930s by $MgTiO_3$ (spinel). It was found that the resistivity of $MgTiO_3$ and its temperature coefficient could easily be adjusted by controlled reduction in hydrogen and by variations in the amount of MgO over and above the stoichiometric amount. CuO was another oxide which was used in the early days of thermistors for electric circuit regulation. Modern thermistors (Sachse, 1975; Hyde, 1971) are almost always mixtures of oxides, rarely stoichiometric, and are manufactured by the sintering of micrometre-sized particles of the components in controlled atmospheres. The presence of an oxidizing or reducing atmosphere during sintering can lead to the existence of an n-type, say, semi-conductor at the surface of a grain turning to p-type in the interior, with all the consequences in the conduction processes that this would entail. In addition to the conductivity of the grain itself, the conductivity of a sintered mixed oxide is strongly influenced by the grain boundaries between regimes of different conductivity. The high-frequency dispersion of thermistors, for example, is a result of the complex network of poorly conducting grain boundaries enclosing relatively highly conducting grains.

The most stable thermistors at temperature below about 250 °C, and hence those of interest for thermometry in this range, appear to be those made from the mixed oxides of manganese and nickel or manganese, nickel and cobalt, having a negative temperature coefficient of resistance. One of the great attractions of thermistors, of course, is the wide range of sizes and shapes that can easily be manufactured. Beads as small as 0.07 mm in diameter mounted on leads of diameter 0.01 mm are commercially available. Such minute temperature sensors lend themselves to such diverse applications as intravenous thermometry and plant-intracellular thermometry.

The two principal types of thermistors are the bead and the disc. Bead thermistors are usually made by first taking a pair of platinum alloy wires, which will eventually form the leads, stretched at the appropriate distance apart and then applying small blobs of the mixed oxide in a binder at intervals along the pair of wires. Subsequent sintering at a temperature near 1300 °C produces a series of thermistors having embedded leads. Individual thermistors may then be coated in glass which, as well as giving strength to the assembly, protects the thermistor from changes in resistivity which would otherwise result from atmospheric oxygen entering the lattice and changing the charge-carrier concentration. Disc thermistors are made by pressing the powder mixture and subsequently heating to about 1100 °C. Electrical contact is made by means of sprayed or screen-printed silver layers on opposite faces of the disc. The fact that disc thermistors are significantly less stable than bead thermistors (see below) is almost certainly a consequence of the inferiority of the surface electrode configuration compared with the embedded electrode configuration of the bead thermistors.

A convenient resistance/temperature relation for thermistors (Wood et al., 1978), is the form

$$R_T = R_0 \exp\left[-B\left(\frac{1}{T} - \frac{1}{T_0}\right)\right] \qquad (5\text{-}39)$$

where B is a constant and R_0 is the resistance at a temperature T_0. The exponential dependence of R_T on T is to be expected in view of the semiconducting nature of the material. Equation (5-39) is, of course, purely empirical since we can make no useful attempt at writing down an explicit relation between R and T for a semiconductor as complicated as a non-stoichiometric mixture of transition metal oxides. The value of B and hence the temperature coefficient α, is given by

$$\alpha = \frac{1}{R}\frac{dR}{dT} = -\frac{B}{T^2} \qquad (5\text{-}40)$$

This can be adjusted during sintering after which it should not change. Commonly employed values of $R(25\ °C)$ are 5 kΩ, 10 kΩ, 15 kΩ and 30 kΩ, although units having a resistance as low as 1 kΩ are available.

The stability of this type of thermistor was the object of an extensive study carried out at the National Bureau of Standards (Wood et al., 1978; but see also La Mers et al., 1982; Zurbuchen and Case, 1982). One of the principal conclusions of the study was that bead thermistors are very much more stable than disc thermistors, and that many bead thermistors have drift rates at the level of only 1 mK per 100 days at temperatures up to 60 °C. Figures 5-39, 5-40 and 5-41 are taken from the results of this study. The disc thermistors showed a much larger change of resistance during the initial hundred days of operation than did bead thermistors. A general conclusion which can be drawn from the NBS work, is that to obtain the best stability from a room temperature thermistor, an initial ageing period of as much as three months is desirable at the maximum temperature of intended operation.

Thermistors for use at temperatures above about 300 °C are made from more refractory oxides than those of manganese or nickel. Not only must the oxide be more refractory but also the activation energy, related to B in equation (5-39), must be higher to maintain adequate sensitivity. The oxides of the rare-earth elements fulfil these requirements, and mixtures of them are employed in thermistors for use up to about 1000 K. At still higher

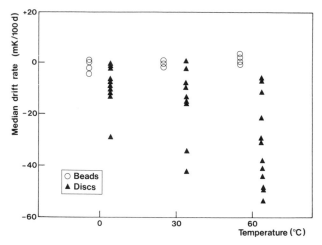

Fig. 5-39. Median drift rate *vs* ageing temperature for bead and disc thermistors (after Wood *et al.*, 1978).

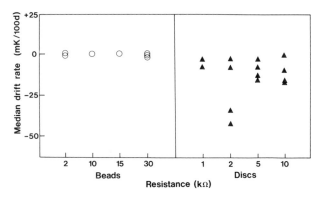

Fig. 5-40. Median drift rate *vs* resistance at 30 °C for bead and disc thermistors (after Wood *et al.*, 1978).

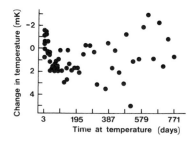

Fig. 5-41. The drift rate of 10 kΩ bead thermistors held at 30 °C. These results are typical of bead thermistors operated in the range 0 °C to 60 °C (after Wood *et al.*, 1978).

temperatures thermistors based upon zirconia doped with small quantities of rare-earth oxides are available. At very low temperatures, the thermistor is of particular interest because of its relative insensitivity to magnetic fields (see Section 5-10). Thermistors for this range, which require a very low activation energy, are usually made from non-stoichiometric iron oxides. Their resistance sensitivity at 20 K is about 15 % per kelvin which can increase to 300 % per kelvin at 4.2 K. The disadvantage of the thermistor as a general-purpose low-temperature thermometer is this same high sensitivity, which implies that the working range of the thermometer is rather narrow. They are mainly of interest above 4.2 K.

5-9   Carbon and carbon–glass thermometers

Since the early 1950s, following the pioneering work of Clement and Quinnell (1952), the simple 1/8 watt carbon resistor has been widely used as a cheap and sensitive thermometer, particularly for temperatures below about 20 K. Although not having the reproducibility of the germanium or rhodium/iron thermometer, it can, with a little care, be used with a reproducibility of about 10 mK in this range. The original work of Clement and Quinnell was carried out using Allen–Bradley resistors, and although these remain in production, changes in manufacturing techniques since then have led to a modified resistance/temperature characteristic (Weinstock and Parpia, 1972) Fig. 5-42. Allen–Bradley resistors are characterized by their room-temperature resistance and their wattage: BB(1/8 W), CB(1/4 W) and EB(1/2 W). For

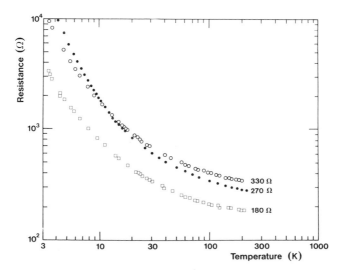

Fig. 5-42. Comparison of calibration curves for 180 Ω and 330 Ω, 1/8 watt Ohmite resistors and a 270 Ω Allen–Bradley resistor (after Weinstock and Parpia, 1972 and Schultz, 1966).

temperatures below about 0.5 K, Speer resistors have been preferred (grade 1002), but are likely soon to become unavailable since manufacture has now ceased (Rubin, 1980).

To take full advantage of the carbon resistor as a thermometer, it is worth taking a number of precautions in its use which help to avoid unwanted changes in calibration (Anderson, 1972; Ricketson, 1975). The resistors are sensitive to changes in humidity, showing an increase in resistance and slope of resistance/temperature with increasing water content. Putting new resistors in epoxy, after drying by warming to 60 °C in vacuum, is therefore to be recommended. Care should be taken not to heat the resistor excessively during connection of the leads, since an excursion to 370 °C irreversible reduces the resistance, Fig. 5-43. On cooling to low temperatures, the resistance slowly approaches an equilibrium value, but is subject to change after even small changes in temperature at low temperatures (Fig. 5-44). Sufficient time should elapse, therefore, before measurements are commenced.

The calibration of carbon thermometers and the fitting of empirical curves to calibration data has been discussed in many monographs on low-temperature techniques. The basic equation, from which refined versions have been developed, is the following:

$$\ln R = AT^{-m} + B \qquad (5\text{-}41)$$

where A, B and m are constants. This equation as it stands allows carbon thermometer calibration data to be fitted over the range from about 3 K to 60 K to within about ± 30 mK. Anderson (1972) and Ricketson (1975) review the

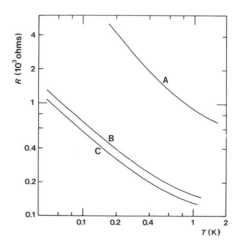

Fig. 5-43. Resistance/temperature characteristics for baked and unbaked Speer (grade 1002) resistors: A, 220 Ω resistor, unbaked; B, 220 Ω resistor baked in air for 1 h at 375 °C; C, 51 Ω resistor, unbaked. These results are typical of this type of resistor (after Anderson, 1972).

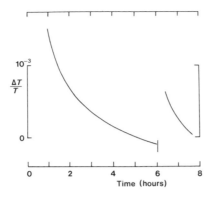

Fig. 5-44. Drift in indicated temperature of a 100 Ω Ohmite resistor, maintained at 1.8 K except for a brief excursion to 4 K at $t = 6$ hours (after Anderson, 1972).

methods of adapting this equation to give better fits for particular resistor values and for more limited temperature ranges.

The thermal anchoring of a carbon thermometer is much less dependent upon the conduction through the leads than is that of a germanium thermometer. This is owing to the much poorer thermal conductivity of carbon compared with that of germanium at low temperatures. Carbon thermometers should therefore be thermally anchored by ensuring good thermal contact between the body of the thermometer and the object whose temperature is being measured. This may be achieved by first grinding away the outer sheath and then inserting the thermometer into a tightly fitting greased hole or by binding the thermometer in copper wire which is itself thermally anchored. An alternative method is to grind the thermometer to a thin wafer which is then glued to the surface of a good conductor. A common use of carbon thermometers is in the detection of cryogenic liquid levels. If a relatively high measuring current is used, thus producing a high heating effect, a sudden drop in resistance is observed on passing from liquid to vapour owing to the much better thermal contact between the thermometer and the liquid.

A modification of the usual carbon thermometer is the carbon-impregnated porous-glass thermometer (Lawless, 1972). The thermometer material is manufactured by first preparing porous glass by leaching out the boron-rich phase of a phase-separated alkali borosilicate glass. The result is a random packed collection of 30 nm diameter silica spheres having a pore size of some 3 to 4 nm. In these pores filamentary carbon can be deposited. Thermometers are made by cutting small rods (about $5 \times 2 \times 1$ mm) from the solid and depositing nichrome–gold electrodes on the ends to which copper leads are silver-epoxied. After baking to remove absorbed water and gases, the units are encapsulated in platinum under a helium atmosphere.

The resistance/temperature relationship for carbon–glass thermometers is

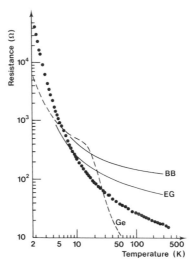

Fig. 5-45. Comparison of the resistance/temperature characteristics of: •••, carbon–glass thermometer; Ge, germanium resistance thermometer; BB, 1/8 watt Allen–Bradley resistor; EB, 1/2 watt Allen–Bradley resistor (after Lawless, 1972).

different from that of ordinary carbon thermometers. The sensitivity is higher, while at the same time the resistance above about 10 K is lower and becomes much lower near room temperature than does that of ordinary carbon resistors, Fig. 5-45. The reproducibility of carbon–glass thermometers is good and they do not appear to suffer from the low-temperature drift observed in ordinary carbon thermometers. A formula of the form already given for ordinary carbon thermometers provides a good fit to carbon–glass experimental data over a temperature range extending from 2 K to 200 K. As would be expected from their construction, the resistance/temperature relationship of samples from the same block of glass differs by only a simple scaling factor. In fact it is found that $R_1 = l\,R_2^m$, where R_1 and R_2 are the resistances of thermometers 1 and 2, and l and m are temperature-independent constants with m being very close to unity. This should allow relatively straightforward calibration of carbon glass thermometers against a reference function determined for a typical example.

5-10 Effects of magnetic fields on resistance thermometers

The phenomenon of magnetoresistance, that is the increase in resistance of a metal observed in the presence of a magnetic field, is not one that is amenable to a simple theoretical description. A magnetoresistance will, in general, be observed (see Ziman, 1972) whenever the Fermi surface is non-spherical, and in particular when it contains contributions from both electrons and holes or electrons from two bands. If there are two such different types of charge carrier having different values of charge, mass or relaxation time, a magnetic field will affect the two carriers in different ways. The net conductivity, which is the vector sum of the two components, will thus be modified. This mechanism leads to a transverse magnetoresistance which is approximately proportional to the square of the field H but which saturates in high fields. A special case is that of a metal in which the groups of charge carriers have the same relaxation time. Under these conditions we can say that the change in resistance $\Delta\rho$ due to the field is given by:

$$\Delta\rho/\rho_0 = f(H/\rho_0)$$

where ρ_0 is the zero-field resistivity (Kohler's Rule). Measurements of magnetoresistance at low temperature (where ρ_0 is low) can be used to predict the magnetoresistance at high temperatures for very high fields. Kohler's Rule is, however, only approximately true because relaxation times are themselves functions of temperature depending upon the proportion of scattering by phonons and impurities. Our interest in magnetoresistance stems from the importance of the ability to measure low temperatures in the presence of strong magnetic fields. It has led to a considerable amount of work being carried out on the behaviour of practical thermometers under these conditions. Much of this has been done at the Francis Bitter National Magnet Laboratory of the Massachussetts Institute of Technology and has been published in a series of papers and reviews (Sample and Rubin, 1977) which have appeared since 1971. In this work, mainly confined to the temperature range 0.5 K to 100 K, it was found convenient to split the range of magnetic field into three parts. The lower part covered fields up to 2.5 T, which is about the upper limit of fields obtainable using iron-core magnets; the middle part included fields up to 8 T, obtainable using medium field superconducting magnets and the third part fields up to 14 T, attainable using high-field superconducting magnets or water-cooled Bitter solenoids. A summary of the results obtained at the National Magnet Laboratory is shown in Table 5-1 taken from Sample and Rubin (1977). In addition to the bare figures given in Table 5-1 for each type of thermometer, the following notes give some further information (Alms et al., 1979; Meaden, 1965).

Carbon resistors
Carbon resistors of the Allen–Bradley type are very widely used because of their relatively small and reproducible magnetoresistance. Figure 5-46 shows the change in resistance for a typical Allen–Bradley resistor over a range of fields at temperatures between 4.2 K and 26 K. There does not appear to be any orientation-dependence for Allen–Bradley resistors, and the effects are similar for similar resistors. Speer resistors, which are used at lower

Table 5-1. (Sample and Rubin, 1977).

Type of sensor	T/K	Magnitude of relative temperature error $\|\Delta T\|/T(\%)$ for fields less than:		
		2.5 Tesla	8 Tesla	14 Tesla
Carbon resistors				
Allen–Bradley (2.7,	0.5	2–4	5–13	7–20
3.9, 5.6 and 10 Ω)	1.0	2–4	6–15	9–25
	2.5	1–5	6–18	10–30
	4.2	1–5	5–20	10–35
Allen–Bradley (47,	4.2	<1	5	10
100 and 220 Ω)	10	<1	3	5
	20	<1	1	2
Speer, Grade 1002	0.5	0–2	0–1	0–6
(100, 220 and 470 Ω)	1.0	1–2	2–4	3–9
	2.5	3–5	1–4	7–14
	4.2	4–9	2–5	4–13
Matsushita (68, 200,	1.5	1–2	10–15	–
and 510 Ω)	2.1	1	10–15	–
	4.2	2–3	4–8	–
Carbon–glass resistors	2.1	0.5	1.5	4
	4.2	0.5	3	6
	15	<0.1	0.5	1.5
	35	<0.1	0.5	1
	77	<0.1	0.5	1.5
Thermistors	4.2	<0.05	1	3
	10	<0.05	0.3	1
	20	<0.05	0.1	0.5
	40	<0.05	0.1	0.5
	60	<0.05	0.1	0.3
Germanium resistors	2.0	8	60	–
	4.2	5–20	30–55	60–70
	10	4–15	25–60	60–75
	20	3–20	15–35	50–80
	70	3–10	15–30	25–50
GaAs diodes	4.2	2–3	30–50	100–250
	10	1.5–2	25–40	75–200
	20	0.5–1	20–30	60–150
	40	0.2–0.3	4–6	15–30
	80	0.1–0.2	0.5–1	2–5

Si diodes	4.2	75	–	–
	10	20	30	50
	20	4	7	10
	30	3	4	5
	77	0.2	0.5	0.5
Platinum resistors	20	20	100	250
	40	<1	5	10
	80	<0.5	1	2

temperatures because of their larger relative sensitivity, show a more complicated magnetoresistance field-dependence than do Allen–Bradley resistors. This is illustrated in Fig. 5-47. Carbon–glass resistors behave in magnetic field in very much the same way as do Allen–Bradley resistors, and in common with them do not show any orientation dependence.

Thermistors

Thermistors show a very small temperature error resulting from magnetoresistance. This is partly due to their having a very high sensitivity ($\Delta R/\Delta T$).

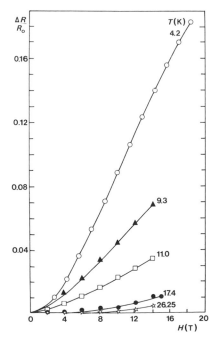

Fig. 5-46. Magnetic field dependence of $\Delta R/R_0$, for Allen Bradley resistors at temperatures from 4.2 K to 26.25 K (after Alms *et al.*, 1979).

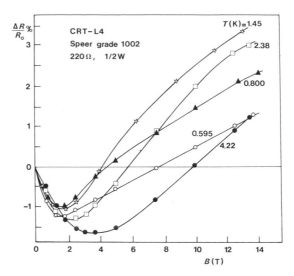

Fig. 5-47. Magnetic field dependence of $\Delta R/R_0$ of a 220 Ω, 1/8 watt Speer (1002) resistor (after Sample and Rubin, 1977).

Germanium resistance thermometers

Germanium resistance thermometers, despite their excellent characteristics for other purposes, are not good thermometers for use in high magnetic fields. Their magnetoresistance is large, Fig. 5.48, and strongly orientation dependent, Fig. 5-49. Their use is not recommended for fields greater than 2.5 T at any temperature.

p–n junction diode thermometers

p–n junction devices are not resistance thermometers in the same sense as platinum or germanium resistance thermometers. Nevertheless, the present discussion of magnetoresistance effects in thermometry provides an opportunity to introduce them and describe their general properties, since they can be useful as thermometers which are relatively insensitive to magnetic fields (Swartz and Swartz, 1974). From Table 5-1, it will be seen that GaAs diodes serve as useful thermometers for fields up to about 5 T for temperatures up to 40 K and at all fields at temperatures above 40 K. The range of figures given in the table indicates the likely spread of magnetoresistance effects due to variations in performance of different samples, orientation of the junction with respect to the field and the dependence upon bias current.

One of the attractions of the GaAs diode as a practical thermometer is that over a wide range of temperature, extending from 1 K up to at least 400 K, the forward voltage at constant current increases very nearly linearly with temperature. This is the case also for the Si diode at temperatures above about

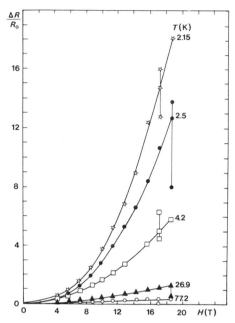

Fig. 5-48. Magnetic field dependence of $\Delta R/R_0$ of a germanium thermometer with the field parallel to the axis of the thermometer. The vertical bars indicate range of values likely to be encountered (after Alms *et al.*, 1979).

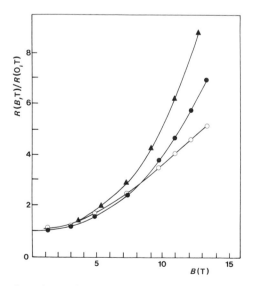

Fig. 5-49. Orientation dependence of magnetoresistance of a germanium thermometer at 7.2 K ●–● parallel to the field, ●–● and ▲–▲ at right angles to the field but in two different positions (after Sample and Rubin, 1977).

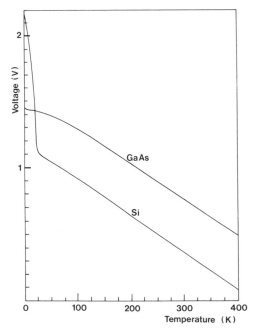

Fig. 5-50. Forward-voltage/temperature characteristics for GaAs and Si diode thermometers (after Swartz and Swartz, 1974).

30 K where the linearity is even better than that of a GaAs diode (Fig. 5-50). Below 30 K, the sensitivity of the silicon diode thermometer greatly exceeds that of the GaAs diode thermometer. The voltage/temperature characteristic of both types of diode is current sensitive, and it is therefore necessary to specify the current. This has an optimum value, for modern low-noise diodes, of between 10 μA and 50 μA. Higher currents are not recommended because of their excessive heating effects, while currents as low as 1 μA lead to non-linear behaviour above 77 K. The reproducibility of GaAs diodes is about $\pm$10 mK above 77 K and below 7 K but with a trough in reproducibility near 20 K, where it is no better than about 30 mK. Below 30 K, however, Si diodes are more reproducible than GaAs diodes but unfortunately are more sensitive to magnetic fields at low temperatures. The silicon diode has a very strong orientation dependence of the magnetic effect, and in Table 5-1, only that for the junction parallel to the field is given since at any other orientation the effect is much larger. At temperatures above 30 K, that is, above the temperature of the sharp drop in sensitivity, the silicon diode is rather less sensitive to magnetic fields than is the GaAs diode.

Glass–ceramic (Sr TiO$_2$) capacitance thermometer
This thermometer is one of the best thermometers for use in high magnetic fields at low temperatures. Its magnetic field dependence is extremely small up

to the highest fields, see Table 5-1 (Rubin and Lawless, 1971). In order to obtain the best results with glass–ceramic thermometers, it is necessary to cool from room temperature in a reproducible way and also to anneal the thermometer at a temperature above its maximum operating temperature for periods which depend upon the temperature but which may be many hours (Swenson, 1977; Lawless et al., 1971).

Platinum resistance thermometers

We remarked in Section 5-1 that at low temperatures, the resistivity of pure platinum depends essentially upon the concentration of defects and impurities. For well annealed pure platinum, the resistivity is extremely low and, since the magnetoresistance is approximately inversely proportional to the resistivity, it becomes very large below about 40 K. Above this temperature, the magnetoresistance is quite small even at high fields. The figures given in the table are for the field parallel to the axis of the thermometer since this leads to a larger effect than when the field is perpendicular to the axis.

Rhodium/0.5 % iron resistance thermometer

The magnetoresistance of rhodium–iron thermometers has not been studied in detail, although measurements have been made at 4.2 K and 2.1 K. These showed that the effect increases on cooling, as expected, but that it is not as large as in platinum or rhodium. Nevertheless at 4.2 K, the increase in resistance of about 6% in a field of 3 T is equivalent to a temperature offset of 1 K (Rusby, 1972; Murani and Coles, 1972).

5-11   Resistance measurement in thermometry

The preceding discussion of resistance thermometry presupposes the ability to make the necessary electrical measurements. In the past, the development of resistance thermometry has been impeded by the inadequacy of available methods of resistance measurement. While this is not so at present, resistance thermometry poses three problems which are absent, or less serious, in the realm of electrical standards. The first is the likelihood of significant stray thermal emfs (commonly of order 1 μV) because of the large temperature gradients likely to be present in the circuit; the second is the need to restrict measuring currents in the interest of acceptable self-heating, and the third is the frequent necessity for long leads. The high resistance of long leads introduces stray admittances in a.c. circuits. Towards the lower limit of IPTS-68, lead resistances may be ten times the resistance of the platinum resistance thermometer they serve. The intrinsically high lead resistance of germanium resistance thermometers are a further aspect of this problem. Although circuits involving compensating leads or three-lead arrangements are still used in

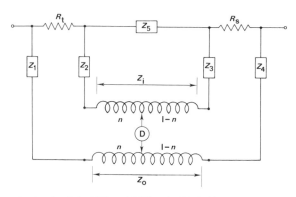

Fig. 5-51. Schematic circuit of the Hill and Miller double bridge.

industrial applications or where simplicity outweighs accuracy, four-lead connection is universal in standards work.

Measuring circuits may be divided into two categories: potentiometric types, in which at balance there is exactly zero current flowing in the potential leads, and bridge circuits where, in principle, a current flows at balance, although it may in fact be negligible. Cutting across this categorization is the distinction between a.c. and d.c. measurement.

Until the late 60s, a.c. methods had no application in temperature standards work. Since then, two factors have altered this situation. Of prime importance is the exploitation of the inductive voltage divider, or ratio transformer, in bridge circuits. This has been aided by the second factor, namely, advances in electronics, which have produced lock-in detectors of high sensitivity and excellent signal-to-noise characteristics. Elaborate self-balancing systems have also become possible.

The overwhelming advantages of the inductive voltage divider are its accuracy and its stability. An accuracy of better than 1 part in 10^6 is easily achieved, and 1 part in 10^8 is possible. In addition, the need for the tiresome periodic re-calibration of resistive bridges (Barber *et al.*, 1955) is absent together with sensitivity to ambient temperature fluctuation.

The earliest bridge of this type to be described is that of Hill and Miller (1963). It is a kelvin double bridge, and the schematic circuit, including lead impedances, is shown in Fig. 5-51. The inner and outer dividers are mechanically coupled to give the same ratio at all times. If Z_i and Z_0 are the input impedances of the inner and outer dividers respectively, the exact balance condition is:

$$R_t = R_s \left[\frac{nZ_0 + z_1}{(1-n)Z_0 + z_4} \right]$$

$$+ z_s \left[\frac{(1-n)Z_i + z_3}{Z_i + z_2 + Z_3 + z_5} \right] \left[\frac{nZ_0 + z_1}{(1-n)Z_0 + z_4} - \frac{nZ_i + z_2}{(1-n)Z_i + z_3} \right] \tag{5-42}$$

If, as is probable, $z_1 = z_2$, $z_3 = z_4$, and $Z_0 = Z_i$, this condition simplifies to:

$$R_t = R_s \left[\frac{n}{1-n} \right] \left[1 + \frac{z_1}{nZ_0} - \frac{z_4}{(1-n)Z_0} \right]. \qquad (5\text{-}43)$$

The effect of lead resistance can be inferred from this. Z_0 is typically of order 5×10^5 Ω, and is nearly pure resistance, z_1 might be 1 Ω and z_4 effectively zero. If R_s is 25 Ω, then at 0 °C, lead resistance results in a correction equivalent to 1 mK. In precise work, the lead resistance must be measured and the corrections applied.

An automatic, seven-decade version of this bridge is commercially available (Automatic Systems Ltd.) and permits rapid, high accuracy measurement. The operating frequency, 375 Hz, is locked to the 50 Hz mains, and is near that at which the divider input impedance is maximum. Rejection of 50 Hz and its harmonics occurs in the lock-in detector. If the full range of IPTS-68 is to be covered, several standard resistors are required. Suitable resistors of 1, 10, 25 and 100 ohms have been made to the Wilkins and Swann (1970) pattern and are commercially available.

To overcome lead resistance errors, bridges using multi-stage transformers have been designed, by Cutkosky (1970, 1982), by Thompson and Small (1971) and Knight (1977). Although the Cutkosky and the Knight bridges employ a three-stage transformer rather than the two-stage transformer of Thompson and Small, the bridges are similar, and can be represented by the schematic circuit of Fig. 5-52. The flux in the transformer core is provided by current flowing in a "second stage" winding, E, having the same number of turns as the fixed winding, N, and connected across R_s. The voltage across E is set very

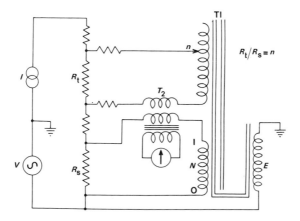

Fig. 5-52. Schematic circuit of the Thompson and Small bridge.

accurately equal to that across N by electronic means, so that essentially no current flows in the potential leads of R_s. At balance the same is true of R_T, and there is then no output from the three-winding detector transformer, $T2$. Various devices are adopted to counteract the effects of admittances between leads and between the windings of $T1$ including, in Cutkosky's and Kibble and Knight's bridges, a third actively energized winding. In all of these bridges, quadrature balance is automatic although the main balance is manual, and all operate near 400 Hz.

In all a.c. thermometer bridges, the design of leads is important. Cutkosky's and Knight's bridges accept two coaxial leads to each resistor, but the bridge of Thompson and Small calls for four coaxial cables per resistor. This entails modification to the heads of stem-type thermometers, and is usually impracticable in cryogenic situations. The most pernicious lead fault is coupling between current and potential leads, which is eliminated in the lead systems already mentioned. Where coaxial leads are inappropriate however, it is essential to twist the leads in pairs, $I + I$ and $V + V$. This reduces coupling between current and voltage leads, and as it minimizes the pick-up of radiated interference, it is desirable in d.c. systems also. For a.c. measuring techniques, it is essential that the lead insulation be polythene or PTFE and not the more common PVC, since the dielectric losses in the latter are much too large.

Another approach to a.c. resistance measurement is that of the Cryobridge which is essentially an a.c. potentiometer and was developed to cope with the high-lead resistances of germanium thermometers. This instrument (Automatic Systems Ltd), which operates at 25 Hz, is automatic and measures sequentially the potentials across standard and unknown resistors, leaving the observer to calculate their ratio.

The ratio transformer principle has been adapted by Kusters and MacMartin (1970) to d.c. resistance-thermometer measurement. The heart of the circuit (Fig. 5-53) is a ratiotransformer having a fixed N_s, and an adjustable N, winding and a flux detector which is able to detect very precisely the condition of zero flux in the transformer core. A servo system actuated by this

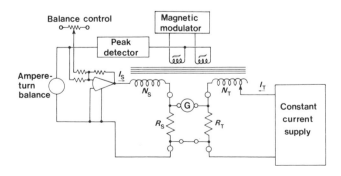

Fig. 5-53. Schematic circuit of the Kusters current comparator bridge.

flux detector, adjustes the current in N_s and R_s so that there is always zero net flux in the transformer. Thus when N_T is changed by the operator, there is a consequent change in I_s. Balance is achieved when the same potential exists across R_s and R_T, and the ratio of currents is then

$$\frac{I_T}{I_s} = \frac{N_s}{R_T} = \frac{R_s}{R_T}.$$

As a further sophistication, the current can be automatically reversed at intervals with a demodulator indicating when the balance is unchanged by reversal. Thus the influence of static thermal emfs is eliminated.

The instrument (Guildline Company) incorporates a photocell-galvanometer amplifier, and is less convenient to use than an equivalent a.c. bridge. It however is immensely superior to traditional d.c. methods, is truly potentiometric, has all the advantages of the ratio transformer and none of the drawbacks of lead admittances and a.c.–d.c. transfer errors.

References

Actis, A. and Crovini, L. (1982). Interpolation equations for industrial platinum resistance thermometers in the range −200 °C to 420 °C. *TMCSI*, **5**, 819–827.

Allen, P. B. and Butler, W. H. (1978). Electrical conduction in metals. *Physics Today*, **31**, 44–49.

Alms, H., Tillmans, R. and Roth, S. (1979). Magnetic-field-induced temperature error of some low temperature thermometers. *J. Phys. E.* **12**, 62–66.

Ambler, E. and Plumb, H. H. (1960). Use of carbon resistors as low temperature thermometers in the presence of strong rf fields. *Rev. Sci. Inst.* **31**, 656–657.

Anderson, A. C. (1972). Carbon resistance thermometry. *TMCSI*, **4**, Part 2, 773–784.

Anderson, M. S. and Swenson, C. A. (1978). Characteristics of germanium resistance thermometers from 1 to 35 K and the ISU magnetic temperature scale. *Rev. Sci. Instr.* **49**, 1027–1033.

Barber, C. R. (1950). Platinum resistance thermometers of small dimensions. *J. Sci. Instr.* **27**, 47–00.

Barber, C. R., Gridley, A. and Hall, J. A. (1955). An improved construction of the Smith bridge type 3. *J. Sci. Instr.* **32**, 213–220.

Barber, C. R. and Hall, J. A. (1962). Progress in platinum resistance thermometry. *Brit. J. Appl. Phys.* **13**, 147–154.

Bedford, R. E. and Kirby, C. G. M. (1969). Notes on the application of the International Practical Temperature Scale of 1968. *Metrologia*, **5**, 83–87.

Berry, R. J. (1963). The relationship between the real and ideal resistivity of platinum. *Can. J. Phys.* **41**, 946–982.

Berry, R. J. (1972). The influence of crystal defects in platinum on platinum resistance thermometry. *TMCSI*, **4**, Part 2, 937–949.

Berry, R. J. (1966). Platinum resistance thermometry in the range 630 °C–900 °C. *Metrologia*, **2**, 80–90.

Berry, R. J. (1975). Control of oxygen-activated cycling effects in platinum resistance thermometers. Temperature-75, 99–106.

Berry, R. J. (1978). Study of multilayer surface oxidation of platinum by electrical resistance technique. *Surface Science*, **76**, 415–442.

Besley, L. M. (1978). Further stability studies on germanium resistance thermometers at 20 K. *Rev. Sci. Instr.* **49**, 1041–1045.

Besley, L. M. (1980). Stability data for germanium resistance thermometers at three temperatures. *Rev. Sci. Instr.* **51**, 972–976.

Besley, L. M. and Kemp, W. R. G. (1977). An intercomparison of temperature scales in the range 1 to 30 K using germanium resistance thermometry. *Metrologia*, **13**, 35–51.

Besley, L. M. and Plumb, H. H. (1978). Stability of germanium resistance thermometers at 20 K. *Rev. Sci. Instr.* **49**, 68–71.

Blakemore, J. S., Schultz, J. M. and Myers, J. W. (1962). Measurements on gallium-doped germanium thermometers. *Rev. Sci. Instr.* **33**, 545–551.

Blakemore, J. S. (1962). Design of germanium for thermometric applications. *Rev. Sci. Instr.* **33**, 106–112.

Bloch, F. (1928). Uber die Quantenmechanik der Elektrone in Kristallgittern. *Z. Phys.* **52**, 555.

Chattle, M. V. (1972). Platinum resistance thermometry up to the gold point. *TMCSI*, **4**, Part 2, 907–918.

Chattle, M. V. (1975). Resistance/temperature relationships for industrial platinum resistance thermometers. NPL Report QU 30.

Chattle, M. V. (1977). Resistance/temperature relationships for industrial platinum resistance thermometers of thick film construction. NPL Report QU 42, 1977.

Clement, J. R. and Quinnell, E. H. (1952). The low temperature characteristics of carbon-composition thermometers. *Rev. Sci. Instr.* **23**, 213.

Cochrane, J. (1972). Relationship of chemical composition to the electrical properties of platinum. *TMCSI*, **4**, Part 2, 1619–1632.

Coles, B. R. (1964). A new type of low-temperature resistance anomaly in alloys. *Physics Letters*, **8**, 243–244.

Compton, J. P. (1972). The realization of low temperature fixed points. *TMCSI*, **4**, Part 1, 195–209.

Curtis, D. J. (1972). Platinum resistance interpolation standards. *TMCSI*, **4**, Part 2, 951–961.

Curtis, D. J. (1982). Thermal hysteresis effects in platinum resistance thermometers. *TMCSI*, **5**, 803–812.

Cutkosky, R. D. (1970). An a.c. resistance thermometer bridge. *J. Res. NBS*, **74C**, 15–18.

Cutkosky, R. D. (1982). Automatic resistance thermometer bridges for new and special applications, *TMCSI*, **5**, 711–713.

Dugdale, J. S. (1977). "The Electrical Properties of Metals and Alloys", Arnold, London.

Evans, J. P. and Burns, G. W. (1962). A study of stability of high temperature platinum resistance thermometers. *TMCSI*, **3**, Part 1, 313–318.

Evans, J. P. and Wood, S. D. (1971). An intercomparison of high temperature platinum resistance thermometers and standard thermocouples, *Metrologia*, **7**, 108–130.

Hill, J. J. and Miller, A. P. (1963). An a.c. double bridge with inductively coupled ratio arms for precision platinum resistance thermometry. *Proc. IEE*, **110**, 453–458.

Hume-Rothery, W. and Raynor, G. V. (1956). "The Structure of Metals and Alloys". The Institute of Metals, London.

Hust, J. G. (1970). Thermal anchoring of wires in cryogenic apparatus. *Rev. Sci. Instr.* **41**, 622–624.

Hyde, F. J. (1971). "Thermistors", Iliffe Press.

IEC (1980). Draft standard for industrial resistance thermometers.

Jackson, J. J. (1965). Point defects in quenched platinum. In "Lattice Defects in Quenched Metals", (R. M. Cotterill, M. Doyama, J. J. Jackson and M. Meshi).

Academic Press, New York and London.

Johnston, J. S. (1975). Resistance Thermometry. Temperature-75, 80–90.

Kemp, R. C., Kemp, W. R. G. and Besley, L. M. (1981). Interpolation methods for platinum resistance thermometers between 13.81 K and 273.15 K. *Metrologia*, 17, 43–48.

Kirby, C. G. M. and Laubitz, M. J. (1973). The error due to the Peltier effect in direct-current measurements of resistance. *Metrologia*, 9, 103–106.

Kittel, C. (1976). "Introduction of Solid State Physics", 5th Ed., Wiley.

Knight, R. B. D. (1977). A precision bridge for resistance thermometry using a single inductive-current divider. IEE Conf. Publ. *Euromeas.-77*, 132–134.

Kunzler, J. E., Geballe, T. H. and Hull, G. W. (1962). Germanium resistance thermometers. *TMCSI*, 3, Part 1, 391–398.

Kusters, N. L. and MacMartin, M. P. (1970). Direct-current comparator bridge for resistance thermometry. *IEEE Trans.*, IM-19, 291–297.

La Mers, T. H., Zurbuchen, J. M. and Trolander, H. (1982). Enhanced stability in precision interchangeable thermistors. *TMCSI*, 5, 865–873.

Lawless, W. N., Radebaugh, R. and Soulen, R. J. (1971). Studies of the glass–ceramic capacitance thermometer between 0.025 K and 2.4 K. *Rev. Sci. Instr.* 42, 567–570.

Lawless, W. N. (1972). Thermometric properties of carbon-impregnated porous glass at low temperatures. *Rev. Sci. Instr.* 43, 1743–1747.

Lindenfield, P. (1962). Carbon and semiconductor thermometers for low temperatures. *TMCSI*, 3, Part 1, 399–405.

Long Guang and Tao Hongtu (1982). Stability of precision high temperature platinum resistance thermometers. *TMCSI*, 5, 783–787.

McAllan, J. V. (1982). Practical high temperature resistance thermometry. *TMCSI*, 5, 789–793.

Macfarlane, J. C. and Collins, H. C. (1979). The low temperature resistivity of platinum rhodium alloys. *Cryogenics*, 18, 668–669.

Mangum, B. W. and Evans, G. A. (1982). Investigation of the stability of small platinum resistance thermometers. *TMCSI*, 5, 795–801.

Marcarino, P. and Crovini, L. (1975). Characteristics of platinum resistance thermometers up to the silver freezing point. Temperature-75, 107–116.

March, J. F. and Thurley, F. (1979). PdPtAu alloys for low temperature precision resistors. *Rev. Sci. Instr.* 50, 616–618.

McLaren, E. H. and Murdock, E. G. (1966). Radiation effects in precision resistance thermometry. *Can. J. Phys.* 44, 2631–2659.

Meaden, G. T. (1965). Chapter 6, in "Electrical Resistance of Metals", Heywood Books, London.

Murani, A. P. and Coles, B. R. (1972). Some magnetic and transport properties of Rh–Mn alloys. *J. Phys. F.* 2, 1137–1144.

Powell, R. L., Hall, W. J. and Hust, J. G. (1972). The fitting of resistance thermometer data by orthogonal functions. *TMCSI*, 4, Part 2, 1423–1431.

Quinn, T. J. (1975). Temperature standards, Temperature-75, 1–16.

Ricketson, B. W. (1975). The 220 Ω Allen–Bradley resistor as a temperature sensor between 2 K and 100 K. Temperature-75, 135–143.

Riddle, J. L., Furukawa, G. T. and Plumb, H. H. (1973). Platinum resistance thermometry: NBS Monograph 126.

Rivier, N. and Zlatic, V. (1972). Temperature dependence of the resistivity due to localized spin fluctuations II. Coles alloys. *J. Phys. F.* 2, L 99-L 104.

Rubin, L. G. (1980). Status of carbon resistors for low temperature thermometry. *Rev. Sci. Instr.* 51, 1007.

Rubin, L. G. and Lawless, W. N. (1971). Studies of a glass–ceramic capacitance thermometer in an intense magnetic field at low temperatures. *Rev. Sci. Instr.* 42,

571–573.

Rusby, R. L. (1982). The rhodium–iron resistance thermometer ten years on. *TMCSI*, **5**, 829–833.

Rusby, R. L. (1975). Resistance thermometry using rhodium–iron 0.1 K to 273 K. Temperature-75, 125–130.

Rusby, R. L. (1972). A rhodium–iron resistance thermometer for use below 20 K. *TMCSI*, **4**, Part 2, 865–869.

Sachse, H. B. (1975). "Semi-conducting Temperature Sensors and their Applications", Wiley, London.

Sample, H. H. and Rubin, L. G. (1977). Instrumentation and methods for low temperature measurements in high magnetic fields. *Cryogenics*, **17**, 597–606.

Sawada, S. and Mochizuki, T. (1972). Stability of 25 ohms platinum thermometer up to 1100 °C. *TMCSI*, **4**, Part 2, 919–926.

Schultz, E. H. (1966). Carbon resistors for cryogenic temperature measurement. *Cryogenics*, **6**, 321–323.

Stickney, T. M. and Dutt, M. (1970). Thermal recovery and accuracy of air total temperature sensors. *Instrumentation in the Aerospace Industry*, **16**, Instrument Society of America.

Swartz, D. L. and Swartz, J. J. (1974). Diode and resistance thermometry: a comparison. *Cryogenics*, **14**, 67–70.

Swenson, C. A. (1977). Time-dependent and thermal history effects in low-temperature glass–ceramic capacitance thermometers. *Rev. Sci. Instr.* **48**, 489–490.

Swenson, C. A. and Wolfendale, P. C. F. (1973). Differences between a.c. and d.c. determinations of germanium thermometer resistances. *Rev. Sci. Instr.* **44**, 339–341.

Thompson, A. M. and Small, G. W. (1971). A.c. bridge for platinum-resistance thermometry. *Proc. IEE*, **118**, 1662–1666.

Thulin, A. (1971). High precision thermometry using industrial resistance sensors. *J. Phys. E.* **4**, 764–768.

Trenkle, F. and Reinhardt, P. (1973). In-flight temperature measurements. AGARDograph No. 160, NATO.

Weinstock, H. and Parpia, J. (1972). A survey of thermometric characteristics of recently produced Allen–Bradley/Ohmite resistors. *TMCSI*, **4**, Part 2, 785–790.

White, G. K. and Woods, S. B. (1959). Electrical and thermal resistivities of the transition elements at low temperatures. *Phil. Trans. Roy. Soc.* **A251**, 273–302.

Wilkins, F. J. and Swann, M. J. (1970). Precision ac/dc resistance standards. *Proc. IEE*, **117**, 841–849.

Wolfendale, P. C. F., Yewen, J. D. and Daykin, C. I. (1982). A new range of high precision resistance bridges. *TMCSI*, **5**, 729–732.

Wood, S. D., Mangum, B. W., Filliben, J. J. and Tillett, S. B. (1978). An investigation of the stability of thermistors. *J. Res. NBS*, 1978, **83**, 247–263.

Ziman, J. M. (1972). "Principles of the Theory of Solids", 2nd Ed., Cambridge University Press.

Zurbuchen, J. H. and Case, D. A. (1982). Aging phenomena in nickel–manganese oxide thermodistors. *TMCSI*, **5**, 889–896.

6

Thermocouples

━━━━━━━━━━━━━━━━━━━━━━━━━━━━

6-1   Introduction

The range of use of thermocouples for the measurement of temperature is so extensive and the problems encountered so diverse that it is possible in this chapter only to touch upon a few of the more important aspects of thermocouple thermometry. The thermocouple remains one of the principal means of temperature measurement in much of industry, particularly the steel and petro-chemical industries. Advances in electronics have, however, allowed a much wider industrial application of resistance thermometers, and it would no longer be true to say that the thermocouple is the single most important industrial thermometer. The advantages of the resistance thermometer over the thermocouple (the measurement of resistance and thermal emf aside) are those that arise from the fundamental difference in the principle of operation of the two devices. The resistance thermometer gives an indication of the temperature in the region of the thermometer element, and the result is relatively independent of the leads and of the temperature gradient therein. The thermocouple, on the other hand, gives a measure of the temperature difference between the hot and cold junctions through a measurement of the potential difference between the two thermoelements at the cold junction. The potential difference is built up in the temperature gradient between the hot and cold junctions. For an ideal thermocouple, it is a function only of the temperature difference between the two junctions, but for a real thermocouple, inhomogeneities of one sort or another, which occur in the wires in the temperature gradient, must be taken into account and indeed constitute the limiting factor in the accuracy with which thermocouples can be used.

To obtain some understanding of these limitations we need to dwell briefly upon the theory of thermoelectricity. It is quite simple to show qualitatively how impurities, phase changes or lattice defects affect the emf of a thermocouple and we can draw certain conclusions as to how to anneal and handle thermocouples to obtain their best performance. Nevertheless, despite a broad understanding of the nature of thermoelectricity, it is not possible to predict with any useful accuracy the detailed thermoelectric behaviour of particular metals and alloys. This is, as we shall see, because thermoelectricity is a result of the energy dependence of the scattering of conduction electrons by the lattice, and for all but the simplest of metals, the shape of the Fermi surface, upon which quantitative calculations depend, is not sufficiently well known. For the multitude of practical thermocouple systems, therefore, the relation between temperature difference and thermal emf is an empirical one. For the most widely used thermocouples, this emf/temperature relation is embodied in internationally agreed standard reference tables based upon the performance of a representative sample of the currently manufactured alloys. In the case of the platinum 10 % rhodium and platinum 13 % rhodium *vs* platinum thermocouples, it is possible to buy wire for which thermoelectric emf up to 1100 °C does not depart from the reference table by more than about 6 μV, equivalent to half of a degree Celsius. We shall see later how the most recent international reference tables were derived and how they may be used to interpolate between fixed-point calibrations of real thermocouples.

In addition to the thermocouples for which there exist internationally agreed reference tables, a total of seven so-called standardized types, there are many other alloy combinations available. There is not the space in this chapter to list them all, in any case there are handbooks which do just this (see for example Kinzie, 1973; ASTM, 1974), but we will mention a few of the specialized or recently introduced alloy combinations designed to overcome limitations of range or performance encountered in the standardized types.

In industrial applications, sealed metal-sheathed thermocouples are very widely used. This construction is essential for those permanently installed thermocouples that would otherwise be destroyed by wear or chemical attack. In the glass industry, Pt 13 % Rh/Pt thermocouples sheathed in Pt 10 % Rh are used, while in the aircraft industry, inconel-sheathed chromel/alumel is common. In the nuclear power industry, the tungsten–rhenium thermocouple sheathed in molybdenum is being adopted as the standard type for use above about 1100 °C. The increasing demand for improved accuracy and long-term stability in industrial measurement has led to a considerable amount of work being undertaken aimed at obtaining a better understanding of the physical and chemical processes which occur inside the closed metal sheath of a sealed thermocouple. This type of thermocouple assembly is known as a "mineral insulated" or "MI" thermocouple.

Of great industrial importance is the whole range of so-called "extension" and "compensating" cables. These were developed to reduce the cost of large thermocouple installations in which many hundreds of thermocouples in a plant are connected to a central data-logging point. The extension or compensating cable is used between the data logger and the point, close to the particular machine or processing furnace, at which the temperature starts to rise significantly above ambient temperature. The thermoelectric properties required of the extension or compensating cable are simply that they match those of the thermocouple over the modest range of temperature in which they are used. The difference between an extension cable and a compensating cable is that the former is made of the same material, but to a much less stringent specification, than the thermocouple it is used with, while the latter is made of quite a different material.

At very low temperatures, the problems of thermocouple temperature measurement are different from those encountered at high temperatures. Chemical reactions between the thermocouple wires and a sheath can usually be ignored. Instead, difficulties arise because of the relatively weak thermopower of those parts of the thermocouple at low temperatures compared with that of those parts near room temperature.

The use of thermocouples in nuclear reactors poses many problems, not the least of which is that it is not uncommon to require a working life of an installed thermocouple of more than twenty years. The design and manufacturing techniques of thermocouples for reactor use are developing rapidly and the short section in this chapter is intended simply as an introduction to the special difficulties of operating thermocouples in a neutron flux. Before embarking upon a description of individual thermocouple types and their applications, however, we begin by introducing some elementary ideas concerning the origin of the thermoelectric effects that are observed when metals and alloys are placed in a temperature gradient.

6-2   Elementary theory of thermoelectricity

The origin of thermoelectricity in a metal can be demonstrated in a qualitative way using the simple free-electron-gas model of the conduction process. A brief introduction to the elementary theory of electrical conduction can be found at the beginning of Chapter 5. Although the electron-gas model does not lead to any useful quantitative results, it can form the basis of an understanding of the mechanism. Upon this can be built the more complex structure of the theory of thermoelectricity which includes the energy dependence of the scattering of the electrons by the lattice, the phenomenon of phonon-drag etc. etc. A very clear account of modern ideas concerning thermoelectricity will be found in Barnard (1972) upon which the very elementary account given in this section is based (see also Blatt et al., 1976).

We begin by imagining a metal to be made up of a rigid lattice of atoms between which can move a gas of free electrons, driven by whatever electric, magnetic or thermal gradients happen to be present. As a result of the temperature gradient in the conductor, the electrons at the hot end diffuse towards the cold end giving up, as they do so, some of their kinetic energy to the lattice. This is the process of thermal conduction. The excess of electrons thereby produced at the cold end leads to a potential gradient being established at the same time. The negative charge at the cold end builds up until a state of dynamic equilibrium is established between: (a) the number of electrons, having relatively high velocity, diffusing towards the cold end of the conductor driven by the temperature gradient, and (b) the number of electrons, having relatively low velocity, diffusing towards the hot end of the conductor driven by the electric potential gradient. This electric potential gradient arises whenever a temperature gradient exists in a conductor and is known as the thermoelectric emf. It follows that no thermoelectric emf can be present in the absence of a temperature gradient.

It is evident that the detailed mechanism of electron scattering must play an important role in thermoelectricity. One might guess, for instance, that electrons having a higher velocity would be scattered through smaller angles by the lattice atoms than those having a lower velocity; in this case an enhanced negative thermoelectric emf would develop. In other words, the mean free path of the electron would depend upon its kinetic energy. This is indeed the case, but the way in which the mean free path is related to the kinetic energy is not a simple one and it depends strongly upon the electronic structure of the lattice.

Despite the simplicity of qualitative explanations, the quantitative prediction of the thermoelectric properties remains elusive because of the complexity of the dependence of the mean free path upon energy. Expressed in another way, we can say that our knowledge of the shape of the Fermi surface (see Chapter 5, Section 1) of real metals remains inadequate for quantitative calculations of thermoelectric effects. It is worth remarking that for semi-conductors the situation is rather better because the number of electrons (and holes) which take part in the conduction process is much smaller. The electron gas model, which implies that the particles obey Maxwell–Boltzmann statistics, is thus much more realistic.

The practical use to which we put the thermoelectric emf is, of course, the measurement of temperature by means of the thermocouple. The complexity of the electron-energy vs scattering relation leads to the thermoelectric emfs in different metals being very different from one another. Were they the same in all metals and alloys, then we would not be able to make use of it for temperature measurement, since the thermocouple is a device in which the difference between the thermoelectric emfs in the two arms of the thermocouple gives an indication of the temperature difference between the hot and cold ends of the wires. This difference emf is known as the Seebeck effect.

The distribution of electric potential $E(T)$ in a pair of conductors of different material, A and B, having a junction maintained at a temperature T_2 and having free ends both maintained at a lower temperature T_0, is shown diagramatically in Fig. 6-1. Both the junction and the free ends of the conductors are shown in regions of uniform temperature and both conductors

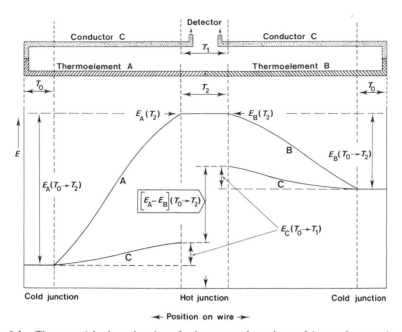

Fig. 6-1. The potentials along the wires of a thermocouple made up of thermoelements A and B having hot junctions in a region of uniform temperature T_2, connected to a pair of identical conductors C at cold junctions in a region of uniform temperature T_0. The conductors C are in turn connected to a detector in a region of uniform temperature T_1. Assuming that $E_C(T_0 \rightarrow T_1)$ is the same in both of the conductors C, the measured emf is $[E_A - E_B]\ (T_0 \rightarrow T_2)$. The thermoelements A and B pass through the same temperature gradient.

pass through the same temperature gradient. To permit measurements to be made of the thermoelectric potential difference between the free ends of A and B, a further pair of identical conductors C is attached at T_0 which lead to a detector at a temperature T_1. It is evident that the Seebeck effect is in no way a junction phenomenon but is, instead, temperature-gradient phenomenon. For a proper understanding of the behaviour of thermocouples, this cannot be overemphasized.

The rate of change of $E(T)$ with temperature is known as the thermoelectric power or thermopower $S(T)$ of the conductor, so that:

$$S_A(T) = \frac{dE_A(T)}{dT} \quad \text{and} \quad S_B(T) = \frac{dE_B(T)}{dT}. \tag{6-1}$$

The thermopower of the thermocouple shown in Fig. 6-1 is given by

$$S_{AB}(T_2) = S_A(T_2) - S_B(T_2) = \frac{d}{dT}[E_A(T_2) - E_B(T_2)]. \tag{6-2}$$

The thermoelectric potential difference between the free ends of the thermoelements, at temperature T_0, is

$$E_{AB}(T_0 \to T_2) = E_A(T_0 - T_2) - E_B(T_0 \to T_2) = \int_{T_0}^{T_2} [S_A(T) - S_B(T)] dT. \qquad (6\text{-}3)$$

$E_{AB}(T_0 \to T_2)$ is a unique function of T_0 and T_2 only if the thermoelements are homogeneous throughout the temperature gradient. The effect of the presence of an inhomogeneity is to add a small additional thermopower at the place in the thermoelement where the inhomogeneity occurs. Since thermopowers are approximately linearly proportional to temperature, an inhomogeneity which happens to be present near the top of a temperature gradient will have a larger effect on the measured thermoelectric emf than the same one near the bottom of the temperature gradient. In consequence, the thermoelectric emf of an inhomogeneous thermocouple will be a function of its position as well as the temperature difference between the two junctions.

In using a thermocouple to explore the temperature gradient in a furnace, there is thus always the problem of deciding whether or not the change in thermoelectric potential observed, as the thermocouple is moved, indicates the presence of a temperature gradient in the furnace or an inhomogeneity in the thermocouple. Fortunately it is usually possible to place an upper limit on the size of the effects of inhomogeneities. Provided the observed variations in emf as a function of position are larger than this upper limit they may safely be interpreted as indicating the presence of temperature gradients in the furnace. Smaller variations in emf cannot be distinguished from those which arise from inhomogeneities in the thermocouple wires and under these circumstances no conclusions can be drawn about the presence of temperature gradients in the furnace.

So far, in our discussion of the mechanism of thermoelectricity, we have mentioned only the Seebeck effect. Although as regards temperature measurement using thermocouples it is the Seebeck effect which is of the most concern to us, it is worth making a few remarks about the other manifestations of thermoelectricity, the Peltier and Thomson effects.

The Peltier effect is exhibited when a current flows across the junction between two dissimilar metals and leads to a release or absorption of heat which is proportional to the current. Thus, in principle, the passage of current across the junction of two dissimilar conductors can be used for cooling as well as heating, depending upon the direction of the current.

The Peltier coefficient Π_{AB} is defined as follows:

$$\Pi_{AB} = \frac{\dot{Q}_{AB}(T)}{J} \qquad (6\text{-}4)$$

where $\dot{Q}_{AB}(T)$ is the heat released per unit time at the junction of conductors A and B at a temperature T for a current density J. The heat released at one junction for a given current in a circuit containing conductors A and B will not equal that released at another unless the two junctions are at the same temperature. It should be remembered that for pure metals and alloys, the Peltier coefficient is very small and the heat released or absorbed at a junction is always much smaller than the Joule heating. This is not the case, however, for semiconductors which have very large Peltier coefficients; they can be used to provide many watts of cooling and the effect is of practical importance.

The third thermoelectric effect is the Thomson effect and is the release or absorption of heat on the passage of an electric current through a single conductor in the presence of a temperature gradient.

The evolution or absorption of this Thomson heat, which is clearly reversible, depends on both temperature and the relative directions of the current and temperature gradient and again in pure metals and alloys it is extremely small. The rate of heat production $\dot{Q}(T)$ per unit volume is given by

$$\dot{Q}(T) = \sigma J^2 - \mu J \nabla T \qquad (6\text{-}5)$$

where J is the current density, σ the resistivity and μ the Thomson coefficient. The first term of (6-5) is evidently the Joule heat, while the second is the reversible Thomson heat.

The general relationship between the thermopower, the Thomson coefficient and the Peltier coefficient is given by the Kelvin relations:

$$\mu = T \frac{dS}{dT} \qquad (6\text{-}6)$$

and
$$ST = \Pi. \qquad (6\text{-}7)$$

These relations allow any two of the thermoelectric effects to be found if one is known and if either μ or S can be determined over a small range of temperature about T. For practical methods of determining S, μ and Π, the reader is referred to Barnard (1972) or Blatt et al. (1976). In establishing the relations given above, Thomson assumed that the reversible processes (the Peltier and Thomson effects) could be treated independently of the simultaneous irreversible processes (Joule heating and thermal conduction). While leading to the correct result, his derivation made certain assumptions not contained in the first and second laws of thermodynamics. It was only in 1931 that a generalized treatment incorporating both the reversible and irreversible processes was developed (Onsager, 1931a,b). For an introduction to the work of Onsager, the reader is once again referred to Barnard (1972) since this is a subject beyond the scope of this book.

The quantitative prediction, from solid-state theory, of the magnitude of the thermoelectric effects has been the aim of a great deal of work, both theoretical and experimental, carried out over the past thirty years or so and which is continuing. At high temperatures, i.e. at temperatures well above the Debye temperature (≈ 300 K for most metals and alloys), the thermopower is given by the expression (Mott and Jones, 1936)

$$S = -\frac{\pi^2 k^2 T}{3e}\left[\frac{\partial(\ln \sigma(E))}{\partial E}\right]_{E=E_F} \tag{6-8}$$

where k is the Boltzmann constant, e the electronic charge and $\sigma(E)$ the electrical resistivity evaluated at the Fermi energy E_F (see Chapter 5). Expressed in terms of the electron mean-free-path l and area of the Fermi surface A, we can write $\sigma(E)$ in the following way

$$\sigma(E) = \left(\frac{e^2}{12\pi^3\hbar}\right)lA \tag{6-9}$$

This is equivalent to equation (5-1).

The thermopower may thus be written

$$S = \frac{-\pi^2 k^2 T}{3e}\left[\frac{1}{A}\frac{\partial A}{\partial E} + \frac{1}{l}\frac{\partial l}{\partial E}\right]_{E=E_F} \tag{6-10}$$

For metals having a spherical Fermi surface, $\partial A/\partial E$ is always positive and $\partial l/\partial E$ would also generally be positive — thus S would be negative. It has long been a problem to explain the positive thermopower of the noble metals, and it remains so despite a considerable body of knowledge built up about the shape of the Fermi surface. Recent work by Bourassa et al., (1978), which included new accurate calculations of $\partial A/\partial E$ based upon de Haas–van Alphen cyclotron mass measurements, led them to conclude that there must be a very strongly negative $\partial l/\partial E$. The final conclusion remains unclear, however, because experimental measurements of $\partial l/\partial E$ lead, in some cases to positive values.

As the temperature falls below the Debye temperature, other transport processes intervene; in particular, lattice waves or phonons, which up to now have been ignored, must be taken into account. It is these lattice waves or phonons that provide the mechanism for heat transport in non-metallic materials in which the free electron gas is absent. At low temperatures in pure metals and alloys thermal conduction by means of phonons becomes significant. There is thus a flux of phonons in which a phonon can interact with other phonons, electrons or impurities, and each species of interaction has its own mean free path. At high temperatures, the mean free path for phonon/electron interaction is very much larger than that for phonon/phonon interactions. Thus, from the point of view of the electron, the lattice is in equilibrium and the diffusion thermopower, which is proportional to T, is the main term in the total thermopower of metals and alloys. At low temperatures, the mean free path for phonon/phonon interaction exceeds that for phonon/electron interaction. The diffusion of the electron gas thus becomes subject to modification by means of the phonon/electron interaction which gives rise to an extra term in the thermopower known as "phonon drag". It produces a peak in the thermopower/temperature curve below room temperature which is present in all metals and alloys. At yet lower temperatures, the phonon drag term decays, because in metals it is proportional to the lattice-specific heat which, in turn, is proportional to T^3 at very low temperatures.

While being unable to predict the magnitude (and sometimes even the sign) of the thermopower and also Peltier and Thomson effects, solid state theory allows us to understand much of the observed behaviour of thermocouples. For example, the fact that the thermopower is pressure-dependent follows from the dependence of the Fermi level on the lattice constant. Similarly, changes in lattice structure, due either to vacancy formation or long- and short-range ordering will affect the thermopower. Since the thermopower is very sensitive to changes in electron scattering mechanism, impurities and strain would also be expected to affect its magnitude as, indeed, is the case.

Any further discussion of the mechanisms of thermoelectricity is beyond the scope of this book since our main purpose is to show how thermocouples can be used to measure temperature. The purpose of this introduction to the theory has been to show how it is that the thermopower of metals and alloys depends so critically upon the composition, homogeneity and state of anneal of the material. It may be remarked that in any well-arranged system for thermocouple temperature measurement, in which the junction of the two elements of the thermocouple is in a region of uniform temperature, the junction itself serves no other purpose than to ensure electrical contact between the two elements. The way in which it is made and whether or not there is alloy diffusion between the two arms in the region of the junction are irrelevant to the thermal emf which is developed in the temperature gradient.

6-3   Thermocouple types

6-3-1   Introduction

In this section we shall consider the various types of thermocouples that are available, give a summary of their more important characteristics, and give a few examples of some special applications. An exhaustive list of thermocouple types can be found in Kinzie (1973).

Although the number of different alloy combinations which have been used is immense, the number which are in widespread use is very much smaller, and these are the ones with which we shall be dealing. For the temperature range

20 K to 2000 K, there are seven different alloy combinations for which internationally agreed reference tables for thermal emf *vs* temperature exist. In addition to these, which make up the vast majority of thermocouples used in science and industry, we shall mention a few others which either have been developed for an important but very specialized use such as nuclear reactor measurements, or have been introduced too recently to have reached

Table 6-1. Compositions, trade names and letter designations for standardized thermocouples.*

Type designation	Materials
	Thermocouple combinations
B	*Platinum*-30 % rhodium/*platinum*-6 % rhodium.
E	*Nickel*-chromium alloy/*a copper*-nickel alloy.
J	Iron/another slightly different *copper*-nickel alloy.
K	*Nickel*-chromium alloy/*nickel*-aluminium alloy.
R	*Platinum*-13 % rhodium/platinum.
S	*Platinum*-10 % rhodium/platinum.
T	*Copper*/a copper-nickel alloy.
	Single-leg thermoelements
... N	Denotes the negative thermoelement of a given thermocouple type.
... P	Denotes the positive thermoelement of a given thermocouple type.
BN	*Platinum*-nominal 6 % rhodium.
BP	*Platinum*-nominal 30 % rhodium.
EN or TN	A *copper*-nickel alloy, constantan: Cupron[†], Advance[§], ThermoKanthal JN[‡]; nominally 55% Cu, 45% Ni.
EP or KP	A *nickel*-chromium alloy; Chromel[‖], Tophel[+], *T*-1[§], ThermoKanthal KP[‡]; nominally 90% Ni, 10% Cr.
JN	A *copper*-nickel alloy similar to but usually not interchangeable with EN and TN.
JP	Iron: ThermoKanthal JP[‡]; nominally 99.5 % Fe.
KN	A *nickel*-aluminium alloy: Alumel[‖], Nial[†], *T*-2[§], ThermoKanthal KN[‡], nominally 95 % Ni, 2 % Al, 2 % Mn, 1 % Si.
RN, SN	High-purity platinum.
RP	*Platinum*-13 % rhodium.
SP	*Platinum*-10 % rhodium.
TP	Copper, usually Electrolytic Tough Pitch.

Registered trade marks: † Wilbur B Driver Co. ‡ Kanthal Corp; § Driver–Harris Co; ‖ Hoskins Manufacturing Co.

In the above table an italicized word indicates the primary constituent of an alloy and all compositions are expressed in percentages by weight.

All materials manufactured in compliance with the established thermoelectric voltage standards are equally acceptable.

* Table from Guildner and Burns (1979).

Table 6-2. Nominal chemical composition of thermoelements.*

	Nominal chemical composition, % by weight									
Element	JP	JN, TN, EN[a]	TP	KP, EP	KN	RP	SP	RN, SN	BP	BN
Iron	99.5	...	...	...	...	...	...	...	...	...
Carbon	...[b]	...	...	...	...	...	...	...	...	...
Manganese	...[b]	...	...	...	2	...	...	...	...	...
Sulphur	...[b]	...	...	...	...	...	...	...	...	...
Phosphorus	...[b]	...	...	...	...	...	...	...	...	...
Silicon	...[b]	...	...	...	1	...	...	...	...	...
Nickel	...[b]	45	...	90	95	...	...	...	...	...
Copper	...[b]	55	100	...	...	...	...	...	...	...
Chromium	...[b]	...	...	10	...	...	...	...	...	...
Aluminium	...	...	...	...	2	...	...	...	...	...
Platinum	...	...	...	...	...	87	90	100	70.4	93.9
Rhodium	...	...	...	...	...	13	10	...	29.6	6.1

[a] Types JN, TN, and EN thermoelements usually contain small amounts of various elements for control of thermal emf, with corresponding reductions in the nickel or copper content, or both.
[b] Thermoelectric iron (JP) contains small but varying amounts of these elements.
* Table from ASTM (1974).

international acceptance. In the latter category are the Nicrosil/Nisil thermocouples which show every promise of being of great importance in the future.

The seven internationally-adopted thermocouples, the so-called "standardized thermocouples", are listed in Table 6-1 and 6-2, together with the nominal composition of each arm, the common trade names of each alloy, and the letter designation of the thermocouples. The letter designation was originally introduced by the Instrument Society of America, but has now gained world-wide acceptance as a convenient shorthand way of referring to the different thermocouple types. The thermocouple arm producing the more positive thermoelectric voltage is denoted by P and the other by N. Conventionally, the more positive alloy is given first when writing down the composition; for example, Pt-10% Rh *vs* Pt is written Pt-10% Rh/Pt.

6-3-2 *Platinum 67 and the absolute thermopower of platinum*

In order to standardize thermoelectric measurements and provide a reference against which the thermoelectric emf of various pure metal and alloy combinations may be judged, it has been found useful to adopt the thermoelectric behaviour of wire made from a batch of very pure platinum as a standard. This practice started in 1922 when work was being carried out at the Bureau of Standards in Washington to compare the behaviour of thermocouples made in different countries. We shall refer to this work again when discussing the origin of the Pt-13 % Rh/Pt thermocouple. At the time, wire from a particular melt of platinum (melt no. 27) was found to be more thermoelectrically negative than any previously obtained. Since impurities in platinum always lead to a more positive thermoelectric emf, it was realized that a particularly pure sample of platinum had been obtained. A small quantity of wire was therefore taken from this melt and subsequently became known as Pt-27. In the early 1960s, by which time all the Pt-27 had long been used up and, in any case, much purer platinum was then available, it was decided to re-establish a platinum reference wire. The result was Pt-67, the number this time referring not to the melt but to the year of fabrication, 1967. The specification of Pt-67 is now available in the NBS report SP 260-56 (1967), and samples of wires are available from the NBS Office of Standard Reference Materials as SRM-1967-Pt-67. The chemical composition of Pt-67 is given in Table 6-3. At temperatures below about 50 K, the thermoelectric power of

Table 6-3. Chemical composition of Pt-67, SRM 1967.[a]

| Element[b] | Concentration in parts per million by weight[c] | |
	Recommended value	Range of values reported
Copper	0.1	(0.087– <1)
Silver	<0.1	(<0.06– <1)
Palladium	0.2	(<0.1– <1)
Lead	<1	(0.6–3)
Iron	0.7	(0.6–2.6)
Nickel	<1	(0.3– <1)
Gold	<1	(<0.1 – 8)
Magnesium	<1	(<0.05–2)
Zirconium	<0.1	(<0.03–0.3)
Rhodium	<0.2	(0.09– <1)
Iridium	<0.01	(0.007–0.01)
Oxygen	4	(3.2–5.2)[d]
Platinum	(remainder −99.999+ %)	

[a] This is the same chemical composition that was certified for SRM 680, High-Purity Platinum, December 28, 1967.
[b] Other elements are also contained in the standards; some of them such as Al, Ca, Na, Si, and Sn may be certified at a later date.
[c] The values listed are based on a consideration of the analytical methods and results reported by co-operating laboratories. For all elements in SRM 1967, either because a single method was used or because of lack of agreement among methods, no estimate of accuracy can be made at this time.
[d] Range from one laboratory only.

platinum becomes too dependent upon trace impurities to be useful and an alternative NBS reference alloy of silver/gold, SRM-733, is recommended instead; this can be used down to 4 K.

In using a thermocouple to measure temperature and, indeed, in using Pt-67 as a reference, it is only the differences between the thermoelectric emfs in different metals and alloys that matter. The absolute value of the thermoelectric emf or the thermopower of a particular material is of secondary interest. Because of the dependence of thermopower upon electron-scattering mechanisms, however, it is of great interest theoretically. An absolute scale of thermopower exists based upon that of lead, a material chosen for its low thermopower. An ideal reference material would, of course, be one having a zero thermopower. Such an ideal reference exists below 20 K in the form of a superconductor, the thermopower of which is strictly zero. Above 20 K, direct measurements of absolute thermopower are not possible; instead, it is obtained through measurements of the Thomson heat $\mu(T)$ using the relation:

$$S(T) = \int_0^T \frac{\mu(T)}{T} \, dT. \tag{6-11}$$

which follows from equation (6-6).

Recent work by Roberts (1977) has shown that the absolute scale of thermopower above 20 K derived for lead by Christian et al. (1958) and based upon measurements of the Thomson heat carried out by Borelius et al. (1932) is wrong. The new scale differs from the old by nearly 0.3 μV/K and is shown in skeleton form in Table 6-4. It is of interest to note that the new scale shows that the phonon-drag component of the thermopower of the noble metals appears to extend to rather higher temperatures than was previously thought. The conclusion, based upon the old scale, that the phonon drag component for the noble metals is negligible at room temperature is probably incorrect.

6-3-3   Pt-10% Rh/Pt and Pt-13% Rh/Pt thermocouples (Types S and R)

The Pt-10 % Rh/Pt thermocouple (the Type S) used to be called the Le Chatelier thermocouple in honour of H. Le Chatelier (1886) who was the first to carry out a systematic investigation of the performance of thermocouples made from various alloys of platinum and rhodium. He concluded that the Pt-10 % Rh/Pt was the alloy composition most likely to lead to stable and reproducible measurements. His conclusions were readily adopted and it soon became widely manufactured and used. During the tests at the Bureau of Standards in 1922 on thermocouples made by a variety of manufacturers (Fairchild and Schmidt, 1922), it was found that those made in Britain

Table 6-4. Absolute thermoelectric power of lead.

T/K	$S/\mu V\ K^{-1}$	$E/\mu V$
0	0	0
7.18	0	0
7.18	−0.2	0.0
8.0	−0.256	−0.186
10	−0.433	−0.873
12	−0.593	−1.904
14	−0.707	−3.213
16	−0.770	−4.700
20	−0.779	−7.827
25	−0.721	−11.592
30	−0.657	−15.032
40	−0.575	−21.142
60	−0.527	−31.978
100	−0.583	−53.878
150	−0.708	−86.072
200	−0.834	−124.672
250	−0.948	−169.293
273.15	−0.995	−191.816
300	−1.047	−219.219
350	−1.136	−273.827

The figures in this table are taken from R. B. Roberts (1977).

contained 0.34 % of iron in the alloy arm. The source of the impurity turned out to be in the rhodium. By this time, however, many instruments had been calibrated in Britain in terms of the Pt-10 % Rh (0.34 % Fe)/Pt thermocouple, and there was considerable resistance to change. Instead, it was decided to increase the rhodium content by the amount required to reproduce the behaviour of the original thermocouple contaminated with iron. In this way the Pt-13 % Rh/Pt thermocouple was born! This thermocouple, now also called the Type R, has remained popular in Britain and certain other parts of the world in preference to the Type S. This has been not without good reason, because the Type R thermocouple appears to be significantly more stable than the Type S, as we shall see.

During an extensive series of measurements made at NPL, NBS and NRC between 1969 and 1971, it became clear that not only was the Type R more stable than the Type S, but the variations between thermocouples made to the same nominal composition by six manufacturers were much less. The reason for this is evident from Fig. 6-2 which shows the emf for a series of temperatures up to 1200 °C for a range of alloys of platinum with rhodium *vs* platinum. The rate of change of emf with rhodium content at a composition of 13 % rhodium is rather less than that at 10 % rhodium. Since most changes in emf of

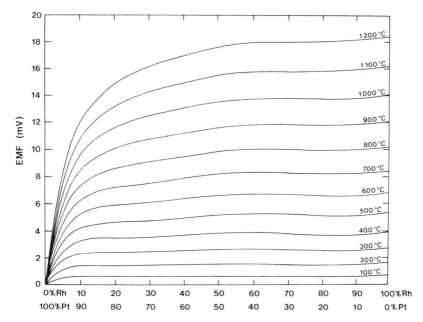

Fig. 6-2. The emf of rhodium–platinum alloys against platinum (after Caldwell, 1933).

platinum/rhodium thermocouples stem from changes in rhodium content of the alloy arm, it is clear why the Type R thermocouple is better than the Type S. More recent work by McLaren and Murdock (1979) indicates that the optimum alloy composition is for a variety of reasons, near 20 % rhodium. The main limitation in these alloys having a higher concentration of rhodium arises because of the danger of contamination of the pure platinum by rhodium carried from the alloy arm by means of the vapour phase. In addition, rhodium-rich alloys of platinum/rhodium become brittle after heating to high temperatures.

Many of the problems encountered with the prolonged use of platinum/rhodium thermocouples arise because of the preferential oxidation of rhodium above about 500 °C. This leads to depletion of the dissolved rhodium in the alloy arm and hence to a reduced thermal emf. The oxide dissociates at high temperatures and often the full thermal emf can be regained by heating for 30 min at 1250 °C that part of the thermocouple which has been in the temperature range 500 °C–900 °C. The oxide is much more volatile than either of the metals and, as we shall see, great care has to be taken if a thermocouple having a closed or tight-fitting sheat is to be used.

At the time the ITS-27 was established, the improvements to be gained by increasing the rhodium content of the alloy arm of the Le Chatelier thermocouple were not recognized. The Pt-10 % Rh/Pt thermocouple was therefore chosen as the interpolation instrument for the scale between 630 °C and the freezing point of gold, then 1063 °C. The present scale in this range is defined in terms of a quadratic equation having constants established by means of calibrations at a temperature of 630.74 °C and the freezing points of silver and gold. The best accuracy that can be obtained using a Type S thermocouple calibrated in this way is ±0.2 °C. The main limitation appears to be due to the effects of rhodium oxide and, apart from a change in rhodium content of the alloy arm, there is little that can be done to improve its performance (Bentley and Jones, 1980; McLaren and Murdock, 1979).

When the criteria for the selection of materials suitable for making thermocouples to establish the IPTS were drawn up, there was no information available on how closely the alloy composition had to be maintained. It was assumed that even a modest change in alloy composition would alter the shape of the emf/temperature curve so that the scale produced between the calibration points would differ from that obtained with the nominal Pt-10 % Rh alloy. The criteria at present appearing in IPTS-68 are as follows:

The thermocouple shall be such that the electromotive forces E (630.74 °C), E $(t_{68}(Ag))$ and E $(t_{68}(Au))$ satisfy the following relations

$$E \ (t_{68}(Au)) = 10\,334 \ \mu V \pm 30 \ \mu V \tag{6-12}$$

$$E \ (t_{68}(Au)) - E \ (t_{68}(Ag)) = 1186 \ \mu V + 0.17[E \ (t_{68}(Au)) -$$
$$10\,334 \ \mu V] \pm 3 \ \mu V \tag{6-13}$$

$$E \ (t_{68}(Au)) - E \ (630.74 \ °C) = 4782 \ \mu V + 0.63 \ [E \ (t_{68}(Au)) -$$
$$10\,334 \ \mu V] \pm 5 \ \mu V \tag{6-14}$$

Recent experimental work has shown that the limitation on the emf at the gold point of ±30 μV is far too restrictive from this point of view. It is equivalent to a restriction of rhodium content of ±0.07 %. Figure 6-3 shows the differences between the temperatures obtained from a quadratic equation fitted to the emfs at 630.74 °C and the silver and gold points of a number of Types S and R thermocouples (Bedford, 1972). In spite of the difference of 3 % in the rhodium content, which is equivalent to 1000 μV in $E(t_{68}(Au))$, the scales realized by the two types of thermocouple do not differ by more than 0.1 °C. Since the overall accuracy of the Type S is not as good as this it does not appear to matter whether Type S or Type R is used! If it had not long been the intention of the Consultative Committee for Thermometry to replace the thermocouple as interpolating instrument of the IPTS by the platinum resistance thermometer, the restrictions on alloy composition would, no doubt, have been modified (Bentley, 1969).

For purposes other than the setting up of IPTS-68 between 630.74 °C and the gold point, the calibration of Types R and S thermocouples has been very much simplified by the introduction of the new International Reference Tables

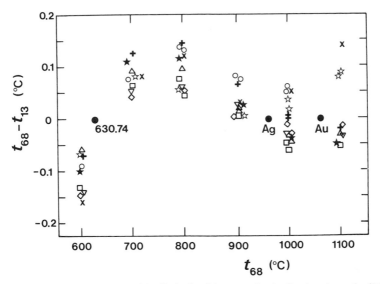

Fig. 6-3. Differences between t_{68} and "t_{13}" obtained from quadratics fitted to the emfs of Types S and R thermocouples respectively calibrated at 630.74 °C, 961.93 °C and 1064.43 °C. The mean results for each of nine different intercomparisons of three Type S and two Type R thermocouples are shown (after Bedford, 1972).

(IEC, 1977). These are discussed in more detail later and given in skeleton form in Appendix IV. Here it suffices to say that currently-manufactured thermocouple material produces an emf/temperature relation which is very close to that in the new Tables and therefore very few calibration points are needed.

The maximum temperature at which the Types R and S can be operated is set by the melting point of the pure platinum arm, 1769 °C. For general use, however, the upper limit is rather lower than this because the pure platinum arm becomes extremely soft at temperatures above 1600 °C. As with all types of thermocouple, the maximum operating temperature depends upon the size of the wire, the environment and the length of life required. For prolonged use and assuming the standard wire diameter of 0.5 mm, an upper limit of 1500 °C is usually recommended for oxidizing atmospheres. For short periods they may be used up to 1650 °C, for example, in the measurement of liquid steel temperatures. For this purpose disposable elements are used and dipped just once into the melt (Schofield, 1942; Oliver and Land, 1944).

6-3-4   The Pt-30% Rh/Pt-6% Rh (Type B) and other noble metal thermocouples

To extend the range of the Types R and S thermocouples, it is necessary to use an alloy negative arm instead of one of pure platinum. A number of different alloy combinations have been tried, but the one which has gained acceptance is Pt-30% Rh/Pt-6% Rh (Type B) and is among the group of seven standardized thermocouples. The melting point of the Pt-6% Rh alloy is well above 1800 °C, and a useful life can be obtained from Type B thermocouples in oxidizing atmospheres up to 1750 °C.

A major advantage of the Type B thermocouple is that up to 100 °C, its thermopower is practically zero and therefore no particular care need be taken in respect of the cold junction. The negative arm containing 6 % Rh is, of course very sensitive to changes in rhodium content and as a result this thermocouple is not quite so stable as either the Type R or the Type S.

Above the range of the Type B thermocouple other alloy combinations are available for specialized applications. The Pt-40 % Rh/Pt-20 % Rh can be used in oxidizing atmospheres up to 1850 °C, for example, although its thermopower is only about 4.5 $\mu V/°C$ between 1700 °C and 1850 °C. Its advantage over the Type B is that in addition to the higher maximum temperature, its mechanical durability and resistance to oxidation are superior. No standardized reference table exists, but a table of emf/temperature has been given by Bedford (1965).

For measurements in oxidizing atmospheres above the upper limits of Pt–Rh alloys, various combinations of Rh–Ir alloys have been proposed. Tables were prepared at NBS in the early 1960s (Blackburn and Caldwell, 1964) for thermocouples of Ir with Ir-40 % Rh, and also for various other Ir–Rh alloys with Ir. Recent experience at NBS with thermocouples made to these nominal compositions, however, does not give an emf/temperature relationship which conforms well with the tables (Burns and Hurst, 1975). The result suggests that modern Ir and Ir–Rh alloys are purer than those which were available twenty years ago. The range of application and life at high temperatures of these thermocouples are also very limited. At 2000 °C, the life of a thermocouple made from 0.8 mm diameter Ir and Ir–Rh alloys is between 10 and 20 h in an oxidizing atmosphere. The mechanical durability is also rather poor which, together with the high cost of the material, leads to their having little to recommend them.

6-3-5 Insulators and sheaths for use with noble metal thermocouples and the problems of contamination

It is not possible to discuss adequately the mounting and choice of refractory insulation for noble metal thermocouples without bringing in the problem of the contamination caused by the insulation or sheath. For temperatures up to the gold point in oxidizing atmospheres, recrystallized alumina, Al_2O_3, provides a perfectly adequate solution. It can be obtained very pure, it is strong and has a high electrical resistance. For higher temperatures, or inert or reducing atmospheres, it is much more difficult to find a suitable material. For industrial use, the choice of sheath is often dictated by the external environment rather than by the thermocouple elements.

A distinction should be made between those thermocouples that are made up by the user from wire supplied by the manufacturer and using whatever insulation and sheath is most convenient and those thermocouples that are made as complete sealed units by the manufacturer. For laboratory use the former are almost always used, largely because of cost and convenience. Sealed thermocouples (the so-called mineral insulated or MI thermocouples) are usually made in the following way: the thermocouple wires are packed in the powder of the insulating material inside a sheath and the whole is then swaged down to whatever diameter is finally required. Provided that proper care is taken in the choice of materials and cleanliness is maintained during assembly, a sealed thermocouple should perform as well as, and sometimes better than, one made in the usual way.

The Pt-10 % Rh/Pt thermocouple used to establish the IPTS-68 and for other precise laboratory applications is usually made as shown in Fig. 6-4. The wire diameter should be between 0.3 mm and 0.5 mm and annealed at 1250 °C in air for half an hour before threading into the twin-bore alumina insulator. The alumina insulator should previously have been passed through a furnace in air at 1200 °C. It is important that the alumina tubing be sufficiently long for there to be no gaps between successive sections in the temperature gradient. Otherwise, migration of rhodium to the pure platinum arm will take place, probably via the oxide phase, at temperatures above about 700 °C. Since twin-bore alumina of the appropriate diameter is available in lengths of 500 mm or more, there is usually no difficulty in meeting this requirement for standards laboratory work. The sheath within which the thermocouple is to be placed must also be made of recrystallized alumina heated in air to 1200 °C to remove grease, etc. Neither the wire nor the sheath should be touched with bare hands after annealing, to avoid the danger of contamination.

Attention should always be paid to ensuring that there is free access of air to thermocouples made in this way. It is well established that the exclusion of oxygen (Darling and Selman, 1972; Glawe and Szaniszlo, 1972), either through inadvertent lack of ventilation or its intentional replacement by an inert gas

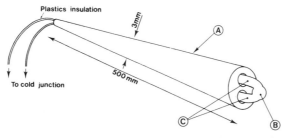

Fig. 6-4. The construction of a platinum/rhodium thermocouple for standards work. *A* is a single length of re-crystallized Al$_2$O$_3$; *B* is the welded junction of the thermoelements *C*.

such as argon or nitrogen, leads to reduction of the alumina and whatever other oxide happens to be present and to contamination of the thermocouple by the metal thus released. For a long time, the contamination which always seemed to result from the exclusion of a free flow of air by means of a close-fitting sheath was taken to imply that a platinum-sheathed fully-sealed platinum/rhodium thermocouple could never be as stable as that made according to the classic design of Fig. 6-4. That this is not necessarily the case was shown by the long series of investigations by Selman (1972) and Darling and Selman (1972), who initially set out to discover the reason for some gross changes in thermoelectric voltage in platinum thermocouples operated in oxygen-depleted atmospheres. Figure 6-5 shows the result of electron-probe microanalysis of the sheath and thermoelements of a magnesia-insulated Type R thermocouple sheathed in Pt-10 % Rh. It had previously been found that MgO is the refractory least likely to react with platinum or platinum alloys. Although almost completely inert with respect to most refractory oxides in air, platinum and its alloys react with them when the partial pressure of oxygen is reduced below a certain level. Alumina, zirconia and thoria dissociate liberating oxygen and the metal which dissolves in the platinum. The thermocouple illustrated in Fig. 6-5 had been kept at 1450 °C for 1400 h, after which time the thermoelectric voltage had dropped by the equivalent of some 200 °C. It is evident that a considerable amount of rhodium has migrated to the pure platinum arm at the expense of both the Pt-13 % Rh arm and the Pt-10 % Rh sheath, although mostly from the latter presumably because it is present in much greater volume. In those regions of the thermocouple where the temperature was below 1200 °C, the transfer of rhodium was very small.

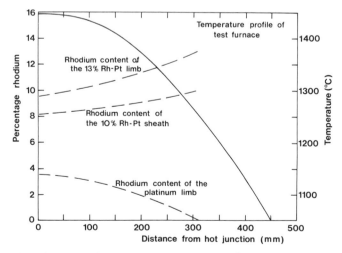

Fig. 6-5. Rhodium distribution in the sheath and elements of a platinum-13 % rhodium/platinum thermocouple clad with a platinum-10 % rhodium sheath, after heating for 1400 h at a hot-junction temperature of 1450 °C (after Selman, 1972).

The rates of loss of platinum and rhodium from the surface of the bulk alloy are a function of the vapour pressures, and these depend upon both the temperature and the composition of the metal. For platinum and rhodium over platinum-10 % rhodium they are shown in Fig. 6-6. It is striking that, in air, material is lost by means of oxide formation and its subsequent evaporation rather than by the direct evaporation of the metal. Whether or not the atmosphere is stagnant above the platinum or the alloy will clearly make a significant difference to the amount of material lost. In this respect, the construction of the thermocouple and the arrangement of insulators and sheath are very important. Figure 6-7 shows, in a schematic way, the various ways of making a thermocouple. That least likely to lead to contamination, but usually the least practical, is Fig. 6-7(a), the in-line thermocouple with the air flow in the direction of the alloy arm. The second configuration, 6-7(b), again not very practical, is that of the bare wires parallel to each other without insulation. The transfer of rhodium to platinum via the rhodium oxide phase will depend upon the direction of the convection currents. Thermocouples of approximately this form, but operated vertically with the junction below, are often used for calibrations by means of the "wire point" method (see Section 6-5). Care should be taken that the two arms are sufficiently well separated for

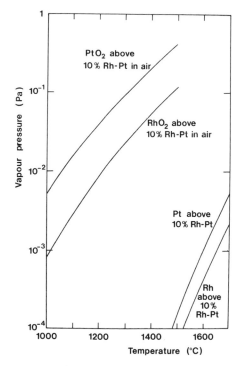

Fig. 6-6. The vapour pressure as a function of temperature of rhodium, platinum and their oxides above a platinum-10 % rhodium surface in air (Selman, 1972).

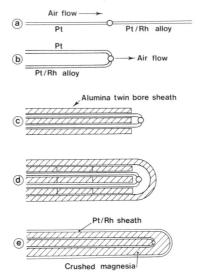

Fig. 6-7. Various constructions of Pt/Rh thermocouples; for description see text.

convection from the alloy arm not to flow over the pure platinum arm. A better arrangement, in fact, might be for the thermocouple to point upwards with the junction at the top. The third example, Fig. 6-7(c) we have already discussed and is that which is commonly used in standards work. The precaution to be taken here is, as we have said, to make sure that the refractory insulator is continuous in those parts of the thermocouple where the temperature is above about 400 °C. Figure 6-7(d) shows a thermocouple similar to that of Fig. 6-7(c), but having a tight-fitting closed-ended sheath and rather short (60 mm) lengths of twin-bore insulation. Measurements by Selman (1972) of the composition along the length of the wires of such a thermocouple showed that above 1100 °C, considerable quantities of rhodium were transferred to the Pt arm, even in those regions between the gaps in the twin-bore insulators. Presumably this was the result of transfer through the vapour phase along the bores of the insulator.

The fifth example, Fig. 6-7(e), is that of a metal-clad, magnesia-filled thermocouple. The composition of the sheath may be chosen in some cases to minimize contamination of the thermocouple. For example, if a Type B thermocouple is being used, the sheath can, with advantage, be made of Pt-5% Rh, although Pt-10% Rh would give better high-temperature strength. If Types R or S are being used, the optimum sheath composition from the point of view of contamination would probably be pure platinum, but in practice this would be too soft, and so Pt-10% Rh is usually chosen. Care must also be taken to avoid cracks and fissures in the insulator. Powder-insulated thermocouples are only satisfactory if the manufacturing techniques and subsequent handling are such as to lead to a continuous barrier between the elements and sheath. In the case of MgO insulation, because of its very low reactivity with platinum and its alloys, the air filling can be replaced by an inert gas such as argon. This will, according to Fig. 6-6, reduce even further the rate of transfer of rhodium to the pure platinum arm of the thermocouple. Thermocouples made in this way show practically the same long-term stability at high temperatures as do those made, for example, according to the design of Fig. 6-7(c).

In all of this it should be remembered that there are two processes operating which can lead to degradation of the thermocouple. The first is contamination by means of metals released through the reduction of oxides present as insulators and sheaths, and the second is the transfer of rhodium via the oxide vapour phase to the pure platinum arm. The former can be eliminated by maintaining an oxidizing atmosphere or, in the case of a low partial pressure of oxygen being necessary, by using MgO insulators. The latter can be suppressed by ensuring that a physical barrier exists between the two arms of the thermocouple or by reducing the partial pressure of oxygen.

6-3-6 Base-metal thermocouples

Base-metal thermocouples (that is, thermocouples made from metals other than the noble metals) have higher thermopowers than noble metal thermocouples. They cannot be used to such high temperatures, however, because of the relatively low melting point of their constituents and the rapidity of failure due to oxidation. The standardized base-metal thermocouples, Types E, J, K and T, are the ones most widely used in industry. They can be mounted and sheathed in any number of ways depending upon the application and environment. For details of the recommended wire sizes, sheath material, insulators, etc. demanded for industrial use, the relevant national standards or specifications should be consulted (see for example ASTM, 1974). The following brief description of the performance of each of the base-metal thermocouples should be supplemented by the more detailed information available in Kinzie (1973) and in the references cited.

Type E, Ni–Cr/Cu–Ni

This thermocouple is known for its very high thermopower, 58.5 μV/°C at 0 °C rising to 81 μV/°C at 500 °C. Despite this, its use in industry is very much less widespread than is that of the other standardized base-metal thermocouples. This is probably due to its having been considered very similar to the Type K but having a lower maximum operating temperature. At temperatures above 800 °C, the negative arm (Cu–Ni) oxidizes much more rapidly than that of the Type K. Nevertheless, at temperatures below 800 °C, the Type E thermocouple has a lower drift rate than the Type K. This is because the changes in thermopower of the two arms of the thermocouple due to oxidation are in the same direction and so there is a measure of cancellation. For example, 1000 hours of operation in air of a Type E thermocouple at 760 °C, having 3 mm wires, should not lead to a change in emf equivalent to more than 1 °C. This is rather better than that to be expected from a Type K. For low-temperature use (down to −250 °C), the Type E is to be recommended since the thermal conductivity of both arms is low and the alloys can be manufactured with good homogeneity without too much difficulty.

Type J, Fe/Cu–Ni

This is one of the most widely used thermocouples in industry, due mainly to its relatively high thermopower and low cost. The temperature range of useful application is from 0 °C to 760 °C. Below 0 °C, the two problems of rust and embrittlement reduce its usefulness, while above 770 °C, the iron undergoes a magnetic transformation. Above 550 °C, traces of sulphur cause large changes in the thermopower of Type J thermocouples.

Type K, Ni–Cr/Ni–Al

This alloy combination is also very widely used in industry. It has a high thermopower and good resistance to oxidation up to 1260 °C, but is rather poor· if used in reducing atmospheres. It can be used successfully at temperatures as low as 4 K and like the Type E, has a low thermal conductivity in both arms. The principal advantage of the Type K over the other base-metal thermocouples is that it has a considerably better resistance to oxidation at high temperatures. Operation in mildly reducing atmospheres leads to the formation of a green oxide of chromium on the positive arm with consequent large changes in thermopower. The effect is most likely to be severe in the range from 800 °C to 1050 °C. This type of thermocouple is also sensitive to attack in sulphureous or carbonaceous atmospheres.

Although the Type K thermocouple withstands oxidation better than the other base-metal thermocouples, and in other respects also it is more versatile, it suffers from two principal defects (Fenton, 1969; Burley 1972). The first is a change in emf due to inhomogeneous short-range ordering in the Ni–Cr alloy which takes place rapidly at temperatures between 250 °C and 550 °C. The second is a long-term drift which takes place at high temperatures and is the result of progressive internal oxidation of the minor reactive solutes in the Ni–Al alloy (see Figs 6-8 and 6-9). Both of these effects can be reduced by relatively small changes in alloy composition and, as we shall see below, considerable improvement in performance can be obtained.

Type T, Cu/Cu–Ni

For general applications, this alloy combination can be used in oxidizing or inert atmospheres over the temperature range −250 °C to about 850 °C. The

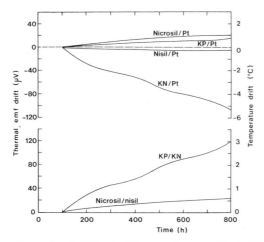

Fig. 6-8. Long-term thermal-emf drifts in 3.3 mm diameter Nicrosil/Nisil and Type K thermocouples and of their individual thermoelements *vs* platinum on exposure to air at 1000 °C (after Burley *et al.*, 1978).

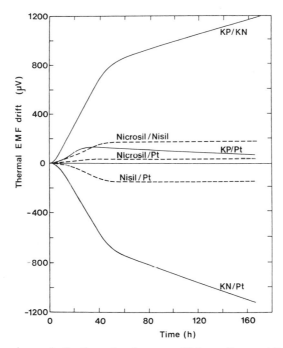

Fig. 6-9. Short-term changes in the thermal emf outputs of 3.3 mm diameter Nicrosil/Nisil and Type K thermocouples and of their individual thermoelements versus platinum on exposure in air at 1250 °C (after Burley *et al.*, 1978).

lower limit for this and for other base-metal thermocouples is simply the temperature below which other alloys have a higher thermopower. The upper limit depends, of course, on wire size but is restricted by the oxidation of the copper element. In reducing or mildly-oxidizing atmospheres, it is possible to use the thermocouple up to nearly 1000 °C, although at these temperatures its useful life is rather restricted. A useful attribute of the Type T thermocouple is its relatively high thermopower, 40 μV/°C, which remains practically constant over a wide range of temperature. At very low temperatures, the high thermal conductivity of copper is often a disadvantage although in some applications its resistance to corrosion under damp conditions more than compensates for this.

Other base-metal thermocouples

There exists a very large number of other alloy combinations for thermocouples which have arisen from the need to develop an alloy composition that will have a better resistance to a particular environment. Many of these are described by Kinzie (1973).

Outside the range of standardized thermocouples, that is, for temperatures below about 20 K and above 2100 K, there are already-existing alloys which are well characterized. In addition, the nuclear power industry requires thermocouples for a wide temperature range having good long-term stability in the presence of a neutron flux. These and other special applications will be mentioned below. First, however, we should consider the recently-developed alternative to the Type K thermocouple: the Nicrosil/Nisil thermocouple.

6-3-7 Nicrosil/Nisil alloys

The problems of the Type K thermocouple, although not apparently inhibiting its widespread use in industry, undoubtedly lead to serious errors in temperature measurement, Figs 6-8 and 6-9. For very critical applications, there has long been a need for a base-metal thermocouple which can be used in air up to about 1200 °C and which is not subject to the large changes in thermoelectric emf shown by the Type K. New alloys have now been produced, which show a much-improved performance over the traditional type K (Burley, 1972; Burley et al., 1982).

The main differences between the old and new alloys are the following: the alloy solute levels are above those required to cause the transition from internal to external oxidation, and minor solutes have been selected which oxidize preferentially to form impermeable diffusion barrier films. These two changes reduce the slow drift at high temperatures. The more rapid changes, which are due to short-range ordering in the Ni–Cr alloy, have been suppressed by increasing the solute levels to those at which the ordering no longer takes place. The composition of the new alloys is as follows:

Nicrosil: Ni – 14.2 % Cr – 1.4 % Si

Nisil: $Ni - 4.4 \% Si - 0.1 \% Mg$.

Experimental work has now shown that these alloys can be used to higher temperatures and are more resistant to oxidation than the Type K alloys and are virtually free from the effects of short-range ordering. Full details of the work which led to the development of Nicrosil and Nisil together with reference tables will be found in Burley *et al.* (1978). The thermopower of Nicrosil/Nisil is rather lower than that of the Type K and thus the two types of thermocouple are not interchangeable. A skeleton table is given in Appendix VI.

6-3-8   *Thermocouples for high temperatures*

For temperatures above the upper limit of the platinum–rhodium types, there are no thermocouples that can withstand oxidizing atmospheres for any length of time. Temperature measurements under these conditions are therefore no longer possible using thermocouples, and recourse must be made to the radiation or perhaps the noise pyrometer. If the atmosphere is non-oxidizing, however, there is a range of thermocouples based upon tungsten–rhenium alloys which have shown good performance up to 2750 °C and which can be used, for short periods, up to 3000 °C.

The alloy combinations that are available for tungsten–rhenium thermocouples have the following compositions:

$$W - 26 \% Re/W$$

$$W^* - 3 \% Re/W - 25 \% Re$$

$$W^* - 5 \% Re/W - 26 \% Re$$

The asterisk indicates that the tungsten contains small quantities of other metals to inhibit grain growth, which otherwise leads to serious room temperature embrittlement after heating to temperatures above 1200 °C (Burns and Hurst, 1972). At temperatures above about 1900 °C, the vapour pressure of rhenium is sufficient for substantial losses to occur if a bare wire thermocouple is being operated in vacuum. For most industrial applications, the mineral insulated (MI) type assembly is used. Both molybdenum and tantalum are commonly used as sheaths with beryllia or magnesia insulation. Figure 6-10 shows the variation of electrical resistivity of these and other insulating materials.

Tungsten–rhenium thermocouples are suitable for use in high-purity inert atmospheres, hydrogen or vacuum with the limitation mentioned above for vacuum use. It is recommended that new thermocouple wires be annealed

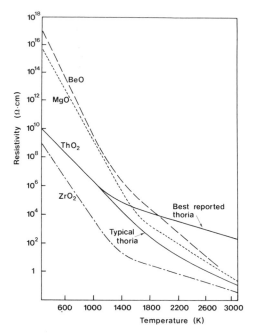

Fig. 6-10. The electrical resistivity of various refractory oxides at high temperature (after Shepard *et al.*, 1972).

before use to stabilize the grain size. This can be accomplished by heating in an inert atmosphere to 2100 °C for periods varying from 1 h for W-3 % Re to a few minutes for W-25 % Re. This annealing procedure also reduces the rate of formation of the σ-phase intermetallic compound in the W-25 % Re alloy which otherwise precipitates out in that part of the wire kept for long periods in the temperature range 800 °C to 1300 °C. A reference table for thermoelectric emf/temperature has been proposed for adoption as an ASTM standard (ASTM, 1974) but has not yet been formally adopted. One of the most important applications of tungsten/rhenium thermocouples is its use in the nuclear power industry for the measurement of temperatures under conditions of strong neutron flux. This will be discussed in more detail below.

6-3-9 Thermocouples for low temperatures

The use of thermocouples at low temperatures requires more care to be taken in the selection of material than is the case at high temperatures (Hurst *et al.*, 1972). The increased role of phonons and their scattering mechanisms leads to the thermopower being even more dependent on impurities and inhomogeneities than is the case at high temperatures, where the thermopower is almost solely a function of electron/lattice scattering. In addition, the thermopower at low temperatures is generally rather low and so the proportion of the total measured thermoelectric emf which originates in the low temperature part of the thermocouple is proportionally low. Inhomogeneities, therefore, in those parts of the wire just below room temperature, where the thermopower may be 50 μV/K, can cause spurious emfs which are very troublesome if the thermopower at, say, 20 K is only 3 μV/K.

As is evident from Fig. 6-11 there are alloys having a thermopower which does not fall quite so much at low temperatures. These are the alloys which exhibit the Kondo effect (see Chapter 5, Section 6) in the scattering of conduction electrons by the magnetic moment of the minor constituent, usually iron or cobalt. Relatively large negative thermopowers can be obtained over a range of temperatures from 1 K to 300 K. The positive arm of the thermocouple is usually an alloy having a low thermal conductivity and small thermopower such as Ni–Cr or Ag–0.3 % Au. The minor constituent of the negative arm now considered best from the point of view of thermoelectric stability is iron. Alloys of cobalt which were used at one time are subject, at room temperature, to structural changes which lead to changes in thermoelectric power. The proportion of iron can be in the range 0.02 atomic percent (At. %) to 0.07 At. %. The thermopowers of various Au–Fe alloys together with those of Type E and K thermocouples are shown in Fig. 6-11. The thermal conductivities of these alloys as a function of temperature is given by Hurst *et al.* (1972).

The choice of iron concentration in the gold/iron alloy depends upon how the thermocouple is going to be used (Berman and Kopp, 1971). Because magnetic fields are frequently present in low-temperature experiments, the dependence of thermopower on magnetic field is an important consideration. Since, at low temperatures, the thermopower is a

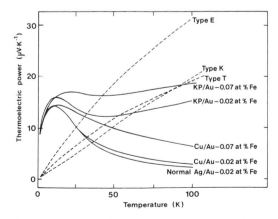

Fig. 6-11. Thermoelectric power of some low temperature thermocouples (after Burns and Hurst, 1975).

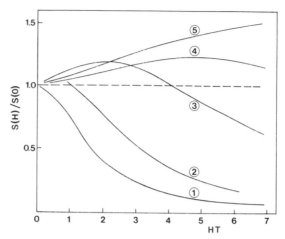

Fig. 6-12. The effect of a magnetic field, H, on the thermopower, S, of Au–Fe thermocouples for various temperatures and iron concentration. (1), 10 ppm, 0.44 K; (2), 110 ppm, 0.40 K; (3), 300 ppm, 1.1 K; (4), 300 ppm, 3.43 K; (5), 1900 ppm, 0.85 K (after Berman, 1972).

result of scattering by the magnetic impurities in the gold, it is clear that it will be strongly influenced by external fields. In fact, the dependence of thermopower upon external fields is much more dependent upon iron concentration than is the thermopower itself. Figure 6-12 shows the effect of a magnetic field upon the thermopower of a number of different alloys of Au–Fe. It shows that alloys having a concentration of iron between 0.03 At. % and 0.1 At. % would be least sensitive to external fields. It should be remembered that, because the thermopower is a result of the temperature gradient a magnetic field can only influence the thermopower of a wire if the wire is in a temperature gradient!

Reference tables for Ni–Cr/Au-0.02 At. % Fe and Ni–Cr/Au-0.07 At. % Fe have been published for the range 1 K to 280 K (Powell *et al.*, 1974). For use in very strong magnetic fields, greater than 8 tesla for instance, Type E thermocouples are recommended. At temperatures below 1 K, the difference between the thermopowers of Ni–Cr and Au–Fe alloys becomes very small and so the thermoelectric voltages developed become small. If, however, the Ni–Cr arm is replaced by a superconducting material such as niobium (the thermopower of which is strictly zero) it is possible to use the thermocouple down to temperatures as low as 0.05 K.

6-3-10 Thermocouples for nuclear reactor applications

Despite the huge investment in nuclear power which has taken place over the past thirty years, it is only in recent years that any serious effort has been made to standardize the techniques necessary to manufacture reliable thermocouples for the high-temperature nuclear reactor environment. The reader is referred to the comprehensive discussion on this in the Proceedings of the Joint Petten–Harwell Colloquium on nuclear reactor thermometry (Petten–Harwell, 1975; and also Thurlbeck (1982). There has been conflicting evidence about the stability of thermocouples used in a neutron flux, and it now seems likely that a lot of this has arisen because of poor control over the manufacture of the thermocouples themselves. This was obscured by the complex nature of the causes of the poor reliability so often observed.

There are two types of thermocouple used extensively in the nuclear power industry, both always in the MI form, the first is the Type K which has an upper limit of about 1100 °C, and the second is the tungsten–rhenium. The latter is either the W*-5 % Re/W-26 % Re or W*-3 % Re/W-25 % Re, both of which have been used up to 2000 °C in nuclear reactors (Cannon, 1982). It is now clear that contamination during manufacture is one of the major causes of failure and excessive drift at high temperatures. Of particular importance is the purity of the refractory, not only in the body of the material but at its surface. The effects of transmutation of the elements of the thermocouple which result from neutron bombardment are severe. They are very difficult to correct for because they lead to a changing composition in the temperature gradient. The thermal emf of the thermocouple will thus be critically dependent upon the position of the temperature gradient with respect to the composition gradient.

Re-calibration of such a thermocouple carried out with the aim of correcting for the drift due to composition changes would need to be done in a furnace having the same temperature gradient as that which the thermocouple experienced in the reactor. The difficulties in verifying that this is the case make the prospect of such a re-calibration being successful rather remote.

In tungsten–rhenium thermocouples, the neutron flux leads to the transmutation of tungsten to rhenium, and rhenium to osmium. Both of these changes have large effects on the thermopower but, fortunately, of opposite sign and so there is a measure of cancellation. An indication of the magnitude of the changes experienced in using a tungsten–rhenium thermocouple for temperature measurement in a neutron flux is given in Fig. 6-13.

The general form of the thermocouple assembly which is increasingly being used in nuclear environments is illustrated in Fig. 6-14 (Coville, 1975; Mason, 1972). The choice of beryllia as insulator and molybdenum as sheath material has been made largely for reasons of compatibility. Thermocouples of this type are not swaged down to size as is usually the case with MI thermocouples. The insulators are made of individual beads of hard-fired BeO threaded onto the thermocouple wire. Before being used, the berrylia beads are heated in air to 2000 °C for 12 h to reduce impurities. Although niobium and tantalum have been used as sheath materials, they have a lower resistance to oxygen permeation at high temperatures than does molybdenum. The twisted, rather than welded, junction leads to less grain growth and therefore fewer failures due to fracture at the junction (Anderson and Ludwig, 1982). For that part of

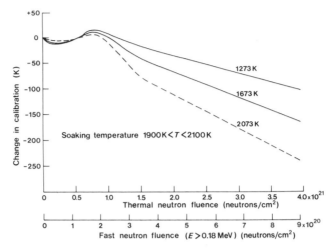

Fig. 6-13. The effect of neutron irradiation on W-3 % Re/W-25 % Re thermocouples (after Heckelman and Kozar, 1972).

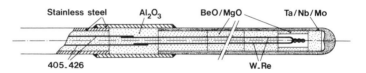

Fig. 6-14. Tungsten–rhenium thermocouple assembly for use in a nuclear reactor at high temperatures (after Coville, 1975).

the thermocouple which is to remain outside the neutron flux and at temperatures below 700 °C, the sheath is stainless steel, the insulant is magnesia and the thermocouple wires themselves are the appropriate extension wires or compensating wires.

In the assembly of high-temperature tungsten rhenium thermocouples, a clean dust-free environment is essential as well as scrupulous exclusion of all organic residues which might result from handling. Vacuum degassing and glove-box assembly under an inert gas atmosphere have been found necessary (McCulloch and Clift, 1982). All of this of course leads to a thermocouple that is expensive to produce.

6-4   Extension and compensating wires

There are many large and extensive thermocouple installations in which the measuring junctions and reference junctions are widely separated. For example, each of a series of furnaces may be instrumented with ten or more thermocouples connected to a data-logging system in a central control room a hundred metres or more away. The thermal emf which is measured is probably almost entirely produced in the first few metres of wire. The remaining hundred metres or so serve mainly to transmit the emf to the measuring system. The thermoelectric properties of the long length of wire near room temperature, or at least at a temperature less than 100 °C, are very much less critical than those of that part of the wire in the steep temperature gradient. Considerable economies can therefore be gained by using, in this less critical section, not the high specification thermocouple wire but another, cheaper, wire whose thermoelectric properties are a reasonable match over the temperature range say from 20 °C to 100 °C.

As mentioned at the beginning of this chapter, there are two types of such extension wires: those having the same nominal composition as the thermocouple wire itself but which are not made to such a high specification and those made of a different alloy altogether. Extension wires of the former category are widely used in industry and are available for Types E, J, K and T. Wires in the second category are usually called "compensating" wires or cables in the UK and are most commonly used with Types R, S and B. Since the thermopower of Type B thermocouples is practically zero up to 100 °C, plain identical copper wires can serve as compensating wires for this type.

Compensating wires for Types R and S are usually made of copper for the positive arm and a copper–nickel alloy for the negative arm and can be used up to 200 °C without leading to an error of more than about 6 °C. Recently, an improved compensating wire has been suggested (Bugden *et al.*, 1975) in which errors of less than 2 °C would be incurred at temperatures as high as 500 °C. The improvement is obtained by using a three-wire system as shown in Fig. 6-15. The negative wire is made from two stranded stainless steels of similar composition braided together, and the positive wire is made from a Ni-20 % Cr-10 % Fe alloy. According to Budgen *et al.*, the criteria for choosing the materials for a successful three-wire compensating system are the following: (1) the temperature/thermal emf characteristic of A *vs* C and B *vs* C (Fig. 6-15) should straddle that of the Pt/Rh thermocouple with which the compensating system is to be used; (2) the ratio of the electrical resistances of A to B should equal the ratio of the deviations of the thermoelectric emfs of the two pairs, A/C and B/C, to that of the Pt/Rh thermocouple. By using stranded wires for the elements A and B, the resistance ratio can easily be adjusted by changing

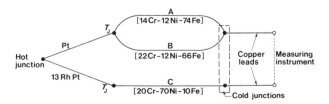

Fig. 6-15. A three-wire compensating lead system for Pt-13 % Rh/Pt thermocouples (after Budgen *et al.*, 1975).

the relative number of strands of each element. There appear to be many advantages in using the three-wire compensating cable, the most important of which is that a much wider range of thermoelectric behaviour becomes accessible.

6-5   The calibration of thermocouples

6-5-1   Standard reference tables

For the calibration of thermocouples, as with most other thermometers, there are a number of possible ways of proceeding. For instance, the emf of the thermocouple may be determined at a relatively small number of fixed points and interpolation carried out by means of agreed formulae or by difference from a standard table. Another method is to compare the emf of the thermocouple being calibrated with that of a standard thermocouple of the same type at a relatively large number of temperatures and then either fit a curve to the differences from the calibration of the standard thermocouple or fit directly the emf *vs* temperature using a least-squares method. For those thermocouple types that do not have a standard reference table, calibration requires comparison with another type of thermocouple, or type of thermometer, which has already been calibrated. The comparison must be made over the whole temperature range of interest and at a sufficient number of points for a proper calibration curve to be fitted to the data.

Standard reference tables play a very important role in thermocouple thermometry and lead to considerable savings in time and effort. A standard reference table represents the behaviour of a typical example of a particular type of thermocouple. The calibration of an individual example of this type of thermocouple is thus reduced to determining the difference between its behaviour and that of the standard embodied in the reference table. Provided that the original work which led to the reference table was well done, and that the manufacturer has maintained the composition of the alloys used in the preparation of the table, it will be found that the differences are very small. The number of calibration points necessary to determine adequately the differences from the standard reference table will be correspondingly small and the whole process will be simple and economic.

Let us take as an example the use of the standard reference table of the Type R thermocouple. The table itself (IEC, 1977) is also given in polynomial form (Appendix V) and requires a seventh order polynomial for the range $-50\,^{\circ}\mathrm{C}$ to 630 °C and a cubic polynomial from 630 °C to 1064 °C. We shall return a little later to the question of simpler, approximate, mathematical expressions for this and other reference tables. Figure 6-16 shows differences from the standard reference table measured for a number of recently manufactured thermocouples. The differences were measured (Coates, 1978a) at the freezing

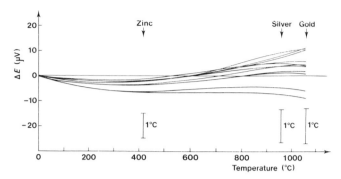

Fig. 6-16. Difference curves between the IEC Standard Reference Table and the emfs of a series of Pt-13 % Rh/Pt thermocouples. The curves are based on the results of calibration of the thermocouples at the freezing points of zinc, silver and gold (Coates, 1978a).

points of zinc ($\sim$419 °C), silver ($\sim$960 °C) and gold ($\sim$1064 °C). The accuracy of the measurement was estimated to be $\pm$0.2 °C. It is clear that a quadratic expression of the differences is perfectly adequate to give a calibration of each thermocouple within the limits of accuracy of measurement. By comparing this quadratic with the expressions required to describe the standard reference table, it is obvious that a considerable amount of information is embodied in the reference table. From Fig. 6-16, one can also deduce that if the accuracy of calibration required is only 0.5 °C, then calibration at the silver point alone with linear interpolation to 0 °C and extrapolation to 1064 °C is all that is strictly necessary. An additional calibration point is of course desirable for checking purposes.

The standard reference tables for all of the common thermocouple Types R, S, B, E, J, K and T, which are given in the national and international standards mentioned above, were unified following experimental work at NBS, NRC and NPL carried out between 1969 and 1971 (Bedford et al., 1972; Powell et al., 1974). Prior to this, the national standards differed from one another by amounts depending upon the origin of the experimental data upon which they were based, and there existed no international standard reference table. It was realized, when the new reference tables were being prepared, that polynomial expressions for the tables would be required for computer use and thus they were given with the tables. Unfortunately, the form in which the polynomials were given was rather cumbersome and, it turns out, unnecessarily complex and polynomials were given only for emf as a function of temperature and not for temperature as a function of emf, the inverse table.

Each of the reference tables was produced by least-squares fitting of experimental data of emf vs temperature. In general, the order of fit was chosen so that the residuals reflected the errors in the original experimental data. In only one case (Type B) was it found possible to use a single polynomial to cover the whole range of use of the thermocouple, and therefore the polynomials were adjusted so that at their junctions both emf and $S(T)$ were continuous. The exceptions to this procedure were the Types R and S, where at 630.74 °C there is a small discontinuity in $S(T)$ of $+0.18\%$ ($S(T)$ is higher above 630.74 °C than it is below). This discontinuity reflects the discontinuity in the IPTS-68 at this temperature (see Chapter 2). Also, the order of fit imposed between 630.74 °C and 1064.43 °C for the Type S thermocouple was of course a quadratic, since this is what is required to realize IPTS-68 in this range.

The polynomials that were obtained using these procedures are not very well suited for many calculators or even computers. This is because the polynomials contain as many as fourteen terms, the coefficients of each of which contains eleven significant figures (truncated to seven for Types R and S). It is not a simple problem to truncate these coefficients, especially where different numbers of digits are required for different coefficients. Nevertheless, for most types truncation is possible (Coates, 1978b) and in Appendix V will be found the coefficients truncated to between five and eight digits for most of the thermocouple types. In addition, there are listed the coefficients of the Chebyshev polynomials which accurately reproduce the tables, and it will be seen that they are a much more efficient way of describing the data. A further step in simplifying the mathematical description of the standard reference tables, is the provision of approximate formulations for both forward and inverse representations (Coates, 1978b).

6-5-2   Experimental methods

The experimental methods used to obtain the data which are needed to make use of the reference tables fall into three categories. In addition to the two mentioned at the beginning of this section, namely, the fixed point and the comparison methods, must be added a third, the "wire-point" method. This third method is a variation of the fixed point method, but it has a number of special features which make it worth treating separately.

The fixed point method is well illustrated by the use of the Type S thermocouple to establish the IPTS-68 between 630.74 °C, and the gold point. An additional fixed point at 961.93 °C, the freezing point of silver, is used and the emfs measured at these three fixed points are used to define the quadratic equation specified in IPTS-68. The metal-freezing point apparatus has been described in Chapter 4 (Fig. 4-26) and the modifications required to accommodate the thermocouple need be no more than the provision of a closed-ended alumina sheath to contain the thermocouple. It should be remembered that to allow free access of air, a tight-fitting sheath should be avoided. The usual dimensions of a Type S thermocouple for work of the highest accuracy are the following: wire diameter between 0.3 mm and 0.5 mm, twin-bore recrystallized alumina insulator 3 mm diameter and 500 mm long, and external closed-ended recrystallized alumina sheath of internal diameter 4 mm to 5 mm and 1 mm wall thickness (Fig. 6-4).

For calibration by means of the comparison method, a liquid bath is usually used up to 600 °C and a furnace with a heavy metal liner or heat pipe for higher temperatures. It is sometimes more convenient to use the same apparatus for the whole temperature range, and if this extends above 600 °C, then the furnace with liner may be used. The designs of stirred-liquid or fluidized-bed baths given in Chapter 4 are quite suitable for thermocouple calibration. If calibration by comparison with a standard thermocouple is being used, the junctions of the thermocouples being calibrated should be bound together with that of the standard to ensure that they are at the same temperature. This applied also to calibration by comparison in a furnace of the type shown in Fig. 4-4. The advantage of calibration by means of the comparison method with the junctions of both test and standard thermocouples bound together, is that much less reliance is placed upon the uniformity of temperature within the stirred-liquid bath or furnace. Automatic methods of thermocouple calibration in which a programmed controller runs a furnace or stirred-liquid bath over a range of temperature while measurements of emf are recorded are much simpler to devise if the junctions are bound together.

A fully-automatic system used for the calibration of thermocouple over the range room temperature to 1100 °C has been described by Jones and Egan (1975). The advent of small computers and microprocessors has had the result

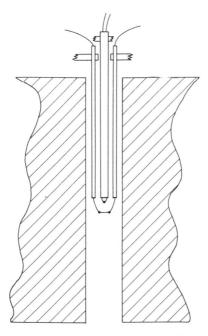

Fig. 6-17. Furnace and thermocouple assembly for wire-point calibration.

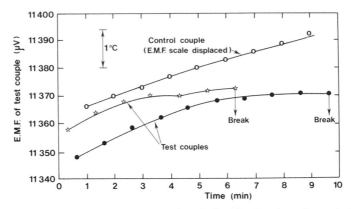

Fig. 6-18. Melting point curves for the gold point obtained by the wire-point method.

that, provided the design of the furnace and switching system is adequate, the whole process, from the moment the thermocouple is mounted in the furnace to the final print out of the calibration report, can be carried out automatically.

For calibration of Types R, S and B at temperatures above 1100 °C, the third method of calibration, the "wire-point" method, is simple and, if care is taken, reliable. The principle of the "wire-point" method is that a small length of wire, usually of gold, palladium or platinum, forms the junction between the two thermocouple wires, Fig. 6-17. As the temperature of the furnace rises through the melting point of the wire, the emf of the thermocouple halts while the wire is melting and finally disappears as it breaks. At the same time, the emf of a control thermocouple is monitored to verify the rise in temperature of the furnace. A typical set of results for a gold wire point for a Type R thermocouple are shown in Fig. 6-18. A precision of measurement of ±0.3 °C at the gold point and ±1.5 °C at the palladium point is possible using this technique. It is widely used at the palladium point for thermocouple wire destined to be used in the quick-immersion method of measuring the liquidus temperature of steel.

6-5-3 Final remarks concerning calibration

In coming to the end of this section on calibration, it is important to recall the discussion, at the very beginning of this chapter, on the effects of inhomogeneities on thermocouples. The measured thermoelectric emf is produced in that section of the thermocouple which is in the temperature gradient. The presence of inhomogeneities will lead to the measured emf being characteristic not only of the temperature difference between the junctions but also of the position of the inhomogeneities in the temperature gradient. This means, in practice, that the calibration of a thermocouple, strictly, only applies to the furnace or bath in which it was calibrated and, even then, only at the time of the original calibration. The removal of the thermocouple from the calibration furnace is usually sufficient to quench-in a sufficient number of lattice vacancies to produce a measurable charge in calibration. Oxidation or phase transformation (as in Type K thermocouples) also leads to a composition gradient being imposed which is a function of the temperature gradient of the calibration furnace (Bentley and Jones, 1980).

It is effects such as these that limit the accuracy with which temperatures can be measured by means of thermocouples. The user must always be aware that such subtle effects are present when he is using a thermocouple and that the possible errors can easily amount to 5 °C or 10 °C; very much more, of course, in Type K thermocouples.

6-6   Reference junctions

The classic cold junction for a thermocouple is an ice and water mixture at 0 °C. For laboratory applications this remains the simplest and most reliable solution, although it is necessary to take a few precautions if it is to be used to full advantage. The temperature of equilibrium between ice and pure water at atmospheric pressure is 0 °C. The effects of dissolved mineral impurities usually found in tap water rarely change the freezing point by more than -0.03 °C but, nevertheless, it is advisable to use ice made from distilled water. The way to produce an ice bath which will last for a long time is to crush ice taken from a freezer at say -18 °C, pack it into a one litre wide-necked vacuum flask and then add distilled water until it just covers the crushed ice. The thermocouple cold junctions should then be immersed some 15 cm into the ice in glass tubes where they will remain at 0 °C within a few millikelvins for up to 40 h without further attention. It is sometimes recommended that the glass tubes be filled, to the level of the top of the ice/water mixture, with a light mineral oil to improve thermal contact. This is not necessary and can lead to problems arising from the capillary rise of the oil in the insulation if the oil reaches the hot parts of the thermocouple. Of course, the number of thermocouple wires, their diameter and thermal conductivity can make a considerable difference to the useful life of the ice bath. An immersion of 150 mm is more than adequate for a pair of 0.45 mm copper wires but for 20 such wires in the same glass tube, an error of about 0.02 °C in the junction temperature would result. Figure 6-19 and Table 6-5 illustrates some of the characteristics of an ice bath.

Although for laboratory use an ice bath is often the best way of maintaining a reference temperature, this is not the case for most industrial applications. There have been many alternative ways of providing a reference temperature, not necessarily at 0 °C. Thermoelectric coolers using the Peltier effect are available and can be useful when large numbers of thermocouples are being used. Their principal defect is that when a reference temperature of 0 °C is being produced, there is rarely provision for sufficient immersion of the thermocouple in the cold well. However, the Peltier cooler is now being

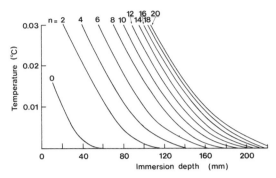

Fig. 6-19. The effect of the number of wires and immersion depth on the junction temperature of an ice-point reference. n is the number of 0.45 mm diameter copper wires in each glass tube (after Sutton, 1975).

Table 6-5. Comparison of different forms of loading.*

Type of wire	No. of 0.45 mm diam Cu for same load
0.45 mm diameter constantan	0.062
0.45 mm diameter iron	0.200
0.45 mm diameter chromel	0.050
0.45 mm diameter alumel	0.077
7/0.2 insulated Cu	1.7
16/0.2 insulated Cu	3.5
1/0.6 insulated Cu	2.8

* From Sutton, 1975.

superseded by electronic systems which inject into the thermocouple circuit the appropriate voltage based upon a measurement of the ambient temperature. These usually work by taking the voltage drop across a calibrated temperature sensor, such as a thermistor or semiconductor junction, to provide the reference for the injected voltage. Developments in miniature electronic systems will undoubtedly lead to reference junctions of this sort displacing all others for large installations.

6-7 The pressure dependence of the emf of thermocouples

The pressure dependence of the thermopower of metals and alloys has already been mentioned at the beginning of this chapter when discussing the theory of thermoelectricity. Although theory leads us to expect a pressure dependence (through the change in Fermi level brought about by the contraction of the

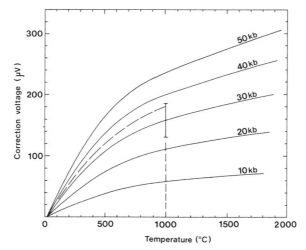

Fig. 6-20. Pressure corrections for Pt-10 % Rh/Pt thermocouples for a pressure-seal temperature of 20 °C. Dashed lines indicate upper limits of experimentally explored regions (after Getting and Kennedy, 1970).

lattice under pressure), it cannot yet make any useful quantitative predictions of the effect. Experimental measurements have, however, been made by many workers starting, of course, with Bridgeman (1918). More recently Bundy (1961), Bloch and Chaissé (1967), and Getting and Kennedy (1970) have measured the effect under a variety of conditions of temperature and pressure. Figure 6-20 shows the results of Getting and Kennedy for Type S thermocouples. Blatt *et al.* (1976) should be consulted for a good discussion on this subject.

References

Anderson, R. L. and Ludwig, R. L. (1982). Failure of sheathed thermocouples due to thermal cycling. *TMCSI*, **5**, 939–951.

ASTM (1974). Manual on the use of thermocouples in temperature measurement, STP-470A, American Society for Testing and Materials (ASTM).

Barnard, R. D. (1972). "Thermoelectricity in Metals and Alloys", Taylor and Francis Ltd., London.

Bedford, R. E. (1965). Reference tables for Pt-20 % Rh/Pt-40 % Rh thermocouples. *Rev. Sci. Instr.* **36**, 1177.

Bedford, R. E., Ma, C. K., Barber, C. R., Chandler, T. R., Quinn, T. J., Burns, G. W. and Scroger, M. (1972). New reference tables for Pt-10 % Rh/Pt and Pt-13 % Rh/Pt thermocouples. *TMCSI*, **4**, Part 3, 1585–1602.

Bedford, R. E. (1972). Remarks on the International Temperature Scale of 1968. *TMCSI*, **4**, Part 1, 15–25.

Bentley, R. E. (1969). The emf criteria for IPTS thermocouples. *Metrologia*, **5**, 26–28.

Bentley, R. E. and Jones, T. P. (1980). Inhomogeneities in Type S thermocouples when used to 1064 °C. *High Temperatures–High Pressures*, **12**, 33–45.

Berman, R. (1972). Gold-iron alloys for low temperature thermocouples. *TMCSI*, **4**, Part 3, 1537–1542.

Berman, R. and Kopp, J. (1971). Thermoelectric power of dilute gold–iron alloys. *J. Phys. F.* **1**, 457.

Blackburn, G. F. and Caldwell, F. R. (1964). Reference tables for thermocouples of iridium–rhodium alloys versus iridium. *J. Res. NBS* **68C**, 41–59.

Blatt, F. J., Schroeder, P. A., Foiles, C. L. and Greig, D. (1976). "Thermoelectric Power in Metals", Plenum.

Bloch, D. and Chaissé, F. (1967). Effect of pressure on emf thermocouples. *J. Appl. Phys.* **38**, 409.

Borelius, G., Keesom, W. H., Johansson, C. H. and Linde, J. O. (1932). Measurements on thermoelectric forces down to temperatures obtainable with liquid or solid hydrogen and with liquid helium. *Proc. Roy. Acad. Sci. Amsterdam*, **35**, 15–24. (Also as Physical Laboratory Leiden Comm. No. 217d.)

Bourassa, R. R., Wang, S. Y. and Lengeler, B. (1978). Energy dependence of the Fermi surface and thermoelectric power of the noble metals. *Phys. Rev. B.* **18**, 1533–1536.

Bridgeman, P. W. (1918). Thermoelectromotive force, Peltier heat and Thomson heat under pressure. *Proc. Amer. Acad. Arts Sci.* **53**, 269–386.

Bugden, W. G., Tomlinson, J. A. and Selman, G. L. (1975). Improved lead compensating systems for platinum-base thermocouples. Temperature-75, 181–187.

Bundy, F. P. (1961). Effect of pressure on emf of thermocouples. *J. Appl. Phys.* **32**, 483–488.

Burley, N. A. (1972). Nicrosil and Nisil: highly stable nickel base alloys for thermocouples. *TMSCI*, **4**, Part 3, 1677–1695.

Burley, N. A., Powell, R. L., Burns, G. W. and Scroger, M. G. (1978). The Nicrosil versus Nisil thermocouple: Properties and thermoelectric reference data. NBS Monograph 161.

Burley, N. A., Hess, R. M., Howie, C. F. and Coleman, J. A. (1982). The nicrosil versus nisil thermocouple system: A critical comparison with the ANSI standard-letter designated base-metal thermocouples. *TMCSI*, **5**, 1159–1166.

Burns, G. W. and Hurst, W. S. (1972). Studies of the performance of W-Re type thermocouples. *TMCSI*, **4**, Part 3, 1751–1766.

Burns, G. W. and Hurst, W. S. (1975). Thermocouple thermometry. Temperature-75, 144–161.

Caldwell, F. R. (1933). Thermoelectric properties of platinum–rhodium alloys. *J. Res. NBS*, **10**, 373–380.

Cannon, C. P. (1982). 2200 °C thermocouples for nuclear reactor fuel center line temperature measurements. *TMCSI*, **5**, 1061–1067.

Christian, J. W., Jan, J. P., Pearson, W. B. and Templeton, I. M. (1958). Thermoelectricity at low temperature VI. A redetermination of the absolute scale of thermoelectric power of lead. *Proc. Roy. Soc. A*, **245**, 213–221.

Coates, P. B. (1978a). These measurements were made at NPL between 1975 and 1978 and the results communicated privately by P. B. Coates to the author.

Coates, P. B. (1978b). Functional approximations to the standard thermocouple reference tables. NPL Report Qu 46 (March, 1978).

Coville, P. (1975). Manufacturing of high temperature thermocouples. *In* Petten–Harwell Colloquium, JRC Petten, Netherlands.

Darling, A. S. and Selman, G. L. (1972). Some effects of environment on the performance of noble metal thermocouples. *TMCSI*, **4**, Part 3, 1633–1644.

Fairchild, C. O. and Schmitt, H. M. (1922). Life tests of Pt–Rh/Pt thermocouples. *Chem. Met. Eng.* **26**, 158–160.

Fenton, A. W. (1969). Errors in thermoelectric thermometers. *Proc. IEEE*, **116**, 1277.

Getting, I. C. and Kennedy, G. C. (1970). The effect of pressure on the emf of chromel–Alumel and platinum–platinum 10 % Rhodium thermocouples. *J. Appl. Phys.* **41**, 4552–4562.

Glawe, G. E. and Szaniszlo, A. J. (1972). Long-term drift of some noble and refractory-metal thermocouples at 1600 °C in air, argon and vacuum. *TMCSI*, **4**, Part 3, 1645–1662.

Guildner, L. A. and Burns, G. W. (1979). Accurate thermocouple thermometry. *High Temperatures–High Pressures*, **11**, 173–192.

Heckelman, J. D. and Kozar, R. P. (1972). Measured drift of irradiated and unirradiated W-3%Re/W-25%Re thermocouples at a nominal 2000 K. *TMCSI*, **4**, Part 3, 1935–1949.

Hurst, J. G., Powell, R. L. and Sparks, L. L. (1972). Methods for cryogenic thermocouples thermometry. *TMCSI*, **4**, Part 3, 1525–1535.

IEC (1977). Thermocouples Part I: Reference Tables International Electrotechnical Commission Standard 584-1. (These reference tables are identical to those in BS 4937 and ASTM E-230-77.)

Jones, T. P. and Egan, T. M. (1975). The automatic calibration of thermocouples in the range 0–1100 °C. Temperature-75, 211–218.

Kinzie, P. A. (1973). "Thermocouple Temperature Measurement", Wiley-Interscience, New York.

Le Chatelier, H. (1886). Sur la variation produite par une élévation de température dans la force électromotrice des couples thermoélectriques. *Compt. Rend. Acad. Sci.* **102**, 819.

Mason, F. (1972). Manufacturing specification for high temperature thermocouples for use in nuclear irradiation experiments, Commission of the European Communities EUR 4807c, J. R. C. Petten, Netherlands.

McCulloch, R. W. and Clift, J. H. (1982). Lifetime improvement of sheathed thermocouples for use in high-temperature and thermal transient operations. *TMCSI*, **5**, 1097–1108.

McLaren, E. H. and Murdock, E. G. (1979). The properties of Pt/Pt Rh thermocouples for thermometry in the range 0–1100 °C. NRCC 17407 and 17408, NRC Ottawa.

Mott, N. F. and Jones, H. (1936). "The Theory of Properties of Metals and Alloys", Clarendon Press, Oxford.

Oliver, D. A. and Land, T. (1944). A thermocouple method for the measurement of liquid steel casting-stream temperatures. *J. Iron Steel Inst.* **149**, 513–521.

Onsager, L. (1931a). Reciprocal relations in irreversible processes I. *Phys. Rev.* **37**, 405–426.

Onsager, L. (1931b). Reciprocal relations in irreversible processes II. *Phys. Rev.* **38**, 2265–2279.

Petten-Harwell (1975). "Colloquium on High temperature in-pile thermometry", JRC Petten, Netherlands.

Powell, R. L., Hall, W. J., Hyinck, C. H., Sparks, L. L., Burns, G. W., Scroger, M. G. and Plumb, H. H. (1974). Reference Tables for thermocouples. NBS Monograph 125.

Roberts, R. B. (1977). The absolute scale of thermoelectricity. *Phil. Mag.* **36**, 91–107.

Schofield, F. H. (1942). Recent work with the quick-immersion thermocouple in the steelworks of Sheffield. *J. Iron Steel Inst.* **145**, 222–243.

Selman, G. L. (1972). On the stability of metal-sheathed noble metal thermocouples. *TMCSI*, **4**, Part 3, 1833–1840.

Shepard, R. L., Hyland, R. F., Googe, J. M. and McDearman, J. R. (1972). Equivalent circuit modelling of insulator shunting errors in high temperature sheathed thermocouples. *TMCSI*, **4**, Part 3, 1841–1853.

Sutton, G. R. (1975). Thermocouple referencing. Temperature-75, 188–194.

Thurlbeck, A. (1982). Temperature measurement in WAGR. *TMCSI*, **5**, 1081–1095.

7

Radiation Thermometry

7-1　Introduction

Optical pyrometry, radiation pyrometry, infrared pyrometry, spectral or total radiation pyrometry are some of the names given to methods of thermometry based upon the measurement of thermal radiation. In this field, the words "pyrometry" and "thermometry" tend to be used synonymously, although the use of "pyrometry", with its connotation of fire, seems out of place when referring to infrared measurements of temperature below 100 °C!

In radiation thermometry, in contrast to resistance thermometry or thermocouple thermometry, we are able to make use of explicit equations relating thermodynamic temperature to the measured quantity: in this case spectral radiance. This is possible because the thermal radiation existing inside a closed cavity (blackbody radiation), depends only upon the temperature of the walls and not at all upon their shape or constitution provided that the cavity dimensions are much larger than the wavelengths involved. The radiation escaping from a small hole in the cavity wall departs from blackbody radiation by only an amount determined by how much the state of equilibrium within the cavity is perturbed by the presence of the hole. By careful design, as we shall see, this perturbation can be made negligibly small, so that equilibrium blackbody radiation is accessible for measurement. Thus, in principle, thermodynamic temperature may be measured very precisely by means of radiation thermometry: an application which we shall mention briefly in Section 7-7.

Another important difference between radiation thermometry and other methods of thermometry, and one that has profoundly influenced its development, is that for the former, there exists a natural sensor: the human eye. The most widely used instrument for radiation thermometry was, until quite recently, the disappearing filament optical pyrometer, here we follow common usage and refer to it as a "pyrometer". In this instrument, the human eye is used to judge equality of brightness between a glowing filament and an image of the source whose temperature is being measured. Under good conditions of illumination, the eye can detect differences in brightness of as little as 1 % between adjacent fields. This is equivalent to a temperature difference of only about 1 °C between blackbodies near 1000 °C. Similarly, the eye is a good detector of colour differences and the often-quoted example of the foundryman using a precious piece of orange peel with which to judge the colour and hence the temperature of his furnace, undoubtedly has some basis in fact.

We have seen in Chapter 1 how the quantitative application of the human eye to temperature measurement first appeared towards the end of the nineteenth century and became codified in ITS-27. It had become apparent at quite an early stage that a convenient, highly reproducible and accurate method of temperature measurement was a monochromatic radiation pyrometer. The availability of first, carbon- and eventually tungsten-filament lamps led to the development of the disappearing-filament optical pyrometer. Although the performance of tungsten-filament lamps was far superior, in most respects, to that of carbon-filament lamps, the latter continued to be used until the 1940s in disappearing-filament pyrometers intended to measure temperatures as low as 650°C. The advantage of the carbon filament in this application was its high emissivity and hence good colour-matching properties when viewed against the image of a blackbody without a colour filter.

The accuracy with which the disappearing-filament pyrometer could be used to measure temperature was adequate for most practical applications. The limiting factor was often, in any case, the uncertainty in the emittance of the object whose temperature was being measured. Nevertheless, despite its convenience, accuracy and reliability, the disappearing-filament optical pyrometer had one great disadvantage: its use required the active participation of a skilled observer. It could not be used for those applications requiring continuous or very rapid measurements, nor for those in inaccessible or dangerous situations. For this reason, from the earliest days some radiation thermometers incorporated thermal, photo-voltaic, photo-conductive or photo-emissive detectors. Among the most successful of these has been the silicon-cell radiation thermometer. Its ruggedness and long-term reproducibility has led to it being used in such diverse situations as the measurement of the temperature of aero-engine turbine-blades and steel-forming furnaces. In standards laboratories, the disappearing-filament optical pyrometer has now been displaced by the photoelectric pyrometer as the instrument used for the realization of IPTS-68 above the freezing point of gold.

Along with the disappearing-filament pyrometer, there was a parallel development of tungsten ribbon lamps designed for maintaining and disseminating the radiation temperature scale. The development of these lamps continues, and they are now used in standards laboratories with photoelectric standards pyrometers. International comparisons of temperature scales are carried out by the circulation of such lamps among national standards laboratories. The agreement now achieved among the radiation temperature scales established by the major national standards laboratories is of the order of ± 0.1 °C in the range 1000 °C to 1700 °C.

Inseparable from the question of the detector of thermal radiation is that of the radiating properties of the emitter of the radiation. The spectral radiating properties of a blackbody are, as we shall see, described by the Planck law. Integrated over all wavelengths, Planck's law leads to the Stefan–Boltzmann law which describes the variation as a function of temperature, of the total radiation emitted by a blackbody. Radiation thermometry would be very much simpler if no more than that need be said about the radiating properties of materials. Unfortunately, real materials do not behave as blackbody radiators, and a correction factor known as the emittance must be applied to the Planck or Stefan–Boltzmann laws. The emittance varies both with temperature and wavelength, and is a function of the detailed electronic structure of the material and also of the macroscopic form of its surface.

We begin this chapter with an outline of the theory of blackbody radiation; this is followed by a discussion of the various methods of calculating the emissivity of near-blackbody cavities and their practical realization. We then come to the tungsten ribbon lamp as a reproducible source of thermal radiation for thermometry. With this as background we shall then deal with radiation thermometry; the realization of IPTS-68 above the gold point, the determination of thermodynamic temperature, the methods aimed at overcoming problems due to lack of knowledge of surface emittance and finally a brief discussion on radiation thermometry of semi-transparent media.

7-2   The properties of thermal radiation

7-2-1   The radiation laws of Wien, Stefan-Boltzmann and Planck

Heat transfer by thermal radiation and radiation thermometry are intimately connected through their common requirement to link thermodynamic temperature to the quantity and quality of thermal energy emitted by a surface. On the basis only of classical thermodynamics and electromagnetic theory, it was possible by the end of the nineteenth century to deduce two important results. The first was Stefan's law (1879), which stated that the energy density inside a cavity was proportional to the fourth power of the temperature of the cavity walls. The second was Wien's displacement law (1893), which stated that as the temperature of a blackbody is increased, the wavelength of maximum emission, λ_m, decreases so as to maintain the product $\lambda_m T$ a constant. The proof of Stefan's law is based upon the treatment of thermal radiation as the working fluid in a thermodynamic engine having a movable mirror as piston and making use of Maxwell's electromagnetic theory to show that the pressure exerted on a surface by isotropic radiation is proportional to the energy density. Wien's law follows if the Doppler effect, arising from the movement of the mirror, is introduced. In both laws there appeared a constant of proportionality about which classical thermodynamics by itself could give no information.

In order to evaluate the constant of proportionality, assumptions had to be made about the radiation process and about the way energy is distributed among the degrees of freedom. Classical electromagnetic theory made it possible to calculate the number of standing waves, dN_v, having frequencies between v and $v+dv$, that can be present in a right parallelepiped of volume V

$$dN_v = \frac{8\pi v^2 V dv}{c^3} \tag{7-1}$$

where c is the velocity of electromagnetic radiation in vacuum. This number dN_v is also known as the average mode density $D_0(v)$ sometimes called the Jeans number. If the energy in each standing wave is u_v, the energy U_v between frequencies v and $v+dv$ is

$$U_v = u_v dN_v = \frac{8\pi v^2 V}{c^3} u_v dv \tag{7-2}$$

or

$$U_v = \frac{8\pi v^2 V}{c^3} \bar{\varepsilon} \tag{7-3}$$

where $\bar{\varepsilon}$ is the average energy of a standing wave. If each standing wave is considered to be a simple harmonic oscillator, classical Boltzmann statistics could be used to derive a simple expression for $\bar{\varepsilon}$ averaged over all frequencies from zero to infinity. The Boltzmann theorem states that the probability of an oscillator having a particular energy ε is $e^{-\beta\varepsilon}$ where $\beta = 1/kT$. Thus the average energy $\bar{\varepsilon}$ of an infinite number of oscillators having energies from $\varepsilon = 0$ to $\varepsilon = \infty$ is given by

$$\bar{\varepsilon} = \int_0^\infty \varepsilon\, e^{-\beta\varepsilon} d\varepsilon \Big/ \int_0^\infty e^{-\beta\varepsilon} d\varepsilon \tag{7-4}$$

$$\therefore \quad \bar{\varepsilon} = -\frac{d}{d\beta}\left(\ln \int_0^\infty e^{-\beta\varepsilon} d\varepsilon \right)$$

$$\bar{\varepsilon} = -\frac{d}{d\beta}\left(\ln\frac{1}{\beta}\right) = \frac{1}{\beta} = kT. \tag{7-5}$$

Thus from (7-3) and (7-5) we have

$$U_v = \frac{8\pi v^2 V}{c^3}kT. \tag{7-6}$$

This is the Rayleigh–Jeans expression for the energy in a cavity, and is the result of applying Boltzmann statistics to the radiation field. For high frequencies, the Rayleigh–Jeans expression fails since it predicts that the energy increases without limit as the frequency increases. At very low frequencies however, it correctly predicts that the energy tends towards a simple proportionality to T.

The way out of this high-frequency impasse in the theory of thermal radiation was provided by the quantum hypothesis of Planck. Instead of allowing the oscillator to take any value of energy ε, Planck postulated that the only energies allowable were integral multiples of an arbitrary energy ε_0, that is to say $\varepsilon_0, 2\varepsilon_0, 3\varepsilon_0 \dots$ where $\varepsilon_0 = hv$ with h being a universal (Planck's) constant.

The calculation of the mean energy $\bar{\varepsilon}$ thus differs slightly from that given above for the classical oscillator, in that we substitute summations for integrals, but the difference in the result is striking.

$$\bar{\varepsilon} = \sum_{n=0}^{\infty} n\varepsilon_0 e^{-\beta n\varepsilon_0} \bigg/ \sum_{n=0}^{\infty} e^{-\beta n\varepsilon_0} \tag{7-7}$$

$$= -\frac{d}{d\beta}\left(\ln\left\{\sum_{n=0}^{\infty} e^{-\beta n\varepsilon_0}\right\}\right)$$

$$= -\frac{d}{d\beta}\left(\ln\left\{\frac{1}{1-e^{-\beta\varepsilon_0}}\right\}\right) = \frac{\varepsilon_0 e^{-\beta\varepsilon_0}}{1-e^{-\beta\varepsilon_0}}$$

$$= \frac{\varepsilon_0}{e^{\beta\varepsilon_0}-1}. \tag{7-8}$$

Substituting for ε_0 and β leads to

$$\bar{\varepsilon} = \frac{hv}{e^{(hv/kT)}-1}. \tag{7-9}$$

Substituting in (7-3) we find

$$U_v = \frac{8\pi v^2 V}{c^3}\frac{hv}{e^{(hv/kT)}-1}. \tag{7-10}$$

This is Planck's law. It relates the energy, per unit frequency interval at a frequency v, in a closed parallelepiped of volume V, to the temperature of the walls. As we would expect, the Planck law tends, in the limit of low frequencies, towards the simple T proportionality of the Rayleigh–Jeans law and in the limit of high frequencies towards the form of the Wien law. The integration of the Planck equation over all frequencies leads to the Stefan–Boltzmann total radiation law. The total energy U in the same cavity is given by

$$U = \int_0^\infty \frac{8\pi V v^2}{c^3} \frac{hv}{e^{(hv/kT)} - 1} \cdot dv \qquad (7\text{-}11)$$

Making the change of variable $hv/kT = x$, and noting that $dx = (h/kT)dv$, we find

$$U = \frac{8\pi V k^4 T^4}{c^3 h^3} \int_0^\infty \frac{x^3}{e^x - 1} dx. \qquad (7\text{-}12)$$

The integral is dimensionless and has the exact value $\pi^4/15$ so that

$$U = \frac{8\pi^5 k^4}{15 c^3 h^3} V T^4. \qquad (7\text{-}13)$$

This is the Stefan–Boltzmann law, relating the total radiation energy in a closed parallelepiped of volume V to the temperature of the walls.

Neither this expression for U nor that in equation (7-10) for U_v tells us the quantity of energy emitted from a small hole in the wall of the parallelepiped. Before coming to this, which will give us expressions for the observable quantities in radiation measurement, we must look a little more closely at the implications and limitations of the Planck and Stefan–Boltzmann laws as expressed in equations (7-10) and (7-13).

7-2-2   Thermal radiation in real cavities

In deriving equation (7-1) for dN_v or $D_0(v)$, a number of important conditions were imposed, chief of which was that the wavelengths of the standing waves were infinitesimally small compared with the dimensions of the cavity. Another was that the walls of the cavity were loss-free, that is to say perfectly reflecting. Yet another was that that cavity was a right parallelepiped.

Since the inception of the quantum theory of blackbody radiation, the question of how well the Planck and Stefan–Boltzmann equations describe the energy density inside real, finite, cavities having semi-reflecting walls, has been the subject of speculation. Much of this took place during the first two decades of this century, but the question was never considered closed and in recent years there has been a revival of interest in this and related problems. Among the reasons for the renewed interest in this, the oldest subject of modern physics, are the growth of quantum optics and the theory of partial coherence and their application to the study of the statistical properties of radiation; the need to understand radiative transfer between closely spaced bodies at low temperatures and the problem of far infrared radiation standards for which the wavelength cannot be considered small, as well as many theoretical problems related to the statistical mechanics of finite systems. A good introduction to modern work in this field is given by Baltes and Hilf (1976), and Baltes (1973, 1976). As early as 1911, it was shown by Weyl that the condition that the cavity be a right parallelepiped could be relaxed provided that $V^{1/3}v/c \rightarrow \infty$. He proved that the Jeans number is valid for any shape of cavity in the limit of large volumes or high temperatures. From Weyl's results, asymptotic expansions were later derived in which $D_0(v)$ was simply the first term in a series for which the sum $D(v)$ was the real average mode density. Recent calculations (Baltes, 1973; Baltes and Kneubühl, 1970) of $D(v)$, using numerical methods to sum the first 10^6 standing waves in simple cavities, have shown that the earlier asymptotic expansions provide good approximations for $D(v)$ over quite wide ranges of $(V^{1/3}v/c)$. As would be expected, the magnitude of $\overline{\Delta U_v}$, defined as $(D(v) - D_0(v)/D_0(v))$, becomes very dependent upon the shape of the cavity as $V^{1/3}v/c$ decreases. For example, in the case of side L, the first order correction is zero and the second order correction is given by

$$\overline{\Delta U_v} \approx -(c/2Lv)^2 = -(\lambda/SL)^2 \qquad (7\text{-}14)$$

where λ is the wavelength.

Figure 7-1 shows $\overline{\Delta U_v}$ for values of $2L/\lambda$ in the range 1 to 100. What is

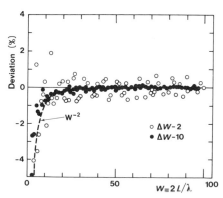

Fig. 7-1. Deviations from Planck's energy density of thermal radiation for a cube of side L. $W = 2L/\lambda$ where λ is the wavelength (after Baltes, 1976).

striking about Fig. 7-1 is that superimposed on a smoothly varying $\overline{\Delta U_v}$ are large fluctuations in ΔU_v. The size of these fluctuations is inversely proportional to the bandwidth dv and therefore decreases with increasing frequency much more slowly than does $\overline{\Delta U_v}$. It is clear from Fig. 7-1 that the departure from Planck's law for the cavity size, wavelength and temperature range encountered in practical radiation thermometry is small. For example at a wavelength of 1 μm and a cavity dimension of 1 mm, we find $\overline{\Delta U_v} = 2.5 \times 10^{-7}$. This is entirely negligible, but the mean square fluctuation $\overline{(\delta U_v)^2}$ is not negligible if a very narrow bandwidth is used. In modern high precision radiation pyrometry it is not uncommon, as we shall see, to use bandwidths of 1 nm or less. This would lead to a value of $\overline{(\delta U_v)^2}$ of 5×10^{-6} or $\delta U_v \approx 2 \times 10^{-3}$, which may not be negligible. Values of $\overline{\Delta U_v}$ and $\overline{(\delta U_v)^2}$ for other shapes can be found in Baltes and Kneubühl (1970).

In view of the link between $\overline{\Delta U_v}$ and the ratio of the wavelength to some characteristic length of the cavity, we would expect the departure, ΔU, from the Stefan–Boltzmann equation to be a function of temperature. This is indeed the case and for a cube of side L:

$$\Delta U \approx -\left(\frac{0.25 \text{ cm K}}{LT}\right)^2 + \left(\frac{0.21 \text{ cm K}}{LT}\right)^3 - \left(\frac{0.14 \text{ cm K}}{LT}\right)^4 \dots \quad (7\text{-}15)$$

This expression is asymptotically correct in the limit of large values of LT, and is in error by about 3 % for $LT = 2$ cm K. For values of LT below 0.2 cm K, more complex expressions have been derived and are to be found in Baltes (1973). From equation (7-15) we can calculate that for $L = 1$ cm and $T = 4.2$ K, $\Delta U \approx -0.3$ %, while for $L = 1$ mm and $T = 1000$ K, $\Delta U \approx 10^{-5}$. While these departures from the Stefan–Boltzmann law are insignificant in terms of the accuracies expected in most practical radiation thermometry, they could become important in the field of far-infrared radiometry.

The condition that the walls of the cavity must be perfectly reflecting is much more difficult to deal with than are the purely geometrical limitations. It is likely that additional correction terms are present if the reflectivity of the walls is less than unity. These correction terms themselves, however, must fall to zero as the reflectivity of the walls tends to zero, since from Kirchhoff's law, the emissivity would at the same time tend to unity, leading once again to perfect blackbody conditions inside the cavity.

A problem we have already mentioned, and one very closely related to those we have just been discussing, is that of calculating the radiative heat transfer between closely spaced, highly reflecting surfaces at very low temperatures. Under these conditions the wavelengths at which most of the thermal energy is transmitted become comparable with the distance apart of the surfaces. It is found experimentally (Domonto et al., 1970) that the apparent radiant heat transfer increases above that to be expected from the application of the Stefan–

Boltzmann law if the mean wavelength exceeds about half the distance apart of the surfaces. Recent theoretical work (Caren, 1974) has correctly predicted the magnitude of this anomalous heat transfer. The calculation is based upon the assumption that the low-temperature radiation field near a metal surface is due to thermal oscillations of electrons in the two dimensional skin at the surface of the metal. These oscillations lead both to travelling waves and quasi-stationary waves. The former make up the classical radiation field observed at large distances from the surface, while the latter are confined to regions close to the surface. At close distances of approach of two such surfaces, the quasi-stationary waves become the dominant means of heat transfer. It is of interest to note that the value of ΔU predicted by equation (7-15) matches very closely the augmentation on the heat transfer calculated using the theory of quasi-stationary waves. Another way of looking at the problem is, therefore, to postulate photon tunnelling across the space between the surfaces. The photons taking part in the tunnelling augment the heat transfer at the expense of the radiation field in the cavity. Although of minor interest for thermometrists, the increase in radiant heat transfer for close distances of approach can be of importance in low temperature calorimetry and in the efficiency of multi-layer metal foil insulation.

7-2-3   Thermal radiation emitted from blackbody cavities

Having looked at some of the restrictions on the application of the Panck and Stefan–Boltzmann laws we shall now come back to their application in the domain in which $D_0(v)$ is a close approximation to $D(v)$. We now also extend the treatment to cavities in which the medium has a refractive index n not necessarily equal to unity. The spectral energy density ρ_v in a cavity of arbitrary shape and for which $(V^{1/3}/vc) \gg 1$ is given by

$$\rho_v dv = \frac{U_v}{V} = \frac{8\pi h n^3 v^3}{c^3} [e^{(hv/kT)} - 1]^{-1} dv. \tag{7-16}$$

In terms of wavelength, equation (7-16) becomes:

$$\rho_\lambda d\lambda = \frac{8\pi hc}{n\lambda^5} [e^{(hv/kT)} - 1]^{-1} dv \tag{7-17}$$

The energy passing through a plane surface from one side to the other is given by $\rho_v \times c/4n$. Thus the rate at which energy leaves a hole in the walls of a blackbody, the spectral exitance M_v, is given by

$$M_v dv = \frac{2\pi n^2 h v^3}{c^2} [e^{(hv/kT)} - 1]^{-1} dv \tag{7-18}$$

where the unit is watts per square metre per unit frequency interval, or in terms of wavelength

$$M_\lambda d\lambda = \frac{2\pi h c^2}{n^2 \lambda^5} [e^{(hc/nk\lambda T)} - 1]^{-1} d\lambda \qquad (7\text{-}19)$$

where the unit is watts per square metre per unit wavelength interval. For the purposes of radiation pyrometry, equation (7-19) is sometimes written

$$M_\lambda d\lambda = n^{-2} c_1 \lambda^{-5} [e^{(c_2/n\lambda T)} - 1]^{-1} d\lambda \qquad (7\text{-}20)$$

where

$$c_1 = 2\pi h c^2 \approx 3.74 \times 10^{-16} \ Wm^2$$

$$c_2 = hc/k \approx 1.439 \times 10^{-2} \ mK$$

In the same way we can write down an expression for the total exitance M from equation (7-13)

$$M = \frac{\rho c}{4} = \frac{U}{V} \times \frac{c}{4} = \frac{2\pi^5 k^4}{15 c^2 h^3} T^4 = \sigma T^4 \qquad (7\text{-}21)$$

where the unit is watts per square metre, where σ is known as the Stefan–Boltzmann constant and has the value of approximately $5.67 \times 10^{-8} \ Wm^{-2} K^{-4}$.

A further manipulation that is sometimes useful is the conversion of spectral exitance to spectral radiance L_λ. The spectral radiance is defined as the radiant flux (i.e. rate of energy flow) propagated in a given direction per unit solid angle about that direction and per unit area projected normal to that direction. It is related to M_λ by

$$L_\lambda = M_\lambda / \pi.$$

Re-writing equation (7-20) in terms of spectral radiance, dividing through by T^5, and taking $n = 1$, we have

$$L_\lambda T^{-5} = c_1 \pi^{-1} (\lambda T)^{-5} [e^{(c_2/\lambda T)} - 1]^{-1} \qquad (7\text{-}22)$$

making it very clear that the spectral distribution of blackbody radiation is a function of λT. This is shown in Fig. 7-2. Equation (7-22) leads easily to

$$(\lambda T)_{max} = 2898 \ \mu m \ K \qquad (7\text{-}23)$$

the Wien displacement law.

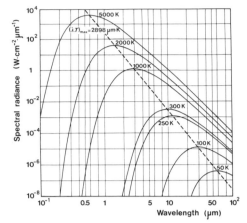

Fig. 7-2. The Planck distribution law; spectral radiance of blackbody radiation as a function of temperature and wavelength.

Differentiating equation (7-22) with respect to T we find

$$\frac{\mathrm{d}L_\lambda}{L_\lambda} = \left(\frac{c_2}{\lambda T}\right)\left[\frac{\mathrm{e}^{(c_2/\lambda T)}}{\mathrm{e}^{(c_2/\lambda T)} - 1}\right]\frac{\mathrm{d}T}{T}. \tag{7-24}$$

This gives the fractional change in spectral radiance as a function of fractional change in temperature. A knowledge of this is useful in radiation thermometry since it allows the accuracy required in spectral radiance measurements, to achieve a given accuracy in temperature, to be calculated. Note the inverse proportionality to wavelength. This is illustrated in Fig. 7-3.

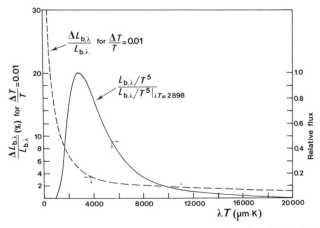

Fig. 7-3. Relative spectral radiance of blackbody radiation as a function of λT, solid curve; relative increase in spectral radiance produced by a one percent increase in temperature as a function of λT, dashed curve.

7-2-4   Thermal radiation emitted by a surface

We have considered in some detail the radiation existing inside a blackbody enclosure and that emitted from a small hole in the walls of such an enclosure. All of this has been done without taking account of the details of the emission/absorption process occurring at the walls. The radiation inside a closed cavity is in thermal equilibrium with the walls, that is to say there must be a balance between the radiation emitted and that absorbed. The processes occurring at the atomic level in the emission and absorption of radiation in a closed cavity were first treated by Einstein in 1917. He considered that the probability of a transition of an atom from a given energy state to a lower energy state with the emission of a photon was

$$\text{Probability of emission} = A + B\rho_v \qquad (7\text{-}25a)$$

where A and B are constants of the transition (Einstein coefficients). A is known as the probability of spontaneous emission, since it is independent of the presence of an external field, and $B\rho_v$ is the probability of induced emission since it is a function of the external field. Similarly the probability of absorption of a photon of the same energy was given as

$$\text{Probability of absorption} = B\rho_v \qquad (7\text{-}25b)$$

Einstein showed, from thermodynamic considerations, that

$$\frac{A}{B} = \frac{8\pi h v^3}{c^3} \qquad (7\text{-}26)$$

if the atoms were in equilibrium with the radiation field.

Although the ratio of the Einstein coefficients was known, the individual values of A and B could not be calculated until the development of wave mechanics. Dirac showed, in principle, how this could be done in 1927. The methods used to carry out such calculations are by no means straightforward and the interested reader is referred to texts on quantum mechanics (see for example, Slater, 1960) for details. The direct calculation of the emitting and absorbing properties of real materials is, in the general case, hopelessly complicated and can give no help whatever to thermometrists. Before leaving the atomic aspect of thermal radiation, however, we can use the relation between the Einstein coefficients to draw a useful distinction between the quantum domain and the classical domain.

The average quantum number n_{av} of an oscillator, such as one at the wall of a blackbody cavity contributing to the thermal radiation, is given by

$$n_{av} = \frac{1}{\mathrm{r}^{(hv/kT)} - 1}. \qquad (7\text{-}27)$$

From equations (7-26), (7-16) and (7-27) we find

$$A + B\rho_v = A(1 + n_{av}) \tag{7-28}$$

In other words the spontaneous emission A is only $1/n_{av}$ times the induced emission $B\rho_v$. Now, n_{av} is very large for low frequency thermal radiation and high temperatures, when $hv/kT \leqslant 1$. Thus for large values of quantum numbers, we find ourselves in the Rayleigh–Jeans domain where radiation density is proportional to T and is explained by classical electromagnetic theory. The radiation in this domain, however, arises almost entirely from induced emission. Thus induced emission behaves as a classical process and can be calculated according to classical mechanics. This is why the emissivities of metals in the far infrared follow very closely the straightforward Drude–Zehner relations. It is also why electronic engineering is so successful using Maxwell's equations!

At high frequencies or low temperatures where $hv/kT \geqslant 1$ and n_{av} becomes small, the spontaneous emission is large compared with induced emission. Spontaneous emission is very much a quantum process and therefore predictions of thermal radiative properties based upon classical methods (the Rayleigh–Jeans law or Drude–Zehner relations) fail.

Unfortunately for radiation thermometry, the sizes of the atomic constants are such that visible radiation is a quantum process and so the radiating properties of materials in this domain cannot be calculated from first principles. As we shall see later on in this chapter, various roundabout ways have to be employed to overcome this difficulty. Even so, the previous discussion might give the impression that the radiation process is so complex and intractable a problem that even experimental measurements are difficult. This is indeed the case; direct measurements of emissive power are difficult, but the way out of the difficulty is provided by Kirchhoff's law.

7-2-5   Kirchhoff's law and the principle of detailed balance

In 1859, very early in the study of thermal radiation, Kirchhoff showed on the basis of general arguments that the absorptive power of a material must equal its emissive power. If this were not so then it would not be possible for bodies of different materials to exist in thermal equilibrium inside a blackbody enclosure. Kirchhoff's law is, however, much stronger than is at first sight apparent. Not only must the overall energy absorbed be balanced by the overall energy emitted, but each induced emission and absorption process must balance. This is called the principle of detailed balance and is a fundamental result based upon statistical mechanics. In a statistical ensemble, representing a system in equilibrium, the probability of occurrence of any process must equal the probability of occurrence of the reverse process.

Thus, for a body in equilibrium in a blackbody enclosure the absorptive power of a given element of surface for a particular wavelength, a particular state of polarization, a particular direction, over a particular solid angle must equal the emissive power for the same radiation.

Before writing down an expression for Kirchhoff's law, we must first establish a convention for the symbols used to express such optical properties of a surface. The whole question of terminology is one that is treated at length in NBS Technical Note 910 (1976), International Lighting Vocabulary (1970), and Touloukian and DeWitt (1972); we shall as far as possible follow the recommendations laid down in these publications.

The symbols we shall use for emittance, reflectance, absorptance and transmittance are ε, ρ, α and τ respectively. The terms emittance, reflectance, etc. refer to real surfaces and include for example the effects of surface geometry. The terms ending in -ivity such as emissivity or reflectivity refer to ideal, smooth surfaces, and here we mostly restrict their use to discussions concerning the aperture of a blackbody cavity. The reflectance factor, R, is also sometimes a useful quantity and is defined as the ratio of the radiant flux reflected from an element of surface, under specified conditions of irradiation and viewing, to that reflected by an ideal, completely reflecting, perfectly diffusing surface irradiated and viewed in the same way.

To each of these quantities we must add symbols to represent the viewing conditions, wavelength and temperature as necessary. With reference to Fig. 7-4, for example, the reflectance of an element x, in the most general case would be written as

$$\rho_x(\theta,\ \varphi,\ \omega;\ \theta',\ \varphi',\ \omega';\ \lambda;\ p)$$

where θ, φ, ω, θ', φ' and ω' are as indicated in Fig. 7-4, λ is the wavelength and p the state of polarization of the radiation. The addition of all of these modifiers all of the time would make most texts incomprehensible, and would obscure

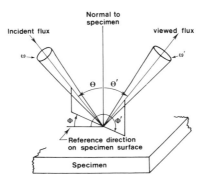

Fig. 7-4. Geometric parameters descriptive of reflection from a surface. θ is the zenith angle, φ the azimuthal angle, ω is the solid angle of the beam, and the superscript refers to viewing conditions.

the train of argument. The convention that must be adopted is to include only those modifiers that are relevant to the discussion. If only one or two are included they are sometimes written as subscripts rather than in parentheses after the symbol for the quantity, or it may be more convenient to place the letter indicating the element inside the parentheses if a different type of modifier is required as a subscript.

The bidirectional reflectance, $\rho_x(\theta, \varphi, d\omega; \theta', \varphi', d\omega')$, is defined by

$$\rho_x(\theta, \varphi, d\omega; \theta', \varphi', d\omega') = \frac{L'_x(\theta', \varphi') \cos \theta' d\omega'}{L_x(\theta, \varphi) \cos \theta \, d\omega} \tag{7-29}$$

where $L_x(\theta, \varphi)$ is the radiance of the beam incident on the element of surface x over the elementary solid angle $d\omega$ from the direction θ, φ, and $L'_x(\theta', \varphi')$ is the reflected radiance over the elementary solid angle $d\omega'$ in the direction θ', φ'. $L_x(\theta, \varphi)$ and $L'_x(\theta', \varphi')$ must not vary over the solid angles $d\omega$ and $d\omega'$.

The relationship between the various types of reflectance and reflectance factors can be deduced from the following identity:

$$\frac{\rho_x(\theta_1, \varphi_1, \omega_1; \theta_2, \varphi_2, \omega_2)}{\cos \theta_2 \omega_2} = \frac{\rho_x(\theta_2, \varphi_2, \omega_2; \theta_1, \varphi_1, \omega_1)}{\cos \theta_1 \omega_1} \tag{7-30}$$

From this we may deduce:

$$R_x(2\pi; \theta', \omega') = R(\theta, \omega; 2\pi) \tag{7-31}$$

where $\theta = \theta'$ and $\omega = \omega'$.

$$\rho_x(2\pi; 2\pi) = R(2\pi; 2\pi) \tag{7-32}$$

$$\rho_x(\theta, \omega; 2\pi) = R(\theta, \omega; 2\pi) \tag{7-33}$$

where θ and ω are the same for ρ_x and R_x.

$$R_x(\theta_1, \omega_1; \theta_2, \omega_2) = R(\theta_2, \omega_2; \theta_1, \omega_1) \tag{7-34}$$

$$\rho_x(\theta, \omega; \theta', \omega') = \frac{\omega' \cos \theta'}{\pi} R_x(\theta, \omega; \theta', \omega') \tag{7-35}$$

where for large solid angles $\omega' \cos \theta' = \int_{\omega'} \cos \theta' \cdot \sin \theta' d\theta' d\varphi'$.

Returning to the question of the Kirchhoff law, we can write down the statement of the law, for an isotropic surface (thus leaving out reference to φ), as follows:

$$\varepsilon_x(\theta', \omega', \lambda, p) = \alpha_x(\theta, \omega, \lambda, p) = 1 - \rho_x(\theta, \omega, 2\pi, \lambda, p) \tag{7-36}$$

where $\rho_x(\theta, \omega, 2\pi, \lambda, p)$ is the fraction of the radiation incident from the direction θ that is reflected into a hemisphere, and where it is assumed that $\tau = 0$. This equation must hold on the basis of conservation of energy and remembering this is often a useful way of verifying that the law is being properly applied; all of the radiation entering or leaving the system must be accounted for.

Kirchhoff's law is more generally valid than just under equilibrium conditions. If it were not, it would be of only limited use since freely radiating surfaces are not in equilibrium in the thermodynamic sense. However in applying Kirchhoff's law to non-equilibrium situations, it is important to define carefully what is meant by emission and absorption. It has been pointed out by Burkhard *et al.* (1972) that there are two ways of stating Kirchhoff's law, only one of which leads to the universally true statement that the emissive power equals the absorptive power.

The first (restricted) definition states that total emission is the sum of the stimulated emission and the spontaneous emission, while the total absorption is just the induced absorption. This statement leads to the ratio of the emissive to the absorptive powers being unity only in a blackbody environment, viz. under equal emission/absorption conditions. The second definition states that, total absorption is induced absorption minus stimulated emission, i.e. stimulated emission is treated as negative absorption. Total emission is just spontaneous emission. This second definition appears to be true under all conditions of thermal emission whether or not equilibrium is present. Further, the second definition is the one more closely corresponding to the operational definition of absorption. Experimentally, it is not possible to separate induced absorption from stimulated emission.

7-2-6   The optical properties of metals

We have already concluded that the calculation from first principles of the emitting properties of real materials is a hopeless task. From Kirchhoff's law, however, the problem may be converted to one of calculating the absorptance. This appears to be more tractable since it concerns the interaction of an external electromagnetic field with the electrons in the solid. A detailed discussion of this problem is not, however, within the scope of this book since it is rare that the results of calculated absorptances are used in thermometry. Although qualitative calculations can be made of the absorptance of metals and dielectrics, particularly in the low-frequency region where classical electromagnetic theory applies, the accuracy of the results for individual materials is not good enough for radiation thermometry. A good review of the optical properties of metals and dielectrics is given by Touloukian and DeWitt (1972).

7-3 The calculation of the emissivity of practical blackbody cavities

7-3-1 Introduction

We have seen that the radiation density inside a closed cavity depends only upon the temperature of the walls. It is quite independent of either their shape or their optical properties. We now come to the practical case of the cavity having a small hole in its wall through which radiation escapes. The problem is to calculate by how much this radiation departs from blackbody radiation for a given cavity geometry and material. This is a very important question, since the only access we have to high-temperature blackbody radiation is by means of a real cavity from which radiation is allowed to escape. In Chapter 2 we showed how IPTS-68 is based upon radiation pyrometry above the freezing point of gold. The scale is established in terms of the ratio of the spectral radiance of the high temperature source to that of a blackbody at the gold point. We therefore need to know how to make a practical blackbody and how to calculate the difference between its spectral radiance and that of an ideal blackbody at the same temperature. We will first of all examine the problem in qualitative terms and then go on to show how numerical solutions are obtained for some representative cavities.

Suppose that we are required to calculate the emissivity of the isothermal cavity shown in Fig. 7-5. The quantity to be calculated is the ratio of the spectral radiance of the element of the wall ΔS viewed by P to that of a blackbody at the same temperature. Now the radiant flux leaving ΔS in the direction of the aperture "a" is made up of two parts: the radiant flux emitted by ΔS and the radiant flux reflected by ΔS. The former depends only upon the emittance of the wall and its temperature and is independent of the presence of

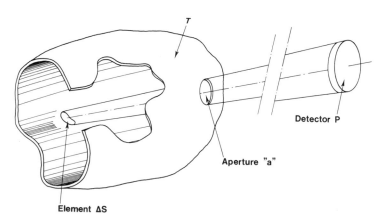

Fig. 7-5. Schematic blackbody cavity emitting radiation through an aperture in its walls which is viewed by a detector at P.

the rest of the cavity. The reflected flux, on the other hand, depends upon the reflectivity of ΔS and the radiant flux arriving at ΔS from the rest of the cavity. This is affected by "a" since the radiant flux which, in a closed cavity would have been coming towards ΔS from "a", is missing. It is thus the effect of the absence of incident flux from "a" on the radiant flux reflected from ΔS that we must calculate. Not only is the radiant flux in the direction "a"$\rightarrow\Delta S$ absent, but also the radiant flux from "a" towards the rest of the walls of the cavity is absent. Thus in turn the radiant flux arriving at ΔS from the whole of the rest of the cavity is slightly depleted. It must by now be clear that the calculation of the emissivity of such a cavity is by no means a trivial operation. An accurate result requires a detailed knowledge of the geometry of the cavity and the viewing system, the angular emitting and reflecting properties of the material of the wall of the cavity and the temperature distribution over the cavity. A non-uniform temperature modifies the radiant flux over the whole of the cavity, as does the presence of an aperture, but with the added complication that the modifications to the radiant flux brought about by temperature gradients can be both positive and negative.

The accuracy with which the calculation represents the real behaviour of an isothermal cavity is almost always limited by lack of knowledge of the reflecting properties of the wall. In practice, the limiting factor is usually non-uniformity in the temperature leading to an uncertainty in the temperature to which the emitted radiation should be ascribed. This is almost always the case for cavities having very high emissivities, i.e. cavities for which the ratio of the size of the hole to the size of the cavity is very small. Under these circumstances the details of the angular reflecting and emitting properties of the walls are not very critical since the total effect of the presence of the hole is small. For the purposes of radiation pyrometry these are the sort of cavities that are of most interest and we do not, therefore, need to know a great deal about the angular optical properties of surfaces. Except for cavities having very unusual geometries, the assumption of diffuse or Lambertian emission leads to very small errors since it is only at very high angles from the normal that this assumption ceases to be valid. The assumption that all materials are diffuse reflectors of thermal radiation is, however, much less justified. In fact all metal and most other surfaces, if they are polished, are specular reflectors of radiation and this we cannot ignore. However, by roughening the surface, it is very easy to render the reflectivity partially diffuse although it is difficult indeed to make the surface into a perfect diffuse reflector. Nevertheless, since at each reflection a significant part of the radiation is diffusely reflected, then in high emissivity cavities where many reflections take place, the assumption of diffuse reflection is reasonable. This statement must be qualified with certain conditions which we will come to later on.

The most successful methods of calculating the emissivities of near-blackbody cavities are based upon the assumption of diffuse emission and reflection. It is found that the results agree with experiment insofar as experimental measurements are possible. Corrections to take into account semi-specular reflection can be made during the course of the calculation.

The emissivities of cavities made from perfect specular reflectors are calculated in quite a different, and usually much simpler, way. The emissivity is calculated directly in terms of the number, n, of reflections undergone by an incident ray before it leaves the cavity. Later on we shall come to some of the simpler geometrical constructions used to arrive at a value for "n" for the more common cavity geometries.

7-3-2 The integral equation method for diffusely reflecting cavities

To illustrate the calculation of the emissivity of a cavity having diffusely reflecting walls, we will consider the cylindrical cavity of Fig. 7-6. This avoids the necessity of writing down the equations in their most general form and allows us to proceed directly to some numerical results. A cavity of the shape shown in Fig. 7-6 is very similar to those used in practice for the realization of blackbodies for the calibration of radiation pyrometers. Although certain modifications are possible to increase its emissivity and to take care of specular reflections (the back wall may be inclined or grooved for example), the simple shape shown in this figure allows the calculation to be demonstrated in detail without too much geometrical complication.

We shall set ourselves the problem of calculating the apparent emissivity ε_a of elements on the back wall and along the cylindrical wall of the cavity. Consider first a cylindrical element dx_0' of area dA_{x_0} at a distance x_0' from the end wall. For the sake of convenience, we shall adopt the dimensionless variables x, r and z defined as $x = x'/D; r = r'/R_2$ and $z = z'/R_2$. Where $D(=2R_2)$ is the diameter of the cylinder and x', r' and z' are co-ordinates on the cylindrical, back and front walls respectively.

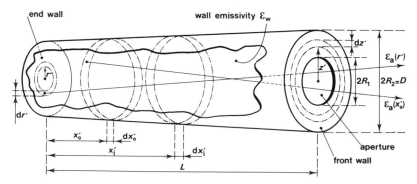

Fig. 7-6. Cylindrical blackbody cavity showing some of the geometrical factors used for calculating the apparent emissivity of an element on the end wall $\varepsilon_a(r)$ and on the cylindrical wall $\varepsilon_a(x_0)$, using the integral equation method.

We define the spectral emittance of the element dA_{x_0} by

$$\varepsilon_a(x_0, \lambda, T) = \frac{M(x_0, \lambda, T)}{M(\lambda, T)} \tag{7-51}$$

where $M(x_0, \lambda, T)$ and $M(\lambda, T)$ are the spectral radiant exitances of the element and a blackbody respectively at a temperature T. Henceforth we shall drop the references to λ and T, but we must remember that they are understood.

The emissivity of dA_{x_0} can now be written down as

$$\varepsilon_a(x_0) = \varepsilon_w + (1 - \varepsilon_w) \int_{x=0}^{L/D} \varepsilon_a(x) dF_{x_0-x} + (1 - \varepsilon_w) \int_{r=0}^{1} \varepsilon_a(r) dF_{x_0-r}$$

$$+ (1 - \varepsilon_w) \int_{z=R_1/R_2}^{1} \varepsilon_a(z) dF_{x_0-z} \tag{7-52}$$

where ε_w is the emittance of the material of the walls of the cavity, $\varepsilon_a(r)$ is the apparent emissivity of an element on the back wall, and $\varepsilon_a(z)$ is the apparent emissivity of an element on the inside of the front wall. This equation simply represents the radiation emitted directly by the element dA_{x_0} (the first term) plus the reflected radiation originating from other elements on the cylindrical wall (the second term) plus the reflected radiation originating from the end wall and the inner surface of the front wall (third and fourth terms respectively).

Similar expressions can be written down for $\varepsilon_a(r)$ and $\varepsilon_a(z)$:

$$\varepsilon_a(r) = \varepsilon_w + (1 - \varepsilon_w) \int_{x=0}^{L/D} \varepsilon_a(x) dF_{r-x} + (1 - \varepsilon_w) \int_{z=R_1/R_2}^{1} \varepsilon_a(z) dF_{r-z} \tag{7-53}$$

$$\varepsilon_a(z) = \varepsilon_w + (1 - \varepsilon_w) \int_{x=0}^{L/D} \varepsilon_a(x) dF_{z-x} + (1 - \varepsilon_w) \int_{r=0}^{1} \varepsilon_a(r) dF_{z-r} \tag{7-54}$$

The angle factors dF_{x_0-x}, dF_{x_0-r}, ... dF_{i-j} are defined as the fraction of the radiation emitted by an element i that is directly intercepted by an element j. There are simple relations between angle factors for diffuse emission, for example

$$dA_i dF_{i-j} = dA_j dF_{j-i} \tag{7-55}$$

where dA_i is the area of element i.

The solution of the three simultaneous integral equations now becomes a problem in mathematics. An iterative method must be used starting from an assumed distribution for $\varepsilon_a(x)$, $\varepsilon_a(r)$ and $\varepsilon_a(z)$ as a first approximation. Successive approximations converge provided that precautions are taken to avoid the difficulties posed by singularities that occur in the integrals at $x = x_0$ and at the junctions of the cylindrical and end walls. The various techniques for overcoming these difficulties have been discussed by a number of authors, notably Sparrow and co-workers (1962) and Peavy (1966). More recently, a rather better method of solving the equations has been demonstrated by Bedford and Ma (1974). Taking advantage of the slowly varying character of $\varepsilon_a(x)$, $\varepsilon_a(r)$ and $\varepsilon_a(z)$, they transformed the integrals of equations $(7\text{-}52) \rightarrow (7\text{-}54)$ into summations over a large number ($n \approx 100$) of zones.

$$\varepsilon_a(x_0) = \varepsilon_w + (1 - \varepsilon_w) \sum_{i=1}^{n} \tfrac{1}{2}[\varepsilon_a(x_{i+1}) + \varepsilon_a(x_i)](dF_{x_0,x_{i+1}} - dF_{x_0,x_i})$$

$$+ (1 - \varepsilon_w) \sum_{j=m}^{n} \tfrac{1}{2}[\varepsilon_a(z_{j+1}) + \varepsilon_a(z_j)](dF_{x_0,z_{j+1}} - dF_{x_0,z_j})$$

$$+ (1 - \varepsilon_w) \sum_{k=1}^{n} \tfrac{1}{2}[\varepsilon_a(r_{k+1}) + \varepsilon_a(r_k)](dF_{x_0,r_{k+1}} - dF_{x_0,r_k}), \qquad (7\text{-}56)$$

$$\varepsilon_a(r) = \varepsilon_w + (1 - \varepsilon_w) \sum_{i=1}^{n} \tfrac{1}{2}[\varepsilon_a(x_{i+1}) + \varepsilon_a(x_i)(-dF_{r,x_{i+1}} + dF_{r,x_i})$$

$$+ (1 - \varepsilon_w) \sum_{j=m}^{n} \tfrac{1}{2}[\varepsilon_a(z_{j+1}) + \varepsilon_a(z_j)](dF_{r,z_{j+1}} - dF_{r,z_j}), \qquad (7\text{-}57)$$

$$\varepsilon_a(z) = \varepsilon_w + (1 - \varepsilon_w) \sum_{i=1}^{n} \tfrac{1}{2}[(\varepsilon_a(x_{i+1}) + \varepsilon_a(x_i)](dF_{z,x_{i+1}} - dF_{z,x_i})$$

$$+ (1 - \varepsilon_w) \sum_{k=1}^{n} \tfrac{1}{2}[\varepsilon_a(r_{k+1}) + \varepsilon_a(r_k)](dF_{z,r_{k+1}} - dF_{z,r_k}), \qquad (7\text{-}58)$$

where

$$dF_{x_0,x_i} = [(x_i - x_0)^2 + 1]^{1/2} - \{2[(x_i - x_0)^2 + 1]^{1/2}\}^{-1} \pm (x_0 - x_i),$$
$$+ (x_0 < x_i < x_{i+1})$$
$$- (x_i < x_{i+1} < x_0)$$

$$dF_{x_0,r_k} = -x_0 \left\{ 1 - \frac{4x_0^2 + r_k^2 + 1}{[(4x_0^2 + r_k^2 + 1)^2 - 4r_k^2]^{1/2}} \right\},$$

$$dF_{x_\theta, z_j} = dF_{L/D - x_0, r_k},$$

$$dF_{r, x_i} = \frac{1}{2}\left\{1 - \frac{4x_i^2 + r^2 - 1}{[(4x_i^2 + r^2 + 1)^2 - 4r^2]^{1/2}}\right\},$$

$$dF_{r, z_j} = \frac{1}{2}\left\{1 - \frac{4(L/D)^2 + r^2 - z_j^2}{[(4(L/D)^2 + r^2 + z_j^2)^2 - 4r^2 z_j^2]^{1/2}}\right\},$$

$$dF_{z, x_i} = dF_{r, L/D - x_i},$$

and

$$dF_{z, r_k} = dF_{r, z_j}.$$

Once again the solution of these equations must be obtained iteratively starting from an assumed distribution for $\varepsilon_a(x)$, $\varepsilon_a(r)$ and $\varepsilon_a(z)$. The simplest initial distribution is evidently $\varepsilon_a(x) = \varepsilon_a(r) = \varepsilon_a(z)$ choosing a value somewhat higher than ε_w. The rate of convergence is only moderately sensitive to the initial values, and it is rarely necessary to proceed beyond four or five iterations to arrive at a solution correct to three significant figures.

The way of avoiding difficulty at the junction of the cylinder and end faces is as follows: when $r_k = 1$, $x_0 \to 0$, and when $z_j = 1$, $x_0 \to L/D$ and we must have $dF_{x_0 - r_k}$ and $dF_{x_0 - z_j} \to 1/2$. Therefore the last term in the second summation of equation (7-56) is replaced by $\varepsilon_a(z_j = 1)/2$, and the last term in the third summation by $\varepsilon_a(r_k = 1)/2$. Similarly the first term in the first summation of equation (7-57) becomes

$$\tfrac{1}{2}\varepsilon_a(x = 0) + \tfrac{1}{2}[\varepsilon_a(0) + \varepsilon_a(x_2)][\tfrac{1}{2} - dF_{1 - x_2}]$$

and the last term of the first sum becomes

$$\tfrac{1}{2}\varepsilon_a(x = L/D) + \tfrac{1}{2}[\varepsilon_a(L/D) + \varepsilon_a(x_n)][\tfrac{1}{2} - dF_{1 - x}].$$

In Figs 7-7 and 7-8 are shown the results of solving equations (7-56)$\to$(7-58) for various values of R_1/R_2, L/D and ε_w. The discontinuity in the value of apparent emissivity at the junction of the cylindrical wall and the base arises because there is a discontinuous change in the angle of the wall with respect to the aperture. This becomes very clear when we look at an alternative reflectivity method of calculating $\varepsilon_a(x)$ and $\varepsilon_a(r)$. The size of the discontinuity falls as the apparent emissivity of the cylindrical and flat walls tends to unity so that in the limit of infinitesimal aperture diameter the discontinuity tends to zero, as must be the case of course.

It should be noted that there is a small increase in apparent emissivity towards the edges of the back wall. This is much more marked in shallow

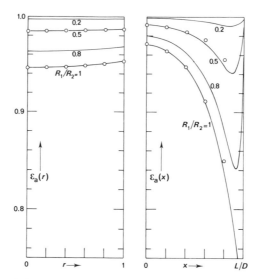

Fig. 7-7. Apparent emissivities of elements on the end wall and cylindrical wall of the cavity shown in Fig. 7-6 for $L/D = 2$, $\varepsilon_\omega = 0.5$ and different aperture sizes, calculated, using the integral equation method, by Bedford and Ma (1974) equations (7-56), (7-57) and (7-58) (solid lines) and using the series reflection method, equations (7-70) and (7-71) (circles).

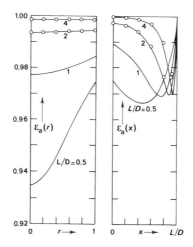

Fig. 7-8. Apparent emissivities as in Fig. 7-7 but in this case for $\varepsilon_\omega = 0.7$, $R_1 = R_2/2$ and different values of L/D.

cavities than in deep ones. Its presence simply represents the fall in the solid angle subtended by the aperture at the element as it moves towards the edge. The presence of the front wall containing the aperture not only increases the emissivity over the whole of the inside of the cavity, but has also another beneficial effect. In calculating the total heat loss to the outside from the inner walls of the cavity, it is found that the largest proportion is lost from those parts of the cylindrical wall at which the aperture subtends the largest angle. Hence for a cylinder having an open end, large amounts of heat are lost from those parts of the wall near the opening. Thus the presence of the front wall not only increases substantially the emissivity of all the elements inside the cavity, as is shown in Fig. 7-7, but reduces the overall heat loss as well.

The emissivity of the front wall has little effect upon the apparent emissivity of the rest of the cavity. The inside of the front wall of the cavity cannot "see" the aperture, and it is irradiated by practically the same radiation flux as streams through the aperture, i.e. practically blackbody radiation. This is made clear in equation (7-54) by the absence of any term containing an integral having as limit R_1. Provided that it is at the same temperature as the rest of the cavity, and since it is irradiated by near blackbody radiation, the sum of its emitted and reflected radiation will be equally black no matter what its value of ε_w. Indeed there is an advantage in having the front face highly reflecting. This stems from the consequent reduction in heat loss from a poorly emitting outside face and therefore a better chance of maintaining a uniform temperature. Even if the front wall is slightly cooler due to heat loss, the fact that it has a low emissivity will decrease the effect on the apparent emissivity of the rest of the cavity. This would be represented by the first term ε_w (in this case not the same as ε_w for the rest of the cavity) in equation (7-54) being small. In the remaining terms the controlling factors are $\varepsilon_a(x)$ and $\varepsilon_a(r)$, which remain high.

The illustrations of the results of calculations of the emissivity of cavities, shown in Figs. 7-7 and 7-8, can give no more than a general impression of the likely values of emissivity in similarly shaped cavities, see also Hattori and Ono (1982) and Ono (1982). In view of the nature of the problem, extrapolation or interpolation of the values calculated for a particular geometry, wall emissivity and temperature distribution, to other values of these parameters is difficult.

7-3-3   Series reflectivity method for diffusely reflecting cavities

In view of the difficulties in extrapolating or interpolating results from the integral equation method, it is useful to examine a simple method of calculating approximate values of $\varepsilon_a(x)$ and $\varepsilon_a(r)$ that avoids the necessity of solving the integral equations for every value of L, D, R_1 and ε_w.

In the past many approximate methods of calculating the emissivities of cavities were tried, for the obvious reason that before the days of easy access to computers, the numerical solution of equations such as (7-52) to (7-54) was not practicable. The most useful of these approximate methods is based upon work of de Vos (1954). In his method, the problem is transformed into one of calculating the absorptivity of the cavity for a ray incident from the direction in which the emissivity is required. From Kirchhoff's law we have

$$\varepsilon_a \text{ (aperture; } \theta', \lambda, T) = a_a \text{ (aperture; } \theta, \lambda, T)$$

$$= 1 - \rho_a \text{ (aperture; } \theta'; 2\pi; \lambda, T). \qquad (7\text{-}59)$$

Thus if we can calculate ρ_a (aperture; θ; 2π; λ, T), we can deduce ε_a (aperture; θ'; λ, T). The quantity to be calculated therefore is the fraction of the radiation entering the aperture from direction θ, that is finally reflected out of the aperture over a hemisphere. Since the aperture itself does not reflect in the ordinary sense of the word, we must consider instead the element of wall directly opposite the aperture in the direction of the entering ray. Thus we re-write equation (7-59) in terms of the reflectance ρ_w of the element of wall x_0; at the same time we drop reference to θ, λ and T, remembering that they are the same for the emissivity, absorptivity and reflectivity. Thus

$$\varepsilon_a(x_0) = a_a(x_0) = 1 - \rho_a(2\pi) \qquad (7\text{-}60)$$

where 2π is left to remind us that radiation reflected from the element x_0 in all directions, and thus finally out of the cavity in all directions, must be accounted for.

The method is sometimes known as the series reflection method since $\rho_a(2\pi)$ is represented as the sum of a series. The first term of the series is the fraction of the incident radiation that is reflected straight out of the aperture after just one reflection, the second term is the fraction reflected out after two reflections and so on up to n reflections. Clearly if n is large enough, this method must lead to results that are just as accurate as the integral equation method. However the mathematical complexity of the reflection method rises very rapidly as n increases, and it is only worth considering when n need be no more than two or at most three.

The series reflection method was originally developed by de Vos to calculate the emissivities of a number of different shaped cavities for both diffuse and semi-specular reflection as well as uniform and non-uniform temperatures. Herein lies the attraction of the reflection method: it is quite easy to apply to a wide range of conditions. The great disadvantage is that it is difficult to estimate the accuracy of the results for a particular set of conditions because it is difficult to show, in the general case, that the number of reflections taken into account is sufficient.

We will demonstrate the reflection method by using it to calculate the emissivity of the cavity shown in Fig. 7-9, the same-shaped cavity as we used for the integral equation method. Here the cavity is of unit radius, length L, has an aperture of radius R and co-ordinates on the cylindrical and back wall of x and r respectively. If we take first the apparent emissivity of the centre of the

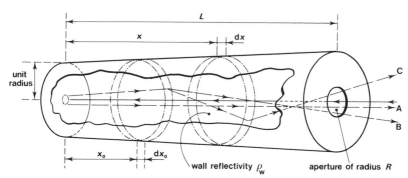

Fig. 7-9. Cylindrical blackbody cavity showing some of the geometrical factors used in the calculation of $\varepsilon_a(r)$ and $\varepsilon_a(x)$ by the series reflection method. "A" represents the term in ρ_w in equation (7-61). "B" the term in ρ_w^2 and "C" the term in ρ_w^3

back wall $\varepsilon_a(r_0)$, we can write down

$$\varepsilon_a(r_0) = a_a(r_0) = 1 - \rho_a(2\pi)$$

$$= 1 - \rho_w I_1(L) - \rho_w^2 I_2(L) - \dots \rho_w^n I_n(L) \qquad (7\text{-}61)$$

where $\rho_w I_1(L)$ is the fraction of the incident radiation that is reflected out of the cavity after one reflection, i.e. that fraction reflected straight back along the incident ray from the element r_0 at the centre of the back wall; $\rho_w^2 I_2(L)$ is the fraction reflected out after two reflections and so on. ρ_w is the diffuse reflectance of the wall and $I_n(L)$ is a geometrical factor.

We can write down the first term to a close approximation as

$$\rho_w I_1(L) = \rho_w \frac{R^2}{L^2 + R^2}. \qquad (7\text{-}62)$$

The second term is of course the result of an integration since we must consider radiation reflected from the centre of the back wall towards elements at all positions along the cylindrical walls. It can be shown (Quinn, 1967) that

$$\rho_w^2 I_2(L) = 2\rho_w^2 R^2 \int_{x=0}^{L} \frac{x(L-x)}{(1+x^2)^2 \{1+(L-x)^2\}^2} \, dx. \qquad (7\text{-}63)$$

This term is also approximate to the extent that we ignore the variation of ρ_w over the solid angle subtended by the aperture at x. Errors due to this are not significant except for cavities having $L < 2$ and $R \to 1$. Numerical integration of (7-63) leads to values of $I_2(L)$ given in Table 7-1.

The problem remains of what to do with the remaining terms for $n > 2$. For the case of the cylindrical cavity, the third term may be found without too much

Table 7-1. Values of the geometrical factor $I_2(L)$ calculated from equation (7-63). L is expressed relative to the unit radius of the cavity (Fig. 7-9).

L	$I_2(L)$	L	$I_2(L)$	L	$I_2(L)$	L	$I_2(L)$
4	0.0488	7	0.0096	10	0.0031	16	0.0007
5	0.0267	8	0.0063	12	0.0017	18	0.0005
6	0.0155	9	0.0043	14	0.0010	20	0.0003

difficulty, but to go any further leads us to a level of complexity approaching that of the integral equation method.

At this point let us digress a little and consider the case of the diffusely reflecting spherical cavity. A sphere has the useful property that the solid angle subtended by an aperture is constant over all elements of the inside surface. Thus for diffuse reflection, the fraction of the radiation remaining in the cavity that leaves through the aperture after each reflection is the same. This fraction is s/S, where S is the surface area of the complete sphere and s is the area of the curved surface removed by the aperture. For a sphere in which the radius of the aperture is R and depth is L, i.e. the distance from the centre of the aperture to the opposite wall, we have

$$s/S = 1/(1 + L^2/R^2). \tag{7-64}$$

It is easy to show that the reflectivity of any element inside the sphere is given by the series

$$\rho_w s/S[1 + \rho_w(1 - s/S) + \rho_w^2(1 - s/S)^2 + \dots] \tag{7-65}$$

which has a sum

$$\frac{\rho_w s/S}{1 - \rho_w(1 - s/S)} \tag{7-66}$$

Thus the emissivity of a spherical cavity is

$$\varepsilon_a(x) = 1 - \frac{\rho_w s/S}{1 - \rho_w(1 - s/S)}. \tag{7-67}$$

This is a result given in slightly different forms by Gouffé (1945), de Vos (1954), Sparrow and Jonsson (1962) and Bedford (1972). Gouffé proposed that this formula be applied to cavities of arbitrary shape. He was apparently unaware that the implicit assumption in the formula is that the reflected radiation becomes uniformly distributed throughout the cavity after the first reflection. For cavities having an approximately spherical or cubic shape his formula

(7-67) indeed gives results that are nearly correct. The further a cavity departs from a spherical shape, however, the greater is the error in applying equation (7-67). In particular for long cylindrical cavities Gouffé's formula gives values of emissivity that are much too low. It is easy to see why this is the case; after the first reflection at the base of the cylinder, most of the reflected radiation impinges upon wall elements near the base and relatively little escapes after the second reflection compared with the predictions of Gouffé's formula. However, after a number of reflections the radiation begins to be more uniformly distributed. In the limit of large n we must have that the fraction of the radiation reflected out of the cavity is $\approx \rho_w s/S$ of that remaining after the nth reflection. After the $(n+1)$th reflection we have $\rho_w^2(s/S)(1-s/S)$ of that remaining after the nth, and after the $(n+2)$th $\rho_w^3(s/S)(1-s/S)^2$ of that remaining after the nth and so on. In other words, as soon as the radiation has become sufficiently diffused throughout the cavity, the Gouffé approach appears valid.

Returning now to the problem of the cylindrical cavity, we can account for the terms in the reflection series, equation (7-61), not explicitly calculated, by asserting that for the third and subsequent terms that radiation is uniformly distributed throughout the cavity. Equation (7-61) thus becomes:

$$\varepsilon_a(r_0) = 1 - \rho_w I_1(L) - \rho_w^2 I_2(L) - \sum_{n=3}^{\infty} \rho_w^n I_n(L) \tag{7-68}$$

where

$$\rho_w^n I_n(L) = \rho_w^3(s/S)\left(1 - \frac{s}{S}\right)^2 [1 + \rho_w(1 - s/S) + \rho_w^2(1 - s/S)^2 + \cdots$$

$$= \rho_w^3(s/S)(1 - s/S)^2/[1 - \rho_w(1 - s/S)] \tag{7-69}$$

in which we write $s/S \approx R^2/(L^2 + R^2)$ for an equivalent sphere of depth L and aperture radius R.

The further modification that can be made to equation (7-61) is to consider values of $\varepsilon_a(r)$ away from the centre of the back surface and $\varepsilon_a(x)$ along the cylindrical wall. This can easily be done by taking account of the cosine relation for diffuse reflectivity and modifying the expression for the angle subtended by the aperture at x or r. For the cylinder shown in Fig. 7-9 we have

$$\varepsilon_a(x) = 1 - \frac{\rho_w R^2(L-x)}{\{1 + (L-x)^2\}^2} - \rho_w^2 R^2 I_2(L) - \rho_w^3 \frac{R^2}{L^2 + R^2}\left\{1 - \frac{R^2}{L^2 + R^2}\right\}$$

$$\left\{1 - \rho_w\left(1 - \frac{R^2}{L^2 + R^2}\right)\right\}^{-1} \tag{7-70}$$

and

$$\varepsilon_a(r) = 1 - \frac{\rho_w R^2 L^2}{(L^2 + r^2)^2} - \rho_w^2 R^2 I_2(L) - \rho_w^2 \frac{R^2}{L^2 + R^2}\left\{1 - \frac{R^2}{L^2 + R^2}\right\}$$

$$\left\{1 - \rho_w\left(1 - \frac{R^2}{L^2 + R^2}\right)\right\}^{-1}. \tag{7-71}$$

Values of $\varepsilon_a(x)$ and $\varepsilon_a(r)$ calculated for various points along the cylindrical wall and across the back wall are given in Table 7-2 and shown in Figs 7-7 and 7-8 along with the results of calculations made by the integral equation method. The agreement is very satisfactory.

It must be remembered that the results obtained using equations (7-70) and (7-71) are in good agreement with the results of the integral equation method because the size of individual terms for $n \geqslant 3$ is small. It follows also that the first two terms are also small, i.e. we are dealing with a cavity having a high emissivity. These conditions obtain if either $R^2/(L^2 + R^2) \ll 1$ or if ρ_w is less than about 0.4. Since these conditions invariably apply to cavities used as spectral radiance standards for optical pyrometry, it can be assumed that equations (7-70) and (7-71) can be used in this case with little risk of significant error.

The application of the modified series reflection method exemplified by

Table 7-2. Values of $\varepsilon_a(x)$ and $\varepsilon_a(r)$ calculated from equations (7-70) and (7-71) with corresponding values from the integral equation (IE) method (Bedford and Ma, 1974). L, r, and R are measured relative to the unit radius of the cavity.

	$\varepsilon_a(x)$			$\varepsilon_a(r)$	
x/L	Eq. (7-70)	IE	r	Eq. (7-71)	IE
$L=8; R=0.5; \rho_w=0.3$					
0.0	0.99957[a]	0.99955	0.0	0.99854[a]	0.99856
0.2	0.99944[a]	0.99946	0.6	0.99855[a]	0.99858
0.4	0.99910[a]	0.99885	1.0	0.99857[a]	0.99862
0.6	0.99781[a]	0.99677			
$L=8; R=0.25; \rho_w=0.3$					
0.0	0.99989	0.99989	0.0	0.99963	0.99964
0.2	0.99986	0.99986	0.6	0.99964	0.99964
0.4	0.99977	0.99971	1.0	0.99964	0.99965
0.6	0.99945	0.99917			

Values shown as circles in Fig. 7-8.

equations (7-70) and (7-71) to other shaped cavities is not difficult provided that $I_2(L)$ can be evaluated. This is straightforward for cones and other simple shapes. In the limit of $\rho_w(R^2/(L^2 + R^2)) \ll 1$ it is not even necessary to take into account terms of order $n \leqslant 2$, the first term is sufficient. Cavities for which this is true would be those having a surface reflectance ρ_w of less than 0.1 and $R^2/(L^2 + R^2) < 0.0005$.

The application of the series reflectivity method for a cylindrical cavity having a re-entrant conical base has been demonstrated by Berry (1981) for the cases of isothermal, non-isothermal diffuse and a mixture of diffuse and specular reflection. The complexity of such a calculation limited the number of terms of the series to two, but nevertheless the results provide useful guidance in designing cavities made of materials having a low reflectivity.

7-3-4   The emissivity of specularly reflecting cavities

The calculations and results we have been discussing up to now refer to cavities having diffusely reflecting and emitting walls. For cavities having specularly reflecting walls, the calculation is, as we have already mentioned, much simpler. In this case the series reflection method is always to be preferred since the problem then becomes one of ray tracing, the radiance being multiplied by ρ_w after each reflection. The emissivity is given, in terms of Kirchhoff's law by

$$\varepsilon_a(x,\ \theta,\ \varphi) = a_a(x,\ \theta,\ \varphi) = 1 - \rho_a \text{ (aperture, } \theta,\ \varphi,\ 2\pi) \qquad (7\text{-}72)$$

where ρ_a (aperture, θ, φ, 2π) is simply ρ_w^n, where n is the number of reflections undergone by a ray incident from the direction θ, φ before it is reflected out of the aperture in any direction.

The most advantageous shape of cavity for specular reflection is a cone. By arranging for an oblique angle of view, "n" can be increased and theoretically tends to infinity at oblique grazing incidence. The geometrical construction for calculating n is shown in Fig. 7-10.

For the general case of a ray entering a cone and striking the surface of the cone at an angle γ to the generating line of the cone and at an oblique angle α, the number of reflections undergone before finally leaving the cone is (Quinn, 1981)

$$n = \text{ent} \left\{ \frac{\pi - 2\gamma + 2\varphi \cos \alpha}{2\varphi \cos \alpha} \right\} \qquad (7\text{-}73)$$

where ent$\{-\}$ represents the greatest integer less than or equal to $\{-\}$ and φ is the half angle of the cone.

If the ray strikes the surface of the cone not at A but at D', between O and A,

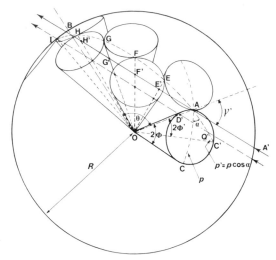

Fig. 7-10. The geometrical construction used for demonstrating the path of an oblique ray A′A entering a specularly reflecting cone OAC and undergoing reflections at A and, in image space at E′, F′, G′ and H′ before leaving the fifth image of the cone at **B**.

an extra reflection may occur. Suppose that the ray entering the cone crosses the projected cone aperture at a point Q' on p'. As Q' moves along p' towards C', a point will be reached for which $AQ' = (1 - f)p'$, where f is the remaining fraction from equation (7-73). For values of $AQ' \geqslant (1 - f)p'$, an extra reflection occurs at I. Only one extra reflection is possible since the point C' will be reached before a further integral value of p' can occur.

It is clear from equation (7-73) that it is preferable to view a specularly reflecting cone at an oblique angle rather than axially. The practical limit to n is set by both the width of the beam, since α varies over the width, and, more particularly, by the small fraction of diffuse reflection that occurs at each reflection.

We shall not pursue this question of the emissivity of specularly reflecting cavities since such conditions rarely occur in thermometry. It is much too difficult to maintain a specularly reflecting surface at high temperatures. A smooth metal surface will always develop grain-boundary grooves and sometimes grain-surface roughness if held at high temperatures for long enough. This is why calculations of the emissivity of cavities are preferably made on the basis of diffuse reflection. Calculations for specular conditions for conical and cylindrical cavities having an inclined or conical end wall, will lead to values of emissivity that are rather higher than those given by the equivalent diffuse calculation. For most real materials at high temperatures such results would be misleading.

For identical viewing conditions the emissivity of a cavity having semi-specularly reflecting walls must lie between the values calculated on the basis of diffuse reflection and specular reflection. It is easy to design a cavity that would have a high emissivity for perfectly specular reflection as well as for diffuse reflection. In a later section we shall see how this can best be done.

7-3-5 Emissivity of non-isothermal cavities

The effect of a temperature gradient along the axis of a cylindrical or conical cavity is quite easy to take into account in the calculation of the emissivity. It is very much more difficult to measure such a gradient in practice. For this reason the application of a correction due to the presence of a temperature gradient is only to be recommended if there is no way of removing the temperature gradient or if its magnitude and distribution are reasonably well known.

Bedford and Ma (1974) have made calculations of the effect of a temperature gradient in various cylindrical, conical and cylindrical/conical cavities.

The presence of a temperature gradient can be taken into account by including in equations (7-52) and (7-54) terms which take account of the variation of emitted radiation from each of the elements i, j and k. This is done by defining $\varepsilon_a(x_0, \lambda, T_{x_0})$ for example, as follows

$$\varepsilon_a(x_0, \lambda, T_{x_0}) = \frac{M(x_0, \lambda, T_{x_0})}{M(\lambda, T_0)} \tag{7-74}$$

where T_0 is a reference temperature. For a cavity of the sort shown in Fig. 7-6, T_0 can conveniently be taken to be the temperature at the centre of the back wall. The results of such a calculation are shown in Fig. 7-11 for a cylindrical

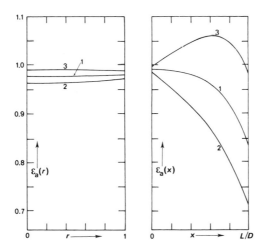

Fig. 7-11. Apparent emissivities as in Fig. 7-7, but in this case for $L/D = 2$, $\varepsilon_\omega = 0.7$, $R_1 = R_2$ and for various temperature distributions along the axis of the cylinder, 1, a uniform temperature; 2, a linear 1 % decrease in T_x from $x = 0$ to $x = L/D$, $T = 1300\,\text{K}$; 3, a linear 1 % increase in temperature from $x = 0$ to $x = L/D$, all for $\lambda = 650\,\text{nm}$ (after Bedford and Ma, 1974).

cavity having $T_0 = 1300$ K and taking a wavelength of 650 nm and for various temperature gradients along the cylinder.

7-4   Practical blackbody cavities

The practical realization of cavities, taking into account the various conditions and limitations mentioned in earlier sections, is illustrated by the following examples.

For the calibration of industrial radiation pyrometers, it is necessary to design a cavity having a relatively large aperture. We shall take this as the example with which to begin our discussion of practical cavities since its design highlights the factors that must be taken into account and which are used in the design of cavities having much higher emissivities. Figure 7-12 shows a cavity designed for this purpose for use up to about 1100 °C. The apparent emissivity required is not as high as is sought for example in the cavity of Fig. 7-13, and a value of 0.99 is often adequate. The cavity is made from stainless steel or nickel, and the surface rapidly oxidizes to give a diffusely reflecting

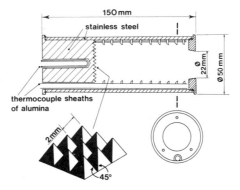

Fig. 7-12. A stainless steel blackbody designed for the calibration of industrial radiation pyrometers requiring a source of relatively large aperture.

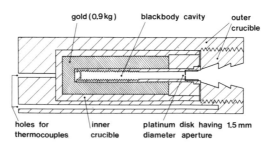

Fig. 7-13. A graphite blackbody used for the realization of the gold point.

surface having an emissivity, in the visible, of about 0.7 rising to 0.85 towards 15 μm beyond which it falls. Provision is made for measuring the temperature of the back wall and the temperature distribution by means of thermocouples. Short lengths of thin-walled alumina thermocouple sheaths are pressed into close fitting holes in the stainless steel or nickel during construction. Subsequent oxidation grips these tubes tightly and they subsequently protect the measuring thermocouples from the oxide and permit them to slide easily in and out.

The principles governing the design of this cavity were drawn from a consideration of the factors which enter into the calculation of the apparent emissivity using the series reflection method. It is clear from equation (7-70) that the fraction of the incident radiation that is reflected out can be substantially reduced by making simple modifications to the cylindrical cavity used as the model in Fig. 7-9. First of all the term in ρ_w can be reduced by grooving the back wall of the cavity. The normal reflectivity ρ_v of a V-groove having diffusely reflecting sides of reflectivity ρ_w and included angle θ is given by Psarouthakis (1963).

$$\rho_v = \frac{\rho_w \sin \theta/2}{1 - \rho_w(1 - \sin \theta/2)} \tag{7-75}$$

A "V" groove made from stainless steel, $\rho_w \approx 0.3$, and having an included angle of 30° would have $\rho_v \approx 0.1$. Thus, if the back wall is made with multiple "V" grooves, ρ_w in the first term of equation (7-70) is simply replaced by ρ_v. The second term, that in ρ_w^2, represents the radiation reflected from the back wall to the cylindrical wall and thence out of the aperture. The second term can be eliminated if the cylindrical walls are lined with a sufficient number of baffles to block completely reflections of this kind. The only reflections of this order which cannot be eliminated are those from the internal edges of the baffles. These can be reduced to a very low level by sharpening the edges. The emissivity of the cavity thus takes the form

$$\varepsilon_a(r) = 1 - \rho_v \frac{R^2 L^2}{(L^2 + r^2)^2} - \rho_v \rho_w^2 I'(L) \tag{7-76}$$

where

$$I'(L) < \frac{R^2}{L^2 + R^2} \left(1 - \frac{R^2}{L^2 + R^2}\right)^2 \left\{1 - \rho_w \left(1 - \frac{R^2}{L^2 + R^2}\right)\right\}^{-1}. \tag{7-77}$$

For $\rho_w = 0.3$, $L = 6$, $R = 0.7$, $\rho_v = 0.1$ we have for the cavity of Fig. 7-12

$$\varepsilon_a(r_0) = 1 - 1.3 \times 10^{-3} - 9 \times 10^{-3} \, I'(L)$$

where $I'(L) < 0.02$. Thus the emissivity, to three significant figures, is governed solely by the second term, namely, the reflectance of the grooved back wall.

It is important to note that the emissivity given here is that of the back wall only. The presence of the baffles modifies considerably the expression for $\varepsilon_a(x)$ given by equation (7-71). In particular the term in ρ_w is increased, since the direct reflection comes from the flat faces of the baffles which are nearly normal to the axis. This design of cavity is, therefore, for use with optical systems that view only the back wall. For wide angle optical systems, the baffles should only begin for values of x outside the marginal rays of the direct viewing cone.

For pyrometers viewing only the back wall, nearly 90 % of the measured radiation is that directly emitted by the back wall, and is thus largely characteristic of the temperature of the back wall alone.

In Fig. 7-13 is shown a cavity made from graphite and used for establishing the freezing point of gold for the primary calibration of photoelectric pyrometers. The temperature uniformity is ensured by immersing the cylindrical cavity directly in the gold. The back face is grooved to avoid direct specular reflection. The front wall is made from a platinum disc having a 1.5 mm diameter aperture. As was mentioned above, the presence of a low-emissivity front wall containing the aperture of the blackbody has two advantages: first, heat loss to the outside is reduced and thus temperature gradients are also reduced; second, it renders the aperture plainly visible from the outside, since it appears as a bright disc against the poorly emitting surroundings. The emissivity of graphite being about 0.9, the apparent emissivity of the cavity viewed along the axis, using a narrow angle optical system, is very high. Calculation leads to a value greater than 0.999 9. This is one example illustrating the point that temperature uniformity is the limiting factor in the realization of blackbody radiation. It is very difficult to show that the cavity of Fig. 7-13 is uniform in temperature to within 10 mK, as it should be to take full advantage of an emissivity of 0.999 9. The only way of demonstrating good uniformity in this case is to carry out a series of melts and freezes at different rates and confirm that the spectral radiance is reproducible on melting and freezing. Examples of good melting and freezing curves obtained using the cavity of Fig. 7-13 are shown in Fig. 7-14.

A rather different design of blackbody is shown in Fig. 7-15. This cavity was designed for a determination of the Stefan–Boltzmann constant (Quinn and Martin, 1983) in which the total radiation from a blackbody at 273.16 K was measured using a calorimeter detector at 2 K (see Section 7-7 and Fig. 7-15). The cavity is large because it was designed to have an effective emissivity $> 0.999\,9$ for 273 K thermal radiation when viewing the lower part of the cylinder and re-entrant cone. The wavelength range covered by radiation at this temperature extends from about 2 μm to 200 μm, beyond which less than 0.1 % of the total energy is radiated. The temperature of the cavity was measured using eight capsule-type high precision platinum resistance

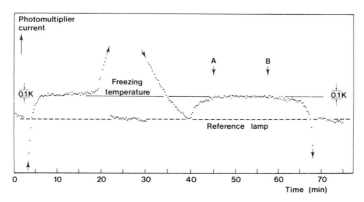

Fig. 7-14. Melts and freezes of the gold surrounding the cavity of Fig. 7-13, observed using the photoelectric pyrometer of Fig. 7-32.

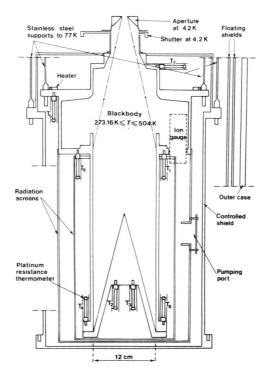

Fig. 7-15. A blackbody cavity designed for use at 273.16 K for total radiation and used in a determination of the Stefan–Boltzmann constant and thermodynamic temperature.

thermometers bolted to various parts of the cavity. The temperature uniformity over the cylindrical and conical parts was better than ± 1 mK. The internal surfaces of the cavity were painted with a 3 M-C-401 black paint, the

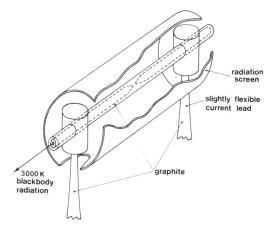

radiation
screen

slightly flexible
current lead

3000 K
blackbody
radiation

graphite

Fig. 7-16. A directly heated graphite blackbody designed for use up to 3000 K.

optical properties of which were known out to wavelengths as long as 300 μm. The reflectivity of the paint is less than 0.06 for wavelengths up to 30 μm. Thus the emissivity of the cavity is given with sufficient accuracy using only the term in ρ_w of equation (7-70), remembering that this black becomes increasingly specular for angles of incidence $> 80°$ for all wavelengths.

A graphite blackbody (Lapworth *et al.*, 1970) for use at temperatures up to 3000 K is illustrated in Fig. 7-16. One of the difficulties associated with a directly heated cavity at temperatures as high as this results from thermal expansion of the cavity itself. Special provision must be made to allow for this expansion. In the case of the cavity shown in Fig. 7-16 the rear support and current lead were made flexible. This cavity was designed to give blackbody radiation over a wide wavelength range so that the presence of a window was to be avoided. The method adopted was to maintain a gentle flow of dry argon through the furnace chamber and out through an open window in the line of sight of the blackbody.

7-5 Tungsten ribbon lamps as reproducible sources for radiation pyrometry

7-5-1 Introduction

A blackbody cavity, although providing an ideal thermal radiator, is not at all convenient for the maintenance or dissemination of IPTS-68. One might argue that there is no need to maintain or disseminate the temperature scale by means of blackbody cavities. For that part of IPTS-68 defined in terms of fixed points and a resistance thermometer, it is the resistance thermometer that is used to maintain and disseminate the scale, not a variable temperature furnace, oil bath or cryostat. The difference between the two parts of the scale is, however, fundamental. In the lower part of IPTS-68 we define T_{68} in terms of the behaviour of the thermometer; i.e. in terms of $W(T_{68})$ or $E(T_{68})$. In the higher part of the scale, T_{68} is defined in terms of the behaviour of a blackbody radiator and not in terms of the instrument used as a thermometer. It would thus be perfectly consistent with the definition of the scale if it were maintained by a reproducible radiator rather than by means of an instrument that measures the radiation. Indeed the maintenance and dissemination of the scale by means of an instrument leads to the additional requirement of showing that the scale thus maintained is independent of the properties of the instrument. As we shall see, both methods are used in practice, but the one that employs a calibrated radiator is simpler and of wider application than the one that employs an instrument.

Although a variable-temperature calibrated blackbody is not easy to employ as a means of disseminating a temperature scale, most of its functions are equally well fulfilled by a carefully-designed tungsten ribbon lamp. The radiation emitted in a given direction at a given wavelength by a small well-defined region on the ribbon may be calibrated in terms of the electric current through the lamp. The current/temperature relation can be made highly reproducible over a wide range of temperature. Vacuum lamps are used from 700 °C to 1700 °C, and gas-filled lamps from 1500 °C to 2700 °C.

The most obvious difference between the radiation emitted by a tungsten ribbon and that emitted by a blackbody stems from the variation with wavelength of the emittance of tungsten (Fig. 7-17). As a consequence of this emittance variation, the spectral radiance temperature is a function of wavelength. We define the spectral radiance temperature, T_R, for a ribbon

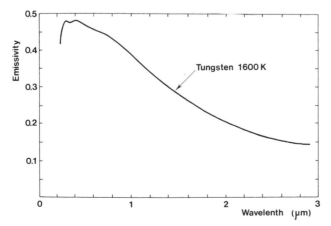

Fig. 7-17. The variation with wavelength of the spectral emissivity of tungsten at 1600 K (after de Vos, 1954).

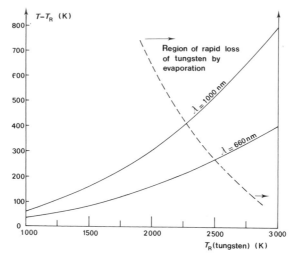

Fig. 7-18. The difference, $T - T_R$, between the true temperature and spectral radiance temperature of tungsten for wavelengths of 0.65 μm and 1 μm.

having an emittance $\varepsilon(\lambda, T)$, viewed through glass of transmittance $t'(\lambda)$, by:

$$\int_{\lambda=0}^{\infty} s(\lambda)t(\lambda)t'(\lambda)\varepsilon(\lambda, T')L(\lambda, T')d\lambda = \int_{\lambda=0}^{\infty} s(\lambda)t(\lambda)L(\lambda, T_R)d\lambda \qquad (7\text{-}77)$$

where $s(\lambda)$ is the spectral response of the detector, $t(\lambda)$ the transmittance of the optical system (including the narrow band filter) and T' the true temperature of the tungsten surface. If the bandwidth, $\Delta\lambda$, of this filter is sufficiently narrow for the variation of $\varepsilon(\lambda, T)$ over $\Delta\lambda$ to be insignificant then, of course:

$$\varepsilon(\lambda, T)L(\lambda, T') = L(\lambda, T_R). \qquad (7\text{-}78)$$

For most purposes, equation (7-78) is quite good enough, since for $T = 2000$ K, $\lambda = 660$ nm and $\Delta\lambda = 10$ nm, $T_R \approx 1854$ K and the difference between T_R calculated from equation (7-78) and from equation (7-77) is only 0.02 K. This error increases as the square of the bandwidth.

Figure 7-18 shows $T - T_R$ for tungsten as a function of T for two wavelengths, 660 nm and 1000 nm. A disadvantage of the tungsten ribbon lamp, which is immediately obvious from Fig. 7-18, is that it is not possible to obtain values of T_R much above 2200 °C without having unacceptably high losses of tungsten by evaporation. In an attempt to overcome this limitation, tungsten lamps were developed that employed a heated cavity as radiating source. We shall come to the properties of these lamps later on in this chapter.

7-5-2   Processes that occur on heating tungsten ribbons

Before coming to the design and performance of tungsten ribbon lamps we shall look briefly at some of the more important physical processes that take place in the interior and on the surface of a heated tungsten ribbon. An appreciation of these processes is useful in understanding both the behaviour of tungsten ribbon lamps and also the various procedures required during their construction in order to produce stable lamps. For the purposes of this discussion the design of lamp shown in Fig. 7-19 is assumed.

For the ribbon to behave as a stable and reproducible emitter of thermal radiation both the interior of the tungsten and its surface must be close to structural equilibrium. We shall consider the principal processes which occur during prolonged heating of a tungsten ribbon mounted in a glass envelope initially connected to a pumping system (Quinn and Lee, 1972). These are: outgassing and loss of tungsten by evaporation, recrystallization, grain boundary grooving and grain surface changes. In addition, in considering the behaviour of the lamp as a whole, it is necessary to take into account effects of outgassing of leads and release of water vapour from the surface of the glass envelope.

Outgassing and loss of tungsten by evaporation

To remove most of the dissolved gas from tungsten it is necessary to heat it in vacuum to a temperature of about 2200 °C and pump for a period of about 2 h (here and in the subsequent discussion on the metallurgical changes taking place in tungsten we are referring to true temperature and not spectral

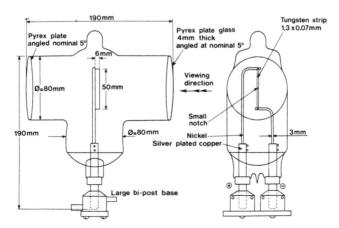

Fig. 7-19. A tungsten ribbon lamp used as a reproducible source of thermal radiation for the calibration of radiation pyrometers and comparison of temperature scales over the range 700 °C–1700 °C (*courtesy GEC Co., London, see Lee et al., 1972*).

radiance temperature). After such treatment most of the remaining gas in the glass envelope of a lamp will originate either from the molybdenum or nickel leads, which remain at a much lower temperature, or from the glass. The gases evolved on heating tungsten are, in order of concentration, nitrogen, carbon monoxide and hydrogen. Their presence in solid solution always increases the electrical resistance of a metal. If, after a lamp has been sealed-off, excessive degassing of the tungsten takes place, a hysteresis in the resistance/temperature relation is usually observed. This hysteresis occurs in the following way. At high temperatures, the gas is released from the body of the metal by diffusion to the surface and evaporation. On cooling, the same gas, if it has not been pumped away or absorbed elsewhere, condenses on the surface of the tungsten and begins to diffuse back into the body of the metal, thereby increasing its resistance. The rates at which all of these processes occur are exponential functions of temperature. For lamps used in the range up to 1800 °C drifts in resistance on cooling to, say, 1200 °C can occur over a period of several days as a result of inadequate initial degassing or subsequent leaks.

At the normal maximum operating temperature of vacuum ribbon lamps, ~1850 °C, the vapour pressure of tungsten is extremely low and can be ignored. For lamps intended for operation at a much higher temperature, however, an inert gas such as argon is introduced into the envelope. The presence of the gas reduces the loss of tungsten by evaporation. Most of the evaporated tungsten atoms re-condense on the tungsten surface after collisions with the gas atoms before they have had a change to diffuse through the gas boundary layer and be swept away by convection. Very large losses of tungsten can occur, however, by the process known as the "water-cycle effect". It is losses from this process that are the most important and can lead to large drifts in calibration at high temperatures.

The generally accepted mechanism for the "water-cycle effect" is as follows. Water vapour, probably originating in the glass, comes into contact with the hot tungsten and is decomposed into oxygen and hydrogen. The oxygen combines with the tungsten to form an oxide which immediately evaporates and condenses on the glass walls, while the hydrogen is converted to atomic hydrogen by the hot tungsten. The atomic hydrogen then reduces the tungsten oxide on the glass to metallic tungsten and recombines with the oxygen released to form water vapour which can return to the tungsten filament to repeat the cycle. In this way a deposit of tungsten can be built up on the walls of the envelope while the temperature of the tungsten is much too low for such a deposit to be formed by normal evaporation.

In the processing of the lamps, some water vapour is expected to be released during the seal-off of the envelope, adding to the small amount remaining even after degassing of the glass by prolonged baking. The clean glass surface will strongly absorb atomic hydrogen, although the amount absorbed before the glass begins to release it as molecular hydrogen is very small. However, since it is possible to produce stable lamps despite the water vapour resulting from sealing-off, it seems likely that there is a certain minimum quantity of water vapour required before the water cycle is self-sustaining.

Recrystallization

On first heating the tungsten ribbon, primary recrystallization, or recrystallization from the cold worked state, begins at about 1200 °C. Grain nuclei form and grow until the grains meet. Little further grain growth occurs after this stage has been reached until the temperature is raised to about 1900 °C. At this temperature secondary recrystallization takes place, which is the growth of some grains at the expense of others. Secondary recrystallization continues until grain-boundary energy is no longer sufficient to overcome the various locking processes that impede the movement of grain boundaries. Subsequent prolonged heating at lower temperatures will have little effect upon grain size.

Grain-boundary grooving and grain-surface changes

The grooves which form on the surface of a heated metal at the junctions of the grains are the result of the adjustment to equilibrium of the grain-boundary and the grain-surface tensions. They arise from plastic deformation under the driving force of the surface and boundary tensions, the material removed from the groove forming a ridge on either side of the groove. After recrystallization, when the grain boundaries have ceased moving, the grooves become deeper and deeper, but at a decreasing rate, due to the driving force remaining essentially constant while the energy required to remove material from the base of the groove increases with depth. The rate of grooving will increase with temperature and in tungsten well-developed grooves appear after only a few hours at about 2200 °C.

The radiance of a grain-boundary groove is greater than that of the flat surface because of inter-reflections occurring between the sides of the groove. The profile of a typical well-developed groove in tungsten is shown in Fig. 7-20. Measurements (Quinn, 1965) indicate that a mean increase in emittance of about 6.5 % is observed at such a grain boundary. Thus a tungsten ribbon having a very fine grain structure should have a higher overall emittance than one having a large grain structure. Figure 7-20 also shows how emittance is related to mean grain size. An obvious result of recrystallization is thus a decrease in surface emittance. This is one of the reasons for the long slow approach to equilibrium observed during the initial annealing at about 1900 °C of a new ribbon lamp.

Of rather more importance than grain-boundary grooving is the growth of grooves or facets over the surface of some grains which takes place in the presence of an electric potential gradient. Tungsten ribbon heated by the passage of an alternating current does not develop grain-surface facets; on the contrary, scratches or other marks tend to disappear, as one would expect.

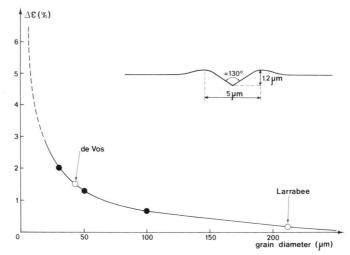

Fig. 7-20. The profile of a typical grain boundary groove on well-annealed tungsten ribbon and the emittance of such a ribbon as a function of grain size. The grain size and calculated grain boundary emittance for two classic measurements of the emissivity of tungsten are shown.

Tungsten ribbon heated by the passage of a direct current, on the other hand, exhibits quite a different structure. This is shown very clearly in Fig. 7-21. The same area of a tungsten ribbon is shown after heating for a long period on a.c. to enable secondary recrystallization to take place (7-21a), and after a period of 70 h heating on d.c. at 2100 °C (7-21b). The structure of the facets, Fig. 7-22, is such as to lead to slight polarization of the emitted radiation (Quinn, 1965).

The equilibrium surface structure depends upon the lattice orientation at the surface with respect to the direction of the potential gradient. The reversal of the polarity of the lamp will thus lead to profound changes in surface structure and should therefore be avoided in a calibrated lamp.

The rate at which the surface structure approaches equilibrium is controlled by factors similar to those that control the rate of recrystallization. In general the higher the temperature, the higher the rate and it follows, as for the recrystallization process, that having achieved a state of near equilibrium after annealing at a high temperature, subsequent annealing at lower temperatures is unlikely to change the structure.

In order to establish a stable surface structure, it has been found necessary to heat new vacuum ribbon lamps to a temperature of about 1900 °C for periods of from 100 to 300 h.

7-5-3 The design and performance of tungsten ribbon lamps

The design of tungsten ribbon lamp that has proved most successful is that shown in Fig. 7-19. In considering the reproducibility of a tungsten ribbon

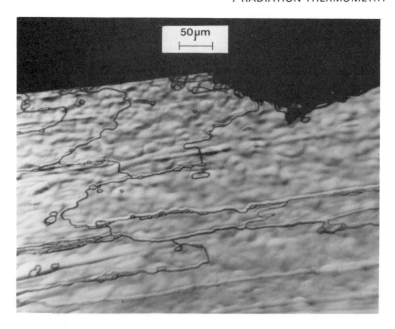

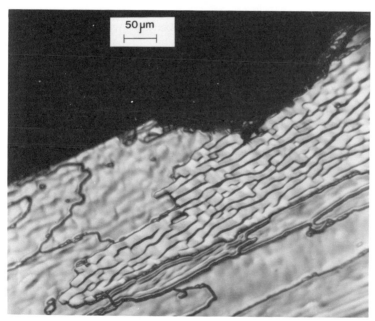

Fig. 7-21. Microphotograph (0.3 × 0.45 mm) showing the grain surface structure of a tungsten ribbon after (a) 100 h heating by a.c. at 2100 °C in vacuum and (b) a further 70 h at the same temperature but on d.c., certain grains show surface restructuring.

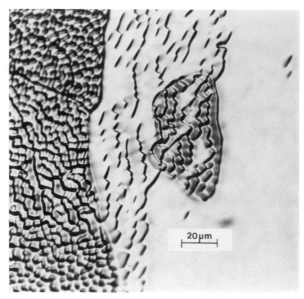

Fig. 7-22. Surface facets on grains of tungsten ribbon after prolonged heating at 2100 °C on d.c.

lamp as a radiance standard, the following factors must be taken into account:
(a) Reproducibility of electrical resistance.
(b) Temperature gradients and emittance variations.
(c) Effects of orientation of ribbon with respect to optic axis of viewing system.
(d) Effects of changes in ambient and pin temperature.
(e) Cleanliness of the windows.

(a) Reproducibility of electrical resistance

The reproducibility of the electrical resistance of a ribbon depends upon many of the factors discussed in the preceding section. Most lamps exhibit a small, repeatable, hysteresis on cycling from 1064 °C (we now revert to spectral radiance temperature at 660 nm) which can amount to as much as 0.05 °C. However this change in radiance temperature is accompanied by an electrical resistance change and therefore a correction may be applied (Jones and Tapping, 1979). The long term drift at radiance temperatures below 1500 °C is insignificant but rises to about 0.02 °C per 100 h at 1600 °C, 0.08 °C at 1700 °C, and 0.15 °C at 1770 °C. It is common to operate tungsten ribbon lamps so that the temperature is expressed as a function of d.c. current only. This is a perfectly adequate method and avoids the difficulties of making precise voltage measurements at the pins in the presence of temperature gradients. The current/temperature relation for a lamp of the design shown in Fig. 7-19 can be expressed as a fourth degree polynomial for a vacuum lamp over the range 1064 °C to 1700 °C and for a gas-filled lamp from 1300 °C to 2200 °C. For lamps of a given design the coefficients of the polynomials follow the same pattern and provide a useful check on the calibration procedures (Andrews and Coates, 1975; Coates, 1981a).

(b) Temperature gradients and emittance variations

For precise measurements of spectral radiance it is necessary to specify both the position and size of the viewed area on the ribbon. This is because it is impossible to avoid both the temperature gradients and the emittance variations from grain to grain which have already been mentioned. Although the details of the temperature distribution along the ribbon depend upon its dimensions, thermal and electrical conductivity and total emittance, the overall distribution near the centre should not be very different from parabolic. Such differences as are observed stem from variations in thickness of the ribbon and are more serious in lamps having wide but correspondingly thin ribbons. In a gas-filled lamp, having a vertical ribbon, the maximum is displaced upwards away from the centre by convection. In a vacuum lamp the Thomson effect leads to a significant asymmetry in distribution about the centre. The highest temperature in a vacuum lamp is always close to the reference notch cut in the side of the ribbon. Figure 7-23 shows temperature gradients measured at two temperatures on a ribbon, having the dimensions shown in Fig. 7-19. Temperature gradients on the ribbon of a gas-filled lamp are somewhat larger than those shown in Fig. 7-23, and are asymmetrical as a result of convection currents. The convection currents are critically dependent upon the shape of the glass envelope and its orientation with respect to the vertical. At certain orientations, the radiance temperature becomes subject to quite large cyclic variations having a period of perhaps 10 sec and amplitude of a few degrees. Before calibration, therefore, the optimum orientation with respect to the vertical must be found to avoid or minimize these convection oscillations.

(c) Orientation of the ribbon with respect to the optic axis of the viewing system

The spectral radiance of vacuum ribbon lamps of the design shown in Fig. 7-19 is fairly insensitive to angle of view. Figure 7-24 shows the results of measurements made of the effects of rotation about the vertical and horizontal axes and also of displacement along the optic axis of the viewing system. These effects are sufficiently small not to lead to any difficulty in setting up the lamp with respect to a pyrometer. In addition, the spectral radiance temperature of the central viewing area is independent of the numerical aperture of the pyrometer over quite a wide range. This is an important property of the lamp since pyrometers vary in their apertures and it is inconvenient if a correction is necessary. It should also be noted that

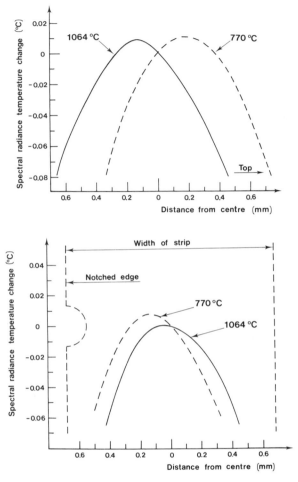

Fig. 7-23. (a) Spectral radiance temperature variations along the tungsten ribbon at 770 °C and 1064 °C for the lamp of Fig. 7-19. (b) The same across the width of the ribbon (after Jones and Tapping, 1979).

pyrometers may also vary in their angular sensitivity if photomultipliers are used as detectors, and if the beam is not diffused before arriving at the photocathode.

(d) Effects of changes in ambient and pin temperature
At temperatures below about 1200 °C, the temperature of the water-cooled pins and the temperature of the room influence the temperature of the centre of the ribbon. At higher temperatures, the ribbon behaves as an infinitely long ribbon, and the temperature of the centre ceases to be affected by the

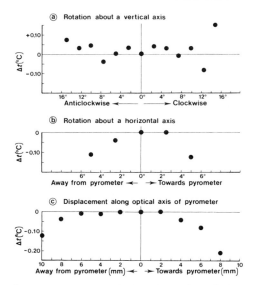

Fig. 7-24. Spectral radiance temperature variations for various displacement of the lamp of Fig. 7-19, (a) rotation about a vertical axis, (b) rotation about a horizontal axis, (c) displacement along the optic axis of the pyrometer.

temperature of the ends. Figures 7-25 and 7-26 show how pin and ambient temperature changes influence the temperature of the centre of the ribbon. For most purposes, the effects of ambient temperature changes can be ignored. It is usual to calibrate the lamp for a conventional pin temperature of 20 °C using the correction curve of Fig. 7-25 for pin temperatures differing from this.

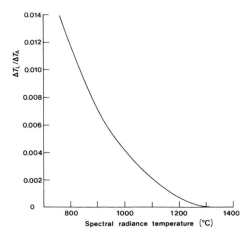

Fig. 7-25. Effect on the spectral radiance temperature of the lamp of Fig. 7-19 of a 1 °C change in ambient air temperature (after Jones and Tapping, 1979).

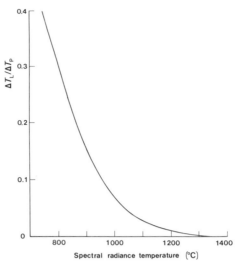

Fig. 7-26. Effect on the spectral radiance temperature of the lamp of Fig. 7-19 of a 1 °C change in pin temperature (after Jones and Tapping, 1979).

(e) Cleanliness of the plane window of the lamp

For the radiance temperature of the lamp to remain reproducible to within 0.05 °C, the window transmission must remain stable to 0.05 %. There are two important ways in which the transmission is likely to change. The first is from deposition of dust over the outer surface, and the second is from changes in transmission of thin grease films as a function of temperature (Jones and Tapping, 1979; Ohtsuka *et al.*, 1979). Dust is easy to remove using a commercial glass-cleaning fluid, but care must be taken to avoid leaving a film of grease. Such films have a transmission that is very temperature-sensitive and, if present, lead to apparent changes in radiance of a few tenths of percent that follow the temperature of the window.

7-5-4   Blackbody lamps

We have already remarked that the low emittance of tungsten leads to a large difference between real and radiance temperatures. This is particularly troublesome if a high radiance-temperature source is required. An alternative to a tungsten ribbon lamp is one having a tungsten element in the form of a blackbody cavity. A high temperature gas-filled version of a commercially available lamp of this sort is shown in Fig. 7-27. The blackbody lamp was originally developed (Quinn and Barber, 1967) as a replacement for the tungsten ribbon lamp for the whole range of temperature from 1064 °C to 2700 °C. Subsequent improvements in ribbon lamp performance have made the blackbody lamp of much less interest in the lower temperature range. At high

Fig. 7-27. A high temperature gas-filled blackbody lamp for use up to 3000 K.

temperatures however it remains a useful device having applications in thermometry and also in radiometry and photometry. Here we restrict our discussion to the gas-filled high temperature version.

The radiating element of the blackbody lamp, Fig. 7-28, is a thin-walled tungsten tube 2 mm in diameter having 1 mm diameter apertures at each end and blocked in the centre by a tightly fitting plug of fine tungsten wires. The emissivity of the tube viewed along its axis is about 0.99, but varies with

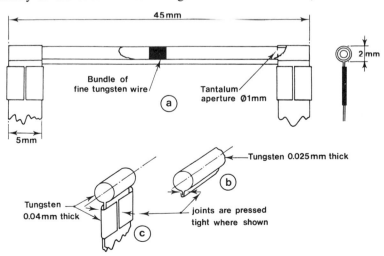

Fig. 7-28. The tungsten radiator of the blackbody lamp showing (a) the mounted tube, (b) the form of the seam along the tube and, (c) the method of holding the end leads in position.

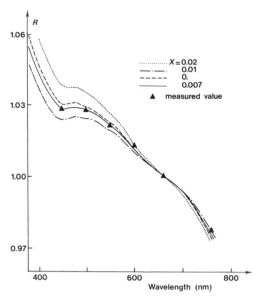

Fig. 7-29. Comparison of blackbody lamp spectral power distribution with that of a blackbody at 2014 K. R = spectral radiance of lamp/spectral radiance of blackbody, normalized at $\lambda = 660$ nm. The various dotted lines are the calculated distributions for a range of emissivities of the lamp. The solid line is the best fit to the measured data and is for a calculated emissivity of 0.992 (after Jones and Gordon-Smith, 1973).

wavelength (Jones and Gordon-Smith, 1973) in the manner shown in Fig. 7-29. The main inconvenience of the blackbody lamp stems from the presence of a temperature gradient along the tube and a small temperature difference between the central plug and the walls of the tube. The emittance of the central plug itself is high, about 0.95. Nevertheless the radiation coming from the wall of the tube remains important even when the angle of the viewing optical systems is small enough for only the central plug to be seen directly. As a result the spectral radiance temperature is dependent upon angle of view. In comparing the temperature of the lamp therefore, using pyrometers of differing numerical aperture, a correction must be made. The radiance increases by about 1.5 % for a viewing angle 2° away from the axis of the tube. For much larger angles it will of course fall as colder regions of the tube come into view.

7-6   Radiation pyrometers

7-6-1   The disappearing-filament optical pyrometer

The disappearing-filament optical pyrometer, at one time universally used in standards laboratories for the realization of the International Temperature

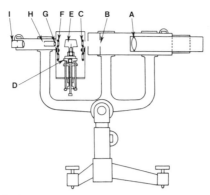

Fig. 7.30a. Sketch of NBS disappearing-filament optical pyrometer, A, objective lens; B, aperture stop (A of Fig. 7.30b); C, neutral filter; E, tungsten-filament pyrometer lamp; F, red glass; G, eye piece objective lens; H, exit stop (C of Fig. 7.30b) (after Kostkowski and Lee, 1962).

Scale, remains widely used in science and industry for practical thermometry. For this reason we shall begin this section with a description of its design and operation.

The principle of the disappearing-filament optical pyrometer is simple and is illustrated in Fig. 7-30a. An objective lens forms an image of the source whose temperature is to be measured, in the plane of the incandescent filament of a miniature lamp. The observer views the filament and the superimposed image of the source through a suitable eyepiece incorporating a red glass filter. The current through the lamp is adjusted until the filament has the same visual brightness as the image of the source. If the optical system has been properly designed, the filament will at this point disappear into the image of the source. The pyrometer is calibrated in terms of the current passing through the miniature lamp. Since the detector of equality of brightness is the human eye, the range of temperature directly accessible without either exceeding a comfortable brightness or becoming too dim to see, is limited. The lower limit depends upon the aperture of the optical system and is usually about 700 °C. The upper limit is about 1250 °C. To measure higher temperatures, a neutral glass filter, C in Fig. 7-30a, is interposed between the objective lens and the filament to reduce the brightness of the source image. The density of the filter is chosen to give a small overlap of ranges so that for a lamp current equivalent to, say, 700 °C on the direct scale, a temperature of 1100 °C will be obtained on the first higher range. In this way, temperature measurements can be extended with a single instrument to any desired maximum temperature. The transmittance, τ, of the filter required to reduce the brightness of a source from a temperature T to, say, that of the gold point T_{Au} is found, using the Wien approximation to the Planck formula, from

$$\ln\left(\frac{1}{\tau}\right) = \ln R(T_{Au}, \; T) = \frac{c_2}{\lambda}\left(\frac{1}{T_{Au}} - \frac{1}{T}\right) \qquad (7\text{-}79)$$

where λ is the wavelength at which the measurements are made, and $R(T_{Au}, T)$ is the ratio of the spectral radiances at temperatures T_{Au} and T. Because the bandwidth of the filter formed by a red glass and the eye is quite large, λ in equations (7-79) and (7-80) (below) is temperature-dependent and needs careful definition. This is discussed in more detail in the next section.

The accuracy of a disappearing-filament pyrometer is limited by the contrast sensitivity of the eye. Under the best conditions of illumination, usually near the middle of the range, a difference of visual brightness of about 2 % can be seen in the red. The repeatability of the mean of a number of measurements by an experienced observer can be as good as ± 0.5 °C. The accuracy attainable using a disappearing-filament pyrometer is thus sufficient for practically all industrial and scientific purposes. The great drawback remains of course the need for the active participation of a human observer. For this reason, the disappearing-filament pyrometer has been replaced in many industrial applications by automatic instruments. Although not necessarily more accurate than a visual disappearing-filament pyrometer, the photoelectric detector opens the way to many other ways of exploiting the properties of thermal radiation for thermometry. We shall be discussing some of these later on.

The conditions that must be met to obtain good disappearance of the filament against the source image relate to the entrance and exit angles of the optical system and stem largely from the consideration of diffraction at the edges of the filament. Unless the entrance angle β in Fig. 7-30b, is more than about three times the exit angle α, dark lines will be seen at the edges of the filament regardless of whether it is a flat or a round filament. In addition, for a round one, it is necessary to degrade the resolution of the eyepiece to avoid seeing bright lines at the edges. These result from the departure from the cosine law of emission at high angles and would, if they were visible, prevent disappearance being obtained. A flat filament is therefore preferable and the ratios of entrance to exit angles given in Table 7-3 should be maintained.

The red glass filter serves, together with the spectral sensitivity curve of the eye V_λ and the Planck distribution for the temperature being measured, to define the wavelength that appears in equation (7-79). The red glass filter is important for two reasons. First of all, from equation (7-79), we need to know

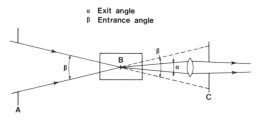

α Exit angle
β Entrance angle

Fig. 7-30b. Schematic diagram showing the entrance angle, β, and exit angle, α, at the pyrometer lamp of 7-30a (after Kostkowski and Lee, 1962).

Table 7-3. Entrance and exit angles of telescope for disappearance of filament.

Exit angle, radian.	Min. entrance angle, radian.
Flat filaments (any width)	
0.02	0.07
0.03	0.10
0.04	0.13
0.06	0.17
0.08	0.21
0.10	0.23
0.12	0.25
Cylindrical filaments (dia. 0.04–0.06 mm.)	
0.04	0.08–0.13
0.02	0.06–0.16
0.01	0.04 and larger

the operating wavelength and second, a good disappearance will only be obtained if the proper red glass is used. The question of the wavelength is a complex one that we shall deal with in the next section. If the source has a spectral distribution that differs markedly from that of the tungsten filament (if, for example, it is a blackbody), good disappearance will only be obtained if a red glass is chosen which, together with the V_λ curve of the eye, gives a relatively narrow bandwidth. This is because it is not possible to obtain visual disappearance of the junction between two fields having a slight colour difference even though their radiance may be the same. A bandwidth of about 30 nm is found to be satisfactory.

There are two types of calibration of a disappearing-filament optical pyrometer. The first is straightforward and consists simply of calibrating the pyrometer-lamp current by observing either a blackbody of known temperature or, more commonly a calibrated tungsten ribbon lamp over the whole range of the pyrometer. The scale over the lower, direct reading, range must be checked in detail at a sufficient number of points for a proper calibration curve to be interpolated. For the higher ranges, the shape of the calibration curve will be nearly the same but the "K" factor of the neutral filters must be confirmed. The "K" factor is defined using equation (7-79) to give

$$K = \frac{1}{T^2} - \frac{1}{T_1} = -\frac{\lambda \ln \tau}{c_2} \qquad (7\text{-}80)$$

For K to be constant, we require that $\ln \tau \approx 1/\lambda$.

The second type of calibration is one in which the scale is built up from a gold point blackbody by direct measurement using a series of filters or sectored discs having known values of τ. Much more is demanded of the wavelength determination for this type of calibration. We shall not enter into the details of the realization of the scale here since it is now rarely carried out using a disappearing filament optical pyrometer. We shall instead deal with the problem of effective wavelength and then pass on to the design and performance of the high-precision photoelectric pyrometer.

7-6-2   The problem of effective wavelength

Above the gold point IPTS-68 is defined by the following equation

$$\frac{L(\lambda,\ T)}{L(\lambda,\ T_{Au})} = \frac{\exp\left[\dfrac{c_2}{\lambda T_{Au}}\right] - 1}{\exp\left[\dfrac{c_2}{\lambda T}\right] - 1}. \tag{7-81}$$

This expression relates, by means of the Planck equation, the spectral radiance $L(\lambda,\ T)$ of a blackbody at a temperature T (here understood to be T_{68}) to that of a blackbody at the gold point, $L(\lambda,\ T_{Au})$. In applying equation (7-81) to practical measurements, the question arises of how to deal with a finite bandwidth, $\Delta\lambda$, which, for a disappearing-filament optical pyrometer is about 30 nm and for a photoelectric pyrometer may be between about 1 nm and 10 nm.

Suppose that we have an ideal neutral filter that has a transmittance τ (the practical realization of such a filter we shall come to later), we can measure a ratio, $R(T_{Au},\ T) = 1/\tau$, in the following way. We arrange that a suitable detector, having a spectral response $s(\lambda)$, views blackbodies, at temperatures T_{Au} and T, in turn through an optical system, which includes a narrow band filter, of which the spectral transmittance is $t(\lambda)$. The temperature, T, of the second blackbody is adjusted until the signal from the detector is the same when viewing the gold-point blackbody directly and the second blackbody through the neutral filter. Under these conditions we can write:

$$R(T_{Au},\ T) \int_{\lambda=0}^{\infty} s(\lambda)t(\lambda)L(\lambda,\ T_{Au})d\lambda = \int_{\lambda=0}^{\infty} s(\lambda)t(\lambda)L(\lambda,\ T)d\lambda. \tag{7-82}$$

If instead of a blackbody at a temperature T, a tungsten ribbon lamp is used, we can write equation (7-82) in the following way

$$R(T_{Au},\ T) \int_{\lambda=0}^{\infty} s(\lambda)t(\lambda)L(\lambda,\ T_{Au})d\lambda = \int_{\lambda=0}^{\infty} s(\lambda)t(\lambda)t'(\lambda)\varepsilon(\lambda,\ T')L(\lambda,\ T')d\lambda \tag{7-82a}$$

but

$$\int_{\lambda=0}^{\infty} s(\lambda)t(\lambda)t'(\lambda)\varepsilon(\lambda,\ T')L(\lambda,\ T')\mathrm{d}\lambda = \int_{\lambda=0}^{\infty} s(\lambda)t(\lambda)L(\lambda,\ T_R)\mathrm{d}\lambda$$

(7-82b)

where $\varepsilon(\lambda, T')$ is the emittance of the tungsten ribbon, $t'(\lambda)$ the transmittance of the lamp window, T' the true temperature of the tungsten ribbon (unknown and of no interest to us) and T_R the spectral radiance temperature of the ribbon lamp as in equation (7-77). In this case, T_R is a function of wavelength and is shown approximately in Fig. 7-31, calculated from measured values of $\varepsilon(\lambda)$ (Latyev *et al.*, 1970) and assuming a constant value for $t'(\lambda)$ of 0.92.

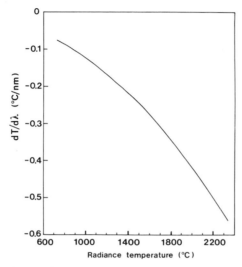

Fig. 7-31. Corrections for wavelength changes in calibrations of tungsten ribbon lamps (after Coates 1981a).

Equations (7-82) and (7-82a) are integral equations for T in terms of the measured or defined quantities $R(T_{\mathrm{Au}}, T)$, $s(\lambda)$, $t(\lambda)$ and $L(\lambda, T_{\mathrm{Au}})$ and may be solved numerically for T. If, instead of using a neutral filter to define $R(T_{\mathrm{Au}}, T)$, a direct ratio of the signals from the detector is measured, the same equation applies, but we must take account of any non-linearity that is present in the detection by including in $s(\lambda)$ a temperature dependence so that it becomes $s(\lambda, T)$.

In order to obtain an accurate value of T from equation (7-82), care must be taken in choosing the numerical methods for carrying out the integration. $s(\lambda)$ and $t(\lambda)$ are always obtained in tabulated form since they result from experimental measurements made at a large number of discrete wavelengths. To carry out the numerical integration there are many ways of fitting an analytical function to the data, and the resulting errors depend upon the way in which such functions are chosen and upon the intervals between the data points. Numerical methods for treating equation (7-82) have been discussed by Tischler (1981) who described a straightforward procedure based upon the fitting of piecewise polynomials to $s(\lambda)$ and $t(\lambda)$. For each interval between data points at wavelengths λ_i and λ_{i+1}, an nth degree polynomial ($4 \leqslant n \leqslant 6$) is fitted to $(n+1)$ points either side of λ_i. A different polynomial is thus used for each interval. Integration is carried out using Simpson's rule with a step size chosen to give an integration error below the desired value. If we define a function $F(\lambda, T)$ such that

$$F(\lambda, T) = \int_{\lambda_a}^{\lambda_b} s(\lambda) t(\lambda) L(\lambda, T) \, d\lambda \tag{7-83}$$

equation (7-82) may be written

$$G(\lambda, T) = F(\lambda, T) - R(T_{Au}, T) F(\lambda, T_{Au}) = 0 \tag{7-84}$$

where $G(\lambda, T)$ is the integration error. By iteration, using Newton's procedure and starting from an initial value for T, a rapidly converging series of solutions is found. The initial value of T may be found from an approximate equation such as (7-87) below, putting λ_e equal to the wavelength of peak transmittance of the filter.

Various other ways of solving equation (7-82) have been used. In the days before computers, when numerical integration was a laborious process, approximate methods were devised to minimize the amount of numerical integration. The best known of these employed the concept of "mean effective wavelength" λ_e defined in the following way for two temperatures T_1 and T_2:

$$R(T_1, T_2) = \frac{\int_{\lambda=0}^{\infty} s(\lambda) t(\lambda) L(\lambda, T_2) \, d\lambda}{\int_{\lambda=0}^{\infty} s(\lambda) t(\lambda) L(\lambda, T_1) \, d\lambda} = \frac{J_2}{J_1} = \frac{L(\lambda_e, T_2)}{L(\lambda_e, T_1)} \tag{7-85}$$

Using the Wien approximation to the Planck equation we can write

$$\ln\left(\frac{J_2}{J_1}\right) = \left(\frac{c_2}{\lambda_e}\right)\left(\frac{1}{T_1} - \frac{1}{T_2}\right) \tag{7-86}$$

so that

$$\lambda_e = \frac{\left(\frac{c_2}{\lambda}\right)(1/T_1 - 1/T_2)}{\ln(J_2/J_1)} \tag{7-87}$$

as $T_1 \to T_2 = T$, the mean effective wavelength, approaches a value known as the "limiting effective wavelength" $\lambda_L(T)$.

There is a value of $\lambda_L(T)$ for every value of T and since $\lambda_L(T)$ is a linear function of $1/T$ we can write

$$\frac{1}{\lambda_e(T_2 \to T_1)} = \frac{1}{2}\left(\frac{1}{\lambda_L(T_1)} + \frac{1}{\lambda_L(T_2)}\right). \tag{7-88}$$

Thus from the calculation of a few values of $\lambda_L(T)$, it is easy to deduce by interpolation λ_e for any pair of temperatures.

Although it is clear that the limiting effective wavelength is inversely proportional to temperature it would be convenient to have an explicit relation linking the two. Various forms have been suggested, among which we find (Kostkowski and Lee, 1962)

$$\frac{1}{\lambda_L(T)} = A + \frac{B}{T} \tag{7-89}$$

or more complicated functions such as

$$\ln\left\{\frac{c_2}{\lambda_L}\left(\frac{1}{T_2} - \frac{1}{T_1}\right)\right\} = B\left(\frac{1}{T_2 - C} - \frac{1}{T_1 - C}\right). \tag{7-90}$$

The constants A, B and C are purely empirical and must be determined for each pyrometer (Bezemer, 1974; Jung and Verch, 1973). It is difficult to make any general statement about the accuracy likely to be achieved using any one of these approximations since it depends so much upon the shape of $t(\lambda)$. Nevertheless we can say that for visual optical pyrometry, equation (7-89) is perfectly adequate in view of the limitations on the accuracy stemming from other sources.

More recent methods which minimize the amount of numerical integration are much more sophisticated than this (Coates, 1977, 1979; Ruffino, 1980). The approach that is adopted is to calculate a pyrometer reference function which is itself the result of a numerical integration of terms of the type J in equation (7-85), evaluated for a reference temperature. For other temperatures, the equivalent J terms are found by difference from that at the reference temperature, a process which is facilitated by the differences being small. Interpolation is made using relatively simple equations involving the pyrometer reference function, T and two or more arbitrary constants. The interested reader is referred to the original papers for details of these methods.

7-6-3   High-precision photoelectric pyrometers and the realization of IPTS-68

Having described the properties of thermal radiation, blackbody cavities, tungsten lamps and effective wavelength, we now have all the elements required to discuss the realization of IPTS-68 using a photoelectric pyrometer.

A photoelectric pyrometer suitable for the realization of IPTS-68 above the gold point is no more than an optical instrument equipped with means of measuring the ratio of the spectral radiance of a tungsten ribbon lamp to that of a blackbody at the gold point. Some of the earlier photoelectric pyrometers (Lee, 1966; Jones and Tapping, 1971) were based upon the disappearing-filament optical pyrometer but with the human eye replaced by a photocell or photomultiplier as detector of equality of brightness.

More recent photoelectric pyrometers dispense with an internal reference lamp, examples are shown in Fig. 7-32a,b (Jones and Tapping, 1982; Quinn and Chandler, 1972). The inherent stability of the photomultiplier is used in the comparison of two external sources: a blackbody at the gold point and a tungsten ribbon lamp for example. The radiance ratios in these pyrometers were measured either by means of sectored discs, direct ratios of photon counts (Coates, 1975a) or photocurrents or by energy doublers.

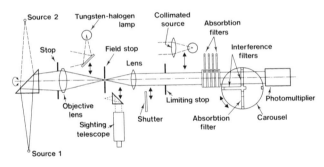

Fig. 7-32a. A photoelectric pyrometer using a refracting optical system (after Jones and Tapping, 1982).

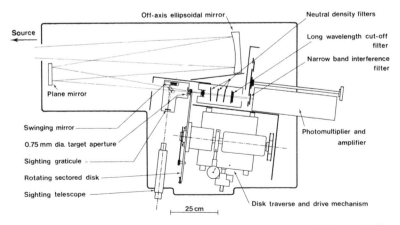

Fig. 7-32b. A photoelectric pyrometer using a reflecting optical system (Quinn and Chandler, 1972).

The measurement of ratios by means of rotating sectored discs has been described in detail by Quinn and Ford (1969). The discs themselves are made with apertures near the periphery defined by pairs of radial knife edges. The axis of rotation of the disc is placed parallel to the beam of radiation to be chopped so that the beam passes through the apertures. The average radiance of a source viewed through the apertures of a rotating sectored disc is given, following Talbot's law, by the product of the radiance of the source and the transmittance of the disc, namely the fraction of the time for which radiation is allowed to pass through the apertures. This fraction is equal to the ratio of the total angle subtended at the centre by all the apertures to 2π. With a carefully made disc, having a transmittance of 1.25 % for example, uncertainties in the measurement of the transmittance can be as low as ± 0.01 % of the transmittance. The transmittance may be measured either mechanically, by direct measurement of the positions of the knife edges, or by timing a light beam passing through the apertures when the disc is rotating *in situ*. For a venetian-blind photomultiplier (EMI 9558 for example) to obey Talbot's law and to take full advantage of this potential accuracy in ratio measurement, it is necessary for the cathode illumination to be low, the mean anode current not to exceed about 0.1 μA, and for the dynode potential to be stabilized.

Using a set of five sectored discs, the pyrometer illustrated in Fig. 7-32b can be used to realize the temperature scale in the following way. Two tungsten ribbon lamps, A and B, are compared in turn with a blackbody cavity, at the freezing point of gold. Using in turn the five sectored discs, one of the ribbon lamps A, say, is calibrated at the temperatures shown in Fig. 7-33 with respect

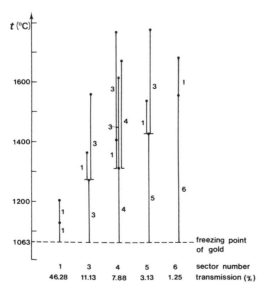

Fig. 7-33. Temperatures obtained with various combinations of five sectored discs working from the gold point (sectored disc No. 2 is not used here).

to *B* which remains at the gold point. This is done by simply comparing *B* with (*A* + sectored disc). Lamp *B* is then raised to a series of higher temperatures, chosen from among those already established on lamp *A*, and these are used in turn as reference temperatures for establishing a further series of temperatures on lamp *A*. This procedure may be repeated until a sufficient number of calibration points are established to enable a polynomial to be derived which adequately describes the current/temperature calibration of the lamps. For a pair of vacuum ribbon lamps operated over the range shown in Fig. 7-33, the standard deviation of the residuals from a fourth degree polynomial should be less than 0.05 °C. As a result of the way in which the calibration temperatures are obtained, there is a good degree of self-checking of the ratio measurement process. For example in Fig. 7-33, two points are indicated, near 1550 °C, that are very close in temperature, but one is obtained using sectored disc no. 6 in one step directly from the gold point while the other is obtained using disc no. 3 twice by means of an intermediate temperature near 1270 °C.

The designs of photoelectric pyrometer shown in Fig. 7-32a,b are but examples of the wide diversity of systems in operation in various national laboratories. It is of course very desirable that the instruments used to establish IPTS-68 differ one from another. In this way, one has confidence that systematic errors originating in the photoelectric pyrometers are small. An example of the results of an intercomparison of temperature scales carried out using photoelectric pyrometers (Lee *et al.*, 1972) is shown in Fig. 7-34. The results of this intercomparison represented the best that could be achieved in photoelectric pyrometry in 1971 and do not reflect the significant improvements that have since been made in the performance of photoelectric pyrometers (Jones and Tapping, 1982; Coates *et al.*, 1982; Coates and

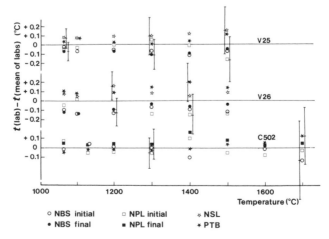

Fig. 7-34. Results of an international comparison of temperature scales between the gold point and 1700 °C among four national laboratories carried out in 1971 (after Lee *et al.*, 1972).

Andrews, 1978). Examples of photoelectric pyrometers in which the ratios are measured from direct ratios of photocurrents without using either sectored discs or beam doublers are given by Kunz (1975) and Jung (1979). We shall not enter into details of any of these photoelectric pyrometers, but rather look more generally at some of the factors affecting the accuracy of such instruments.

In designing a photoelectric pyrometer, one of the most important choices to be made, and one constrained by the temperature range over which the pyrometer is intended to operate, is that of detector.

For photoelectric pyrometry from about 700 °C upwards, the photomultiplier having an S-20 type photocathode is the preferred detector. Its competitor is the silicon photodiode which, although offering some advantages, is not widely used because its optimum operating wavelength lies at about 1 μm, just in the infrared (Ruffino, 1971) whereas that of the photomultiplier lies in the visible region of the spectrum. The optimum wavelength of operation of a detector is related to the signal-to-noise ratio and the rate of change of signal with temperature.

There are two main sources of the noise that appears in the output signal of the detector: the detector itself and the fluctuations present in the thermal radiation arriving at the detector (Lovejoy, 1962). Neither of these are limiting factors in the sensitivity of photoelectric pyrometers in the range above about 700 °C. Both photomultiplier and silicon-cell detectors can be used with averaging times sufficiently long to reduce the random uncertainty, due to detector noise and fluctuations in the radiation, to levels equivalent to a few millikelvins in temperature sensitivity.

The noise and other properties of photomultipliers for radiation thermometry have been extensively studied by Coates (1970, 1971, 1972, 1975b,c) and Coates and Andrews (1981). The choice of operating a photomultiplier in the d.c. mode (Jones and Tapping, 1982) or the photon-counting mode (Coates and Andrews, 1978) is very much a matter of the personal preference of the user. There are no overriding advantages of one method compared with the other. In both cases, it is necessary to ensure that the photomultiplier does not suffer from excess noise, fatigue or non-linearity. The photon-counting method has, however, the advantage that the dependence of signal amplitude upon gain is reduced, as are the effects of leakage currents inside the photomultiplier and around its base. Also the signal is in digital form which facilitates the direct connection to digital data-handling and computer-control systems. In both d.c. and photon-counting modes, temperature control of the photomultiplier is critical, since the spectral response (particularly near the long-wavelength cut-off) as well as the dark current, is a function of temperature. Photomultipliers having an extended-red S-20 photocathode such as the EMI-9558 (a plug-in replacement for the EMI 9658 S-20 photomultiplier) must be operated at a temperature of about -25 °C to reduce the dark current. The use of the extended-red photocathode allows wavelengths up to about 800 nm to be used, although if the instrument is destined solely for the realization of IPTS-68 above the gold point, such long wavelengths are rarely required.

The wavelength at which IPTS-68 realizations are made is almost always near 660 nm, although this is not specified in the definition of the scale. This wavelength is used for two reasons: it is close to the effective wavelength of the disappearing-filament optical pyrometer, defined by a red glass and the long wavelength cut-off of the eye, and it happens also to be close to the wavelength at which maximum signal is obtained from an S-20 photomultiplier illuminated by a blackbody at the gold point.

The requirements called for in the interference filter which defines the bandwidth of a photoelectric pyrometer are severe. In particular the transmittance at wavelengths well away from the principal peak must be less than about 10^{-5} times that of the peak. If not, the calculation of the temperature using equation (7-82) becomes critically dependent upon out-of-peak transmittance, and this is likely to lead to errors. If any of the approximation methods are used to solve (7-82), out-of-peak transmittance becomes very difficult to deal with and errors are sure to arise. Transmittance curves of three typical filters studied by Coates (1979) are shown in Fig. 7-35,

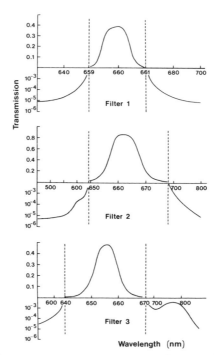

Fig. 7-35. The transmittances of three interference filters: filters 1 and 2 are suitable for high precision photoelectric pyrometry but filter 3 is not, due to its high transmittance outside the peak (after Coates, 1979).

filters 1 and 2 would be considered suitable for high precision photoelectric pyrometry but filter 3 not, because its out-of-peak transmittance is too high. The rapid fall in response of an S-20 photocathode at wavelengths beyond 700 nm is useful in compensating for the long-wavelength transmittance of filters, which otherwise would be troublesome in view of the exponential rise in spectral radiance of a blackbody in this range.

The calibration of a filter for high-precision photoelectric pyrometry should always be done *in situ* to ensure that illumination during use matches exactly that during calibration. The typical angular sensitivity of an interference filter is 0.1 nm per angular degree rotation and the temperature coefficient is likely to be ≈ 0.2 nm/$^\circ$C.

A final point to note on the use of filters is that reflections between filters can be a source of error and one that is very difficult to discover. It is most likely to arise if neutral filters are used to maintain the cathode illumination low when comparing two high-temperature sources.

In assessing the uncertainties of photoelectric pyrometry we find that there exists a category of such uncertainties comprising those related to the way in which the optical system interacts with the source. Uncertainties in this category turn out to be rather difficult to deal with and are often the result of subtle combinations of different effects. One of the most important of these effects is related to the size of the source being viewed and its distribution of radiance outside the geometrical viewing area. For an object of finite size in the plane of the source, the radiant flux falling within the geometrical image in the plane of the field stop is, as a result of diffraction, less than would be predicted by geometrical optics. It would be necessary to increase the size of the source so that it subtended an angle of 2π steradians at the aperture stop for this loss to fall to zero. Thus, if the pyrometer is measuring in turn two sources of different size, the comparison will be subject to an error due to diffraction. There will be an additional error resulting from scattering at the objective lens or mirror. This also will be dependent upon the size of the source since the scattering is proportional to the irradiance of the objective element.

The combination of these two effects is called simply the "size of source" effect, and its magnitude amounts to a few tenths of a percent under normal conditions when viewing in turn a blackbody in a furnace and a tungsten ribbon lamp. This is illustrated in Fig. 7-36. The size of the component due to diffraction is quite easy to calculate (Blevin, 1970) and is shown dotted in Fig. 7-36. In comparing a tungsten ribbon, 2 mm wide but very long, with a blackbody in a furnace, the size of source effect would amount to about 0.2 %. In comparing two blackbodies, the size of source effect would depend upon the difference in the radiance distribution in the two furnaces. As with all diffraction and scattering processes, the effect increases very rapidly at small angles, but dies away only very slowly at large angles, as is clear from Fig. 7-36.

Since diffraction accounts for only about half of the size-of-source effect, with the remainder being due mostly to scattering, it is necessary to measure

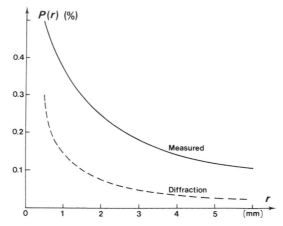

Fig. 7-36. Size-of-source effect where $P(r)\%$ is the fraction of the measured radiant flux originating from outside a circle of radius r.

the effect for each optical system. This may be done by using one of the arrangements shown in Fig. 7-37. In the first method (Quinn and Ford, 1969), Fig. 7-37a, the change of response of the pyrometer is measured as the size of the source is increased. The disadvantage of this method is that a small change in a large signal is being sought with all the attendant problems that this brings. In addition, the source does not provide a good replica of the spatial radiance distribution of a furnace containing a blackbody. A better method (Jones and Tapping, 1982) is shown in Fig. 7-37b; a furnace is used to provide off-axis irradiation and an aperture is placed at the centre of the furnace

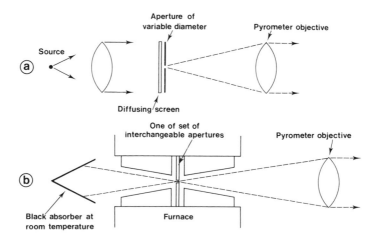

Fig. 7-37. Methods for measuring the size-of-source effect. (a) using a source of variable diameter illuminated by a supplementary lamp; (b) using a range of apertures placed in the centre of a furnace.

through which a cold black surface outside can be viewed by the pyrometer. In the absence of a size-of-source effect, the response of the pyrometer is, of course, zero so that the whole of the observed signal is a measure of the size-of-source effect. A method similar in principle but not using a furnace was described by Coates and Andrews (1978).

Closely related to the size-of-source effect is the change in irradiance of the field stop due either to a change in transmission or reflection of the objective element, or a change in the size of the aperture stop, resulting from heating on exposure to radiation from a furnace. The largest effects of this nature arise if organic films remain on the outer surface of the objective element. This has already been mentioned (Ohtsuka et al., 1979) in relation to the stability of transmission of the windows of tungsten ribbon lamps. If mirror optics are used, a substantial quantity of heat may be focussed on the field stop. Adequate means of thermal dissipation must be provided to avoid dimensional changes taking place, since this would be indistinguishable from an increase in radiance of the source.

7-7   Radiation thermometry in the measurement of thermodynamic temperature

In Chapter 3, we described in detail how thermodynamic temperatures are measured by means of a gas thermometer and other primary thermometers. We showed that for temperatures above about 30 K practically all numerical values of thermodynamic temperature are based upon gas thermometry. Improvements in radiation thermometry, however, are likely to change this. Already, measurements of temperature intervals in the range 630 °C to the gold point have shown that IPTS-68 is in error near 800 °C by about 0.4 °C (Bonhoure, 1975; Quinn et al., 1973). Photoelectric pyrometry, by itself, is not primary thermometry since it is not possible to measure the spectral radiance of a source absolutely, but only the ratio of the spectral radiances of two sources, and it is not possible for one of them to be at the triple point of water. Nevertheless, photoelectric pyrometry can give very precise values for the temperature difference between two sources at temperatures above about 400 °C. The accuracy with which this can be done is increasing and is likely to attain a level of ±5 mK (Coates and Andrews, 1982). The methods used are extensions of those described above for the realization of IPTS-68 above the gold point. The principal differences are that wavelengths out to 1000 nm are used, measurements are exclusively based upon blackbody radiation i.e. tungsten ribbon lamps as transfer sources are not used, and extreme care is taken to reduce uncertainties.

At temperatures below 100 °C, Quinn and Martin (1982) have measured thermodynamic temperatures directly by measuring the ratio of the total radiant powers of the thermal radiation from two blackbodies, one of which is

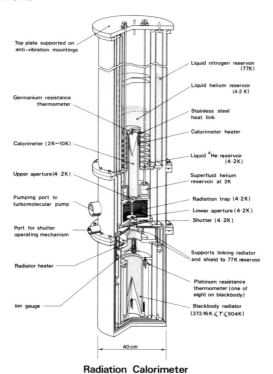

Top plate supported on
anti-vibration mountings

Liquid nitrogen reservoir
(77K)

Liquid helium reservoir
(4·2 K)

Germanium resistance
thermometer

Stainless steel
heat link

Calorimeter heater

Calorimeter (2K–10K)

Liquid ^{4}He reservoir
(4·2K)

Upper aperture (4·2K)

Superfluid helium
reservoir at 2K

Pumping port to
turbomolecular pump

Radiation trap (4·2K)

Lower aperture (4·2K)

Shutter (4·2K)

Port for shutter
operating mechanism

Supports linking radiator
and shield to 77K reservoir

Radiator heater

Platinum resistance
thermometer (one of
eight on blackbody)

Ion gauge

Blackbody radiator
(273.16 K $\leqslant T \leqslant$ 504K)

40 cm

Radiation Calorimeter

Fig. 7-38. Overall view of radiation calorimeter used for the determination of thermodynamic temperature between 0 °C and 100 °C and the Stefan–Boltzmann constant.

at the temperature of the triple point of water, 273.16 K, and the other at the unknown temperature T. The apparatus used for these measurements is shown in outline in Fig. 7-38. The principle features of the method are as follows: radiation from a blackbody at 273.16 K (see also Fig. 7-15) passes through a pair of apertures at 4.2 K to a heat-flow calorimeter in the form of a blackbody at about 2 K. At equilibrium, the calorimeter has risen in temperature to about 4 K, the radiation is then interrupted by means of a shutter at 4.2 K and electrical power applied to the calorimeter to maintain the same temperature. The electrical power is a direct measure of the thermal radiative power. This is repeated when the temperature of the radiating blackbody has been changed to an unknown temperature T. The electrical power required to balance the thermal radiative power is once again measured and the ratio of the measured electrical powers equals $T^4/(273.16)^4$, from the Stefan–Boltzmann law. The accuracy so far achieved appears to match that of classical gas thermometry, namely about ± 3 mK in the range -40 °C to 100 °C, and the preliminary results confirm those obtained by Guildner and Edsinger (see Chapter 3) up to 100 °C. A value of the Stefan–Boltzmann constant has also been obtained (Quinn and Martin, 1983) from absolute measurements of the thermal radiative power at 273.16 K together with the measured geometry of the apertures at 4.2 K.

7-8 Practical radiation thermometry aimed at overcoming the emittance problem

7-8-1 Introduction

The most common problem in radiation thermometry outside the standards laboratory is that of using a radiation thermometer to measure the temperature of a body for which the emittance is not known. It is the exception rather than the rule in most industrial applications to find good blackbody conditions present. It is much more likely that the object, whose temperature is to be measured, is either a clean freely-radiating metal surface, a partially-oxidized metal surface, a mixture of molten metal and slag partially obscured by smoke, or even a semi-transparent object such as molten glass. Both highly specular and nearly diffuse surfaces are encountered. The former, although simpler in many ways as we shall see, brings the added difficulty of reflected radiation from other hot bodies in the vicinity.

There is, unfortunately, no single method of radiation pyrometry that can cope with all of these diverse situations. There are, however, a number of different techniques each of which can overcome one or perhaps two of the difficulties mentioned above. The choice of method depends very much on the details of the particular application, and all that can be given here are some general guidelines concerning the principles of the various methods. The details of the many instruments developed for particular applications will be found in the Proceedings of the more important Symposia on thermometry cited in the bibliography.

7-8-2 Two-colour or ratio pyrometry

Even though the emittance of a particular surface may not be very well known, or the transmittance of the intervening atmosphere or window or even the size of the source itself may be varying, it is sometimes the case that all of these effects are relatively wavelength-independent. If this is so, then a two-colour or ratio pyrometer is useful. The principle of the method is simple; using the Wien approximation for the Planck function, which is perfectly adequate for these purposes, we can write

$$Q = \frac{L(\lambda_2 T)}{L(\lambda_1 T)} = \frac{\varepsilon_2 \lambda_2^{-5}}{\varepsilon_1 \lambda_1^{-5}} \exp\left[\frac{c_2}{T}\left(\frac{1}{\lambda_1} - \frac{1}{\lambda_2}\right)\right] \qquad (7\text{-}91)$$

This can be written

$$Q = R(\varepsilon) \left(\frac{\lambda_1}{\lambda_2}\right)^5 \exp\left[\frac{c_2}{T}\left(\frac{1}{\lambda_1} - \frac{1}{\lambda_2}\right)\right].$$

(7-92)

Thus a measurement of the ratio of the spectral radiances at two wavelengths enables T to be calculated provided that $R(\varepsilon)$ is known. Although $R(\varepsilon)$ is here defined as the ratio of the emittance, it can also be considered as the ratio of any other wavelength-dependent but temperature-invariant quantity such as atmospheric transmission, spectral sensitivity of the detector and so on. Note that parameters which are at the same time wavelength and temperature-independent do not appear in equation (7-92) and can be ignored. The size of the source is one such parameter. The sensitivity of the method increases as the separation of the wavelengths increases. Unfortunately, knowledge of $R(\varepsilon)$ decreases the further apart are the two wavelengths λ_1 and λ_2. One of the most useful applications of the two-colour pyrometer is in the measurement of the temperature of very small objects, too small to be resolved by ordinary pyrometers. Examples of such objects are high-temperature incandescent tungsten filaments (Lechner and Schob, 1975) or clouds of incandescent particles.

An alternative way of writing the equation for a ratio pyrometer is to consider it as a device that measures the spectral radiance temperatures T_{R1} and T_{R2}, at two wavelengths λ_1 and λ_2. In this case

$$Q = R(\varepsilon) \left(\frac{\lambda_1}{\lambda_2}\right)^5 \exp\left[\frac{c_2}{T}\left(\frac{1}{\lambda_1} - \frac{1}{\lambda_2}\right)\right] = \left(\frac{\lambda_1}{\lambda_2}\right)^5 \exp\left[c_2\left(\frac{1}{\lambda_1 T_{R_1}} - \frac{1}{\lambda_2 T_{R_2}}\right)\right]$$

(7-93)

$$\therefore \qquad \ln R(\varepsilon) + \frac{c_2}{T}\left(\frac{1}{\lambda_1} - \frac{1}{\lambda_2}\right) \equiv c_2\left(\frac{1}{\lambda_1 T_{R_1}} - \frac{1}{\lambda_2 T_{R_2}}\right)$$

(7-94)

from which

$$\frac{1}{T} = \left(\frac{1}{\lambda_1} - \frac{1}{\lambda_2}\right)^{-1}\left[\left(\frac{1}{\lambda_1 T_{R_1}} - \frac{1}{\lambda_2 T_{R_2}}\right) - \frac{\ln R(\varepsilon)}{c_2}\right].$$

(7-95)

This is sometimes written

$$\frac{1}{T} = \frac{1}{T_{\text{ratio}}} - \left(\frac{1}{\lambda_1} - \frac{1}{\lambda_2}\right)^{-1}\frac{\ln R(\varepsilon)}{c_2}$$

(7-96)

in which T_{ratio}, the ratio temperature, is defined as

$$\frac{1}{T_{\text{ratio}}} = \left(\frac{1}{\lambda_1 T_{R_1}} - \frac{1}{\lambda_2 T_{R_2}}\right)\left(\frac{1}{\lambda_1} - \frac{1}{\lambda_2}\right)^{-1}$$

(7-97)

In order to look more closely at the performance to be expected of a ratio pyrometer we shall differentiate equation (7-92) with respect to T

$$\frac{dQ}{dT} = -R(\varepsilon)\left(\frac{\lambda_1}{\lambda_2}\right)^5 \frac{c_2}{T^2}\left(\frac{1}{\lambda_1} - \frac{1}{\lambda_2}\right)\exp\left[\frac{c_2}{T}\left(\frac{1}{\lambda_1} - \frac{1}{\lambda_2}\right)\right]$$

(7-98)

for incremental changes in Q and T we may deduce from (7-92) and (7-98)

$$\frac{\Delta Q}{Q} = -\frac{c_2}{T^2}\left(\frac{1}{\lambda_1} - \frac{1}{\lambda_2}\right)\Delta T$$

(7-99)

or

$$\Delta T = -\frac{T^2}{c_2}\left(\frac{1}{\lambda_1} - \frac{1}{\lambda_2}\right)^{-1}\frac{\Delta Q}{Q}.$$

(7-100)

This equation demonstrates how the sensitivity of the ratio pyrometer, ΔT, for a given discrimination in Q improves the greater the difference between λ_1 and λ_2. For example in order to achieve an accuracy of 1 K at 1200 K using wavelengths of 650 nm and 750 nm, an accuracy in the measurement of Q of 1% is called for, which is not too difficult. It follows however that an equal accuracy in $R(\varepsilon)$ is required and this is much more difficult to achieve. There are few real surfaces for which the relative emittance is known at these two wavelengths to anything like 1 %. There is an application however, for a cavity whose emissivity is high but not known exactly, for under these conditions the emissivity variation with wavelength is very much less.

To find the accuracy with which the two wavelengths must be known, we differentiate equation (7-92) with respect to wavelength, λ_1 assuming that $dR(\varepsilon)/d\lambda = 0$,,

$$\frac{dQ}{d\lambda_1} = R(\varepsilon)\left(\frac{\lambda_1}{\lambda_2}\right)^5 \exp\left[\frac{c_2}{T}\left(\frac{1}{\lambda_1} - \frac{1}{\lambda_2}\right)\right]\left(\frac{5\lambda_1 T - c_2}{\lambda_1^2 T}\right)$$

(7-101)

Combining (7-101) with (7-98) we find

$$\frac{dT}{d\lambda_1} = \frac{T}{c_2\left(\frac{1}{\lambda_1} - \frac{1}{\lambda_2}\right)}\left[\frac{5\lambda_1 T - c_2}{\lambda_1^2}\right]$$

whence for incremental changes in T and λ_1

$$\Delta T = - \left[\frac{T}{\lambda_1^2 c_2} \left(\frac{1}{\lambda_1} - \frac{1}{\lambda_2} \right)^{-1} (c_2 - 5\lambda_1 T) \right] \Delta \lambda. \qquad (7\text{-}102)$$

For the same temperature and wavelengths as before, we find that for an accuracy of 1 K, we need to know λ_1 to an accuracy of 0.1 nm, which is rather difficult.

The general conclusion must be that although ratio pyrometry has certain attractions (Gardner et al., 1982), it is not a suitable method for making high-precision temperature measurements unless special steps are taken to measure $R(\varepsilon)$ and very high precision is available in the determination of wavelength.

A novel method of measuring the ratio of the emittances in situ is due to de Witt and Kunz (1972) and Kunz and Kauffmann (1975). They used a laser to measure the ratio of the absorptances of the material at the two wavelengths used in the ratio pyrometer. This was done by using a spectral pyrometer working at a third wavelength to measure the temperature rise of the sample on irradiation by the laser at the ratio pyrometer wavelengths λ_1 and λ_2 in turn. The optical system of their pyrometer is illustrated in Fig. 7-39. The

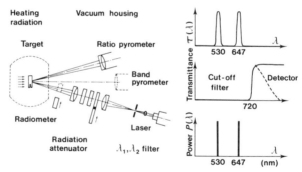

Fig. 7-39. A simple arrangement for laser-absorption two-wavelength ratio pyrometry (after Kunz and Kauffmann, 1975).

spectral pyrometer is operated at a wavelength λ_3, greater than either λ_1 or λ_2, so that a cut-off filter can easily be used to eliminate interference by laser radiation. The temperature rise observed in a tungsten ribbon of the type used in standard lamps amounts to a few degrees at 1100 °C.

7-8-3   Methods employing an auxiliary source of thermal radiation

A method useful for measuring the temperature of a wide range of surfaces in the range up to about 300 °C is due to Kelsall (1963). His method is illustrated in Fig. 7-40. The unknown temperature T_3 is found by adjusting the

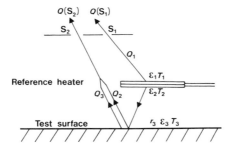

Fig. 7-40. Kelsall's method using an auxiliary heater (after Kelsall, 1963).

temperature T_2 of a reference source in close proximity to the surface. A balance is obtained when the signal from a detector is the same when viewing in turn the top of the reference source T_1 and the test surface. Under these conditions, the radiant flux Q_1 emitted by the upper surface equals the sum of Q_3, the radiant flux emitted by the test surface, and Q_2, that reflected from the test surface but originating from the lower surface of the reference source:

$$Q_1 = g_1 \varepsilon_1 L(\lambda T_1)$$

$$Q_2 = g_2 \varepsilon_2 \rho_3 L(\lambda T_2)$$

$$Q_3 = g_3 \varepsilon_3 L(\lambda T_3)$$

where ε_1, T_1; ε_2, T_2; and ε_3, T_3 are the emittances and temperatures of the upper and lower surfaces of the reference heater and the test surface respectively. The constants g_1, g_2 and g_3 represent the geometrical factors which we will set equal for the purposes of this discussion. It will also be assumed that $\varepsilon_3 + \rho_3 = 1$, an assumption that is justified if the test surface is a specular reflector as in Fig. 7-40. If it is not specular, then, from our earlier discussion of the Kirchhoff law, some of the radiation from the lower surface Q_2 is lost by scattering and Q_2 no longer equals $\varepsilon_2 \rho_3 L(\lambda T_2)$. This is likely to be the main source of error in the method.

If we now set $\varepsilon_1 = \varepsilon_2$ and $T_1 = T_2$, we can write

$$\varepsilon_1 L(\lambda T_1) = \varepsilon_1 (1 - \varepsilon_3) L(\lambda T_1) + \varepsilon_3 L(\lambda T_3) \tag{7-103}$$

whence
$$\varepsilon_1 L(\lambda T_1) = L(\lambda T_3). \tag{7-104}$$

Thus if ε_1 is known, $T_1 = T_3$. The instruments that have been made to take advantage of this method have usually incorporated a blackbody so that $\varepsilon_1 = 1$. Even so, errors are likely to arise if the emittance of the test surface ε_3 is low. From equation (7-103) it is clear that if ε_3 is small, the significant term, $\varepsilon_3 L(\lambda T_3)$, makes up only a small fraction of the total signal detected. Losses due

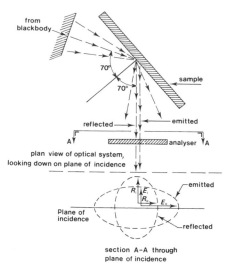

Fig. 7-41. The polarization method (after Murray, 1972).

to scattering, which influence the first term on the right-hand side of equation (7-103), will therefore be of great importance.

A modification of this method by Toyota and others (1972), allows some compensation to be made for the radiation lost by scattering. A shutter obscures some of the radiation emitted by the lower reference heater so that at balance T_2 is rather higher than it would otherwise be. Provided that $(1/\rho_3)(1 - \varepsilon_3/\varepsilon_1)$ is roughly constant for different materials, this method leads to somewhat better results. For most common metals, whether clean or partially oxidized, in the temperature range up to 350 °C, an accuracy of about ± 10 °C can be expected from the modified version of the method. The original version of Kelsall was liable to errors of as much as 30° for some low-emissivity materials such as polished brass.

A quite different method of compensating for lack of knowledge of the emittance is based upon the polarization of thermal radiation emitted and reflected at angles away from the normal. The method is based upon a proposal of Tingwaldt (1952) later developed by Murray (1972) and Berry (1980), the principle of the method is as follows: blackbody radiation is unpolarized; if, therefore, blackbody radiation is reflected at a high angle of incidence from a hot metal surface, the sum of the emitted and reflected radiation will be polarized unless the temperature of the reflecting surface equals that of the blackbody. This is illustrated in Fig. 7-41.

The radiation emitted by the test surface at an unknown temperature T_2 and transmitted by a polarizer at an arbitrary angle φ to the principal plane of polarization I_2 is given by

$$I_2 \infty [\tfrac{1}{2}\varepsilon_p L(T_2) \cos^2 \varphi + \tfrac{1}{2}\varepsilon_n L(T_2) \sin^2 \varphi) \tag{7-105}$$

where ε_p and ε_n are the emittances of the surface for radiation polarized parallel and normal to the surface.

If the surface is also irradiated by blackbody radiation of a known temperature T_1, the sum of the reflected and emitted radiations I transmitted by the polarizer is

$$I \infty [I_2 + \tfrac{1}{2}(1 - \rho_p)L(T_1) \cos^2 \varphi + \tfrac{1}{2}(1 - \rho_n)L(T_1) \sin^2 \varphi]. \qquad (7\text{-}106)$$

The reflectances ρ_p and ρ_n for the two planes of polarization are related to the emittances by

$$\rho_p = 1 - \varepsilon_p - \Delta\rho_p \qquad \text{and} \qquad \rho_n = 1 - \varepsilon_n - \Delta\rho_n$$

where $\Delta\rho_p$ and $\Delta\rho_n$ represent the fractions of the reflected radiation that are diffusely scattered and therefore lost. Substituting for ρ_p and ρ_n in (7-106) and combining with (7-105), we find that the signal, S, from a detector is proportional to

$$L(T_2)\{\varepsilon_p \cos^2 \varphi + \varepsilon_n \sin^2 \varphi\} + L(T_1)\{(1 - \varepsilon_p - \Delta\rho_p)\cos^2 \varphi$$

$$+ (1 - \varepsilon_n - \Delta\rho_n)\sin^2 \varphi\}. \qquad (7\text{-}107)$$

The form of S as a function of φ is shown in Fig. 7-42 for a range of values of $T_1 - T_2$. S shows maxima and minima as φ passes through the angles normal and parallel to the plane of the test surface. If $L(T_1)$ is adjusted until the

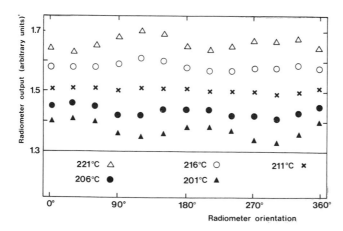

Fig. 7-42. Sum of the reflected and emitted radiation from a stainless steel surface at 212 °C for blackbody temperatures from 201 °C to 221 °C (after Berry, 1980).

oscillatory term disappears, then from (7-107)

$$L(T_2) = L(T_1)\{1 + (\Delta\rho_p - \Delta\rho_n)(\varepsilon_p - \varepsilon_n)^{-1}\}. \tag{7-108}$$

Thus if $\Delta\rho_p = \Delta\rho_n$, the condition for the disappearance of the oscillatory terms is that $T_1 = T_2$. This is a much less strict requirement than was the case for the method of Kelsall. Here we do not require $\Delta\rho$ to be equal to zero. The requirement is only that whatever radiation is scattered must be unpolarized, i.e. scattered diffusely. In practice, this means that errors are liable to be encountered only in measuring the temperature of surfaces having a preferential direction, such as might result from grinding or polishing in one direction. The sensitivity of the method depends upon the magnitude of the difference between ε_p and ε_n. This is greatest at large angles from the normal and in the instrument developed by Murray an angle of $70°$ to the normal was chosen. Figure 7-43 shows results obtained by Murray on a variety of surfaces at temperatures in the range 150 °C–450 °C.

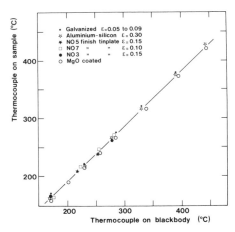

Fig. 7-43. Experimental results obtained using the polarization method on a variety of materials (after Murray, 1972).

Both of the methods just described require the presence of a supplementary source of thermal radiation. Another method described by Drury *et al.* (1951) and widely used in industry is much simpler. Instead of a separate supplementary source, the surface itself is used by bringing a hemispherical gold-plated mirror into contact with, or very close proximity to, the surface. A silicon-cell detector is used to provide a measure of the radiation density inside the hemisphere. If the hemisphere is a perfect reflector (and gold has an infrared reflectivity greater than 99 %), and the surface area of the hemisphere occupied by the silicon cell is negligible, the method will give accurate results. The maximum temperature of operation is of course limited by the heating of the device and also the method cannot be used on small or inaccessible objects nor can it be used for continuous measurements. It is, nevertheless, one of the most widely used techniques in industry to overcome the lack of knowledge of the emittance of the object the temperature of which is to be measured (Becker and Wall, 1981).

7-8-4 Multi-wavelength pyrometry

Considerable interest has been shown in recent years in the application of methods of measurement which make use of the normally redundant information contained in the spectrum of the radiation from hot bodies. The principle of the new methods is contained in the statement that if it is known that the emittance of a material varies as the nth power of the wavelength, then the temperature can be derived from measurements of the relative spectral radiance at $n+2$ wavelengths. For $n=0$ we have the case of two colour or ratio pyrometry in which the emittance is independent of wavelength. The method has been extended to the case of $n=1$, in which the emittance varies linearly with wavelength, and three wavelengths are used. The problem with the two-colour pyrometer, as we have seen, is that for the condition of equality of emittance at the two wavelengths to be satisfied, the wavelengths must in practice be close together. Our elementary analysis showed that the sensitivity on the other hand, improves as the wavelengths move further apart. For the three-colour pyrometer, a similar analysis shows that even small departures from the assumed linear relation between emittance and wavelength can lead to large errors. Svet (1972) has pointed out, however, that there are advantages to be gained in using modern computing methods to arrive at the true temperature from measurements at m wavelengths on the assumption that the emissivity is a function of only the nth power of the wavelength when $m \geq n$. In this way, the redundant information contained in the $[m-(n+2)]$ measurements should compensate for the lack of precision in the measurements of relative radiance at the m wavelengths. The difficulty in achieving a useful accuracy has been demonstrated by Coates (1981b), who concluded that none of these methods are likely to lead to a more accurate value of T than that obtainable from a single-wavelength pyrometer using an estimated value for the emittance.

7-9 Radiation thermometry of semi-transparent media

Important applications of radiation thermometry occur in the glass industry where the temperature of glass in various conditions must be measured: thin solid or liquid sheets, thick slabs or large molten volumes. The transfer of heat through glass by radiation is a very complex process (Condon, 1968; Isard, 1980). It is akin, in many respects, to the transfer of heat or momentum through a gas in the intermediate regime between molecular and viscous states. The mean free path of the molecules of a gas can be likened to the distance travelled by a ray of glass before it is significantly absorbed, viz. α^{-1} where α is the absorption coefficient. The value of α is strongly dependent upon wavelength, and increases from a very small value at wavelengths below about 2.5 μm to a very large value, > 10 cm^{-1}, for wavelengths greater than about 4 μm. An intermediate region exists, between about 2.7 μm and 4 μm, within which α is strongly temperature-dependent and varies between 4 cm^{-1} to 6 cm^{-1}. These large variations in absorption occur just in the wavelength range where the bulk of the thermal radiation exists during glass working from 1000 K to 2000 K.

The thermal radiation emitted by glass contains information not only about the surface temperature, but also about temperature distribution within it over a depth of the order of α^{-1}. The temperature distribution is often a matter of great importance to the glass technologist, particularly during the cooling process. The toughening of plate glass by rapid cooling of the outer skin is a well known and widely used procedure. Accurate temperature measurement of the skin and inner body of the glass is thus essential in the development and quality control of such industrial processes.

A different situation is that in which a layer of hot glass, or any other solid or liquid material, is supported on a hot liquid or solid substrate. The radiation emitted by the glass is a mixture of the radiation from the glass and that from the hot substrate. This is illustrated in Fig. 7-44, where we consider the normal radiation leaving a layer of glass, of thickness d, within which we imagine two blackbody cavities to be placed, one of which protrudes from the surface while the other is submerged.

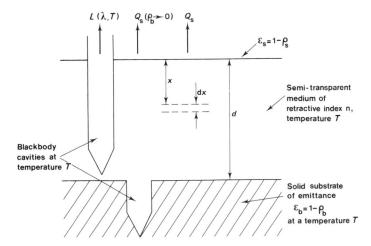

Fig. 7-44. Radiation emitted normal to a layer of semi-transparent material of thickness d resting upon an opaque substrate.

We define the spectral volume radiance $J(\lambda T)$ as the radiant flux omitted per unit volume of glass at temperature T per unit solid angle:

$$J(\lambda T) = \frac{\alpha_\lambda n^2}{\pi} M(\lambda T) = \alpha_\lambda n^2 L(\lambda T) \tag{7-109}$$

in which α_λ is the wavelength-dependent absorption coefficient. The radiant flux emitted per unit area from a layer of thickness dx is thus

$$dQ = J(\lambda T) dx. \tag{7-110}$$

The radiant flux arriving at the surface a distance x from dx is thus

$$dQ_s = J(\lambda T) e^{-\alpha_\lambda x} dx \tag{7-111}$$

of which a fraction $(1 - \rho_s)$ is emitted from the surface. Thus the radiant flux Q_d emitted into air from a layer of thickness d is

$$Q_d = \int_{x=0}^{d} \frac{(1 - \rho_s)}{n^2} J(\lambda T) e^{-\alpha_\lambda x} dx. \tag{7-112}$$

Assuming that the layer is isothermal, we can write

$$Q_d = (1 - \rho_s) \alpha_\lambda L(\lambda T) \int_{x=0}^{d} e^{-\alpha_\lambda x} dx = (1 - \rho_s) L(\lambda T)(1 - e^{-\alpha_\lambda d}). \tag{7-113}$$

If either $d \to \infty$ or $\alpha \to \infty$ equation (7-113) reduces to

$$Q_d = (1 - \rho_s) L(\lambda T) = \varepsilon_s L(\lambda T). \tag{7-114}$$

We can now write down an expression for Q_s, the radiant flux emitted from the surface of the layer of glass resting upon a substrate of emittance $\varepsilon_b = 1 - \rho_b$ at the same temperature as the glass:

$$Q_s = (1 - \rho_s) \alpha_\lambda L(\lambda T) \int_{x=0}^{d} e^{-\alpha_\lambda x} dx$$

$$+ \rho_b e^{-\alpha_\lambda d} (1 - \rho_s) \alpha_\lambda L(\lambda T) \int_{x=0}^{d} e^{-\alpha_\lambda x} dx$$

$$+ (1 - \rho_s)(1 - \rho_b) L(\lambda T) e^{-\alpha_\lambda d} \tag{7-115}$$

in which the first term represents the radiation from the body of the glass emitted upwards through the surface. The second term is the radiation from the body of the glass emitted downwards, reflected from the substrate and emitted through the surface in an upward direction. The third term represents the radiation emitted by the substrate and transmitted up through the glass and out through the top surface. This expression simplifies, using the result shown in equation (7-113), to

$$Q_s = (1 - \rho_s)L(\lambda T)[(1 - e^{-\alpha_\lambda d})(1 + \rho_b e^{-\alpha_\lambda d}) + (1 - \rho_b)e^{-\alpha_\lambda d}]$$

$$= (1 - \rho_s)L(\lambda T) \cdot A. \tag{7-116}$$

If we wish to take into account the radiation that undergoes multiple reflections between the substrate and the upper glass surface we transform (7-116) into

$$Q_s = (1 - \rho_s)L(\lambda T)[A + \rho_s\rho_b e^{-2\alpha_\lambda d}A + (\rho_s\rho_b e^{-2\alpha_\lambda d})^2 A + \dots \tag{7-117}$$

The series sums to

$$Q_s = (1 - \rho_s)L(\lambda T)A\left[\frac{1}{1 - \rho_s\rho_b e^{-2\alpha_\lambda d}}\right]. \tag{7-118}$$

Thus

$$Q_s = (1 - \rho_s)L(\lambda T)\left[\frac{(1 - e^{-\alpha_\lambda d})(1 + \rho_b e^{-\alpha_\lambda d}) + (1 - \rho_b)e^{-\alpha_\lambda d}}{1 - \rho_s\rho_b e^{-2\alpha_\lambda d}}\right] \tag{7-119}$$

which simplifies to

$$Q_s = (1 - \rho_s)L(\lambda T)\left[\frac{1 - \rho_b e^{-2\alpha_\lambda d}}{1 - \rho_s\rho_b e^{-2\alpha_\lambda d}}\right]. \tag{7-120}$$

If ρ_b is close to unity, i.e. the substrate is a highly reflecting clean metal surface such as a liquid metal surface for example, equation (7-120) becomes very close to equation (7-113) except that the thickness appears to be $2d$.

$$Q_{s(\rho_b \to 1)} \approx (1 - \rho_b)L(\lambda T)[1 - e^{-2\alpha_\lambda d}]. \tag{7-121}$$

If on the other hand the substrate is close to a blackbody $\rho_b \to 0$, and we have

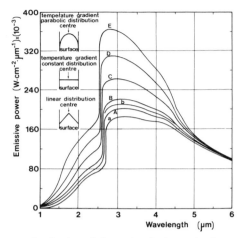

Fig. 7-45. Spectral energy distribution of thermal radiation emitted normal to the surface of a glass plate 6 mm thick having a surface temperature of 600 °C for: (a) uniform internal temperature distribution; (b) linear temperature gradient with maximum at centre of 650 °C; (c) parabolic distribution, maximum 625 °C; (d) parabolic, maximum 650 °C; (e) parabolic, maximum 700 °C; (f) parabolic, maximum 750 °C; (g) parabolic, maximum 800 °C (after Beattie and Coen, 1959).

$$Q_{s(\rho_b \to 0)} \approx (1 - \rho_s) L(\lambda T) \qquad (7\text{-}122)$$

we thus have a means of determining the surface emissivity of glass or any other semi-transparent liquid or solid medium (Keene and Quinn, 1979) by measuring the ratio of $Q_{s(\rho_b \to 0)}$ to $L(\lambda T)$. The latter is obtained by arranging a blackbody at the same temperature as the glass while the former is obtained by arranging a blackbody within the layer of glass.

The treatment given above is considerably simplified due to the initial assumption of a uniform temperature within the glass. For non-uniform temperatures, equation (7-115) must be modified by the inclusion of $L(\lambda T)$ within the integral. For particular temperature gradients, the equation must then be solved numerically (Beattie and Coen, 1959), and no simple solution is possible. Fortunately the absorption coefficient and surface reflection coefficients are usually such that even for a layer only 5 mm thick, the higher order internal reflections are very small and can usually be ignored.

Radiation thermometry of glass is carried out using wavelengths either below 3 μm or beyond 5 μm depending upon whether the internal temperature or the surface temperature is required. Figure 7-45 shows the spectral distribution of the thermal radiation emitted by a 6 mm layer computed for three internal temperature profiles, across its thickness; constant temperature, parabolic and linear increase towards the centre. Radiation pyrometry at both 5 μm and 3 μm can thus give information on the surface temperature and also the profile of the internal temperature distribution of a semi-transparent medium.

References

Andrews, J. W. and Coates, P. B. (1975). Standard lamps and their calibration. Temperature-75, Inst. of Physics Conf., Series No 26, 256–263.

Baltes, H. P. (1973). Deviations from the Stefan–Boltzmann law at low temperatures. *Applied Physics*, **1**, 39–43.

Baltes, H. P. (1976). Planck's radiation law for finite cavities and related problems. *Infrared Physics*, **16**, 1–8.

Baltes, H. P. and Kneubühl, F. K. (1970). Spectral density, thermodynamics and temporal coherence of non-Planckian blackbody radiation for small cavities. *Optics Comm.* **2**, 14–16.

Baltes, H. P. and Hilf, E. R. (1976). "Spectra of Finite Systems", Bibliographisches Institut Manheim/Wein/Zurich.

Beattie, J. R. and Coen, E. (1959). Spectral emission of radiation by glass. *Brit. J. Appl. Phys.* **11**, 151–157.

Becker, H. B. and Wall, T. F. (1981). Effect of specular reflection of hemispherical surface pyrometer on emissivity measurement. *J. Phys. E.* **14**, 998–1001.

Bedford, R. E. (1972). Effective emissivities of blackbody cavities — a review. *TMCSI*, **4**, Part 1, 425–434.

Bedford, R. E. and Ma, C. K. (1974). Emissivities of diffuse cavities: Isothermal and non-isothermal cones and cylinders. *J. Opt. Soc. Amer.* **64**, 339–349; *See also, ibid.*, (1975), **65**, 565–572; *ibid.*, (1976), **66**, 724–730.

Berry, K. H. (1980). New ways to save energy — proceedings of the international seminar held in Brussels, Oct. 1979. Reidel, Dortrecht Holland, 880–889.

Berry, K. H. (1981). Emissivity of a cylindrical blackbody cavity with a re-entrant cone end face. *J. Phys. E.* **14**, 629–632.

Bezemer, J. (1974). Spectral sensitivity corrections for optical standards pyrometers. *Metrologia*, **10**, 47–52.

Blevin, W. R. (1970). Diffraction losses in photometry and radiometry. *Metrologia*, **6**, 31–44.

Blevin, W. R. (1972). Corrections in optical pyrometry and photometry for the refractive index of air. *Metrologia*, **8**, 146–147.

Bonhoure, J. (1975). Détermination radiométrique des températures thermodynamiques comprises entre 904 et 1338 K. *Metrologia*, **11**, 141–150.

Burkhard, D. G., Lochhead, J. V. and Peuchina, C. M. (1972). On the validity of Kirchhoff's law in a non-equilibrium environment. *Amer. J. Phys.* **40**, 1794–1798.

Caren, R. P. (1974). Thermal radiation between closely spaced metal surfaces at low temperatures due to travelling and quasi-stationary components of the radiation field. *Int. J. Heat Mass Transfer.* **17**, 755–765.

Coates, P. B. (1970). The edge effect in electron multiplier statistics. *J. Phys. E.* **4**, 1290–1296.

Coates, P. B. (1971). Noise sources in the C31000D photomultiplier. *J. Phys. E.* **4**, 201–207.

Coates, P. B. (1972). Photomultiplier noise statistics. *J. Phys. D.* **5**, 915–929.

Coates, P. B. (1975a). The NPL photon counting pyrometer. Temperature-75, 238–243.

Coates, P. B. (1975b). Fatigue and its correction in photon counting experiments. *J. Phys. E.* **8**, 189–193.

Coates, P. B. (1975c). Limitations of the magnetic defocusing technique for photomultipliers. *J. Phys. E.* **8**, 614–617.

Coates, P. B. (1977). Wavelength specification in optical and photoelectric pyrometry. *Metrologia*, **13**, 1–5.

Coates, P. B. (1979). The direct calculation of radiance temperatures in photoelectric pyrometry. *High Temperatures — High Pressures*, **11**, 289–300.

Coates, P. B. (1981a). The calibration of radiation pyrometers, blackbody sources and standard lamps. NPL Report QU 61.

Coates, P. B. (1981b). Multi-wavelength pyrometry. *Metrologia*, **17**, 103–109.

Coates, P. B. and Andrews, J. W. (1978). A precise determination of the freezing point of copper. *J. Phys. F. Metal Physics*, **8**, 277–285.

Coates, P. B. and Andrews, J. W. (1981). Measurement of gain changes in photomultipliers. *J. Phys. E.* **14**, 1164–1166.

Coates, P. B. and Andrews, J. W. (1982). Measurement of thermodynamic temperature with the NPL photon-counting pyrometer. *TMCSI*, **5**, 109–114.

Condon, E. U. (1968). Radiative transport in hot glass. *J. Quant. Spectrosc. Radiat. Transfer*, **8**, 369–385.

de Vos, J. C. (1954). Evaluation of the quality of a blackbody. *Physica*, **20**, 669–689.

De Witt, D. P. and Kunz, H. (1972). Theory and technique for surface temperature determinations by measuring the radiance temperatures and absorptance ratio for wavelengths. *TMCSI*, **4**, 599–610.

Domonto, G. A., Boehm, R. F. and Tien, C. L. (1970). Radiative transfer between metallic surfaces at cryogenic temperatures. *J. Heat Transfer*, **92** c(3), 412.

Drury, M. D., Perry, K. P. and Land, T. (1951). Pyrometers for surface temperature measurement. *J. Iron Steel Inst.* **169**, 145.

Gardner, J. L., Jones, T. P. and Davies, M. R. (1982). A broadband ratio pyrometer. *TMCSI*, **5**, 409–412.

Gouffé, A. (1945). Corrections d'ouverture des corps noirs artificiels compte tenu des diffusions multiples internes. *Rev. d'Optique*, **24**, Nos 1–3, 1–10.

Hattori, S. and Ono, A. (1982). The effective temperature to express radiant characteristics of non isothermal cavities. *TMCSI*, **5**, 521–527.

International Commission on Illumination (CIE) (1970). "International Lighting Vocabulary", 3rd Edition. Bureau Central de la CIE, 4 Ave. du Recteur Poincaré, F-75016 Paris.

Isard, J. O. (1980). Surface reflectivity of strongly absorbing media and calculation of the infrared emissivity of glasses. *Infrared Physics*, **20**, 249–256.

Jones, O. C. and Gordon-Smith, R. C. (1973). Absolute radiometry by means of a blackbody source. *Proc. Roy. Soc.* A**335**, 369–386.

Jones, T. P. and Tapping, J. (1971). The realization of IPTS-68 above 1064.43 °C using the NSL photoelectric pyrometer. *Metrologia*, **8**, 4-11.

Jones, T. P. and Tapping, J. (1979). The suitability of tungsten strip lamps as secondary standard sources below 1064 °C. *Metrologia*, **15**, 135–141.

Jones, T. P. and Tapping, J. (1982). A precision photoelectric pyrometer for the realization of IPTS-68 above 1064.43 °C. *Metrologia*, **18**, 23–31.

Jung, H. J. (1979). Spectral nonlinearity characteristics of low-noise silicon detectors and their application to accurate measurement of radiant flux ratios. *Metrologia*, **15**, 173–181.

Jung, H. J. and Verch, J. (1973). Ein Rechenverfahren zur Auswertung pyrometrischer Messungen. *Optik*, **38**, 95–109.

Keene, B. J. and Quinn, T. J. (1979). An apparatus to determine total normal emissivity of opaque and diathermanous liquids at high temperatures. *High Temperatures — High Pressures*, **11**, 693–702.

Kelsall, D. (1963). An automatic emissivity-compensated radiation pyrometer. *J. Sci. Instr.* **40**, 1–4.

Kostkowski, H. J. and Lee, R. (1962). Theory and methods of optical pyrometry. NBS Monograph 41.

Kunz, H. (1969). Representation of the temperature scale above 1337.58 K with photoelectric direct current pyrometers. *Metrologia*, **5**, 88–102.

Kunz, H. (1975). On the state of ratio pyrometry with laser absorption measurements.

Temperature-75, 273–277.

Kunz, H. and Kaufmann, H. J. (1975). Photoelectric direct current standard pyrometers and their calibration at PTB. Temperature-75, 244–255.

Lapworth, K. C., Quinn, T. J. and Allnutt, L. A. (1970). A blackbody source of radiation covering a wavelength range from the ultraviolet to the infrared. *J. Phys. E.* **3**, 116–120.

Latyev, L. N., Chekhovskoi, V. Ya. and Shestakov, E. N. (1970). Monochromatic emissivity of tungsten in the temperature range 1200–2600 K and in the wavelength range 0.4–4 μm. *High Temperatures–High Pressures*, **2**, 175–181.

Lechner, W. and Schob, O. (1975). Temperature measurement of filaments above 2500 K applying two-wavelength pyrometry. Temperature-75, 297–305.

Lee, R. D., Kostkowski, H. J., Quinn, T. J., Chandler, T. R., Jones, T. P., Tapping, J. and Kunz, H. (1972). Intercomparison of the IPTS-68 above 1064 °C. *TMCSI*, **4**, Part 1, 377–393.

Lee, R. D. (1966). The NBS photoelectric pyrometer and its use in realizing the International Practical Temperature Scale above 1063 °C. *Metrologia*, **2**, 150–162.

Lovejoy, D. R. (1962). Detection limits in radiation and optical pyrometry. *J. Opt. Soc. Amer.* **52**, 1387–1398.

Murray, T. P. (1972). The polarized radiation method of radiation pyrometry. *TMCSI*, **4**, Part 1, 619–623.

NBS (1976). Technical Note 910-1 and 910-2. "Self-study Manual on Optical Radiation Measurements", (F. Nicodemus, ed.).

Ohtsuka, M., Bedford, R. E. and Ma, C. K. (1979). A note on the temperature dependent transmission of residual films of glasses. *Metrologia*, **15**, 165–166.

Ono, A. (1982). Apparent emissivities of cylindrical cavities with partially specular conical bottoms. *TMCSI*, **5**, 513–516.

Parker, J. W. and Abbott, G. C. (1965). Total emittance of metals, Symposium on Thermal Radiation of Solids, NASA SP-55.

Peavy, B. A. (1966). A note on the numerical evaluation of thermal radiation characteristics of diffuse cylindrical and conical cavities. *J. Res. Nat. Bur. Stds.* **70C**, 139–147.

Psarouthakis, J. (1963). Apparent thermal emissivity from surfaces with multiple V-shaped grooves. *AIAA Journal*, **1**, 1879–1882.

Quinn, T. J. (1965). The effects of thermal etching on the emissivity of tungsten. *Brit. J. Appl. Phys.* **16**, 973–980.

Quinn, T. J. (1967). The calculation of the emissivity of cylindrical cavities giving near blackbody radiation. *Brit. J. Appl. Phys.* **18**, 1105–1113.

Quinn, T. J. (1981). The absorptivity of a specularly reflecting cone for oblique angles of view. *Infrared Physics*, **21**, 123–126.

Quinn, T. J. and Barber, C. R. (1967). A lamp as a reproducible source of near blackbody radiation for precise pyrometry up to 2700 °C. *Metrologia*, **3**, 19–23.

Quinn, T. J. and Chandler, T. R. (1972). The freezing point of platinum determined by the NPL photoelectric pyrometer. *TMCSI*, **4**, Part 1, 295–309.

Quinn, T. J. and Ford, M. C. (1969). On the use of the NPL photoelectric pyrometer to establish the temperature scale above the gold point (1063 °C). *Proc. Roy. Soc.* **A312**, 31–50.

Quinn, T. J. and Lee, R. D. (1972). Vacuum tungsten strip lamps with improved stability as radiance temperature standards. *TMCSI*, **4**, Part 1, 395–412.

Quinn, T. J. and Martin, J. E. (1982). Radiometric measurements of thermodynamic temperature between 327 K and 365 K. *TMCSI*, **5**, 103–107.

Quinn, T. J. and Martin, J. E. (1983). A radiometric determination of the Stefan–Boltzmann constant. PMFC-2, NBS.SP-617.

Quinn, T. J., Chandler, T. R. D. and Chattle, M. V. (1973). The departure of IPTS-68 from thermodynamic temperatures between 725 °C and 1064.43 °C. *Metrologia*, **9**, 44–46.

Ruffino, G. (1971). Comparison of photomultiplier and Si photodiode as detectors in radiation pyrometry. *Applied Optics*. **10**, 1241–1245.

Ruffino, G. (1980). Primary temperature measurement above the gold point. *High Temperatures — High Pressures*, **12**, 241–246.

Slater, J. C. (1960). "Quantum Theory of Atomic Structure", Vol. 1. McGraw-Hill.

Sparrow, E. M., Ulbers, L. U. and Eckert, E. R. (1962). Thermal radiation characteristics of cylindrical enclosures. *J. Heat Transfer*, **C84**, 73–81.

Sparrow, E. M. and Jonsson, V. K. (1962). Absorption and emission characteristics of diffuse spherical enclosures. *J. Heat Transfer*, **C84**, 188–189.

Svet, D. Ya. (1972). Optimal utilisation of redundant information in thermal radiation in thermophysical measurements. *High Temperatures — High Pressures*, **4**, 715–722.

Tingwaldt, C. P. (1952). Uber die Messung von Reflexion, Durchlässigkeit und Absorption an Prüfkorpen beliebiger Form in der Ulbrichtschen Kugel. *Optik*, **9**, 232.

Tischler, M. (1981). High accuracy temperature and uncertainty calculation in radiation pyrometry. *Metrologia*, **17**, 49–58.

Touloukian, Y. S. and De Witt, D. (1972). Thermal radiative properties of metallic solids. In "Thermophysical Properties of Matter", Vol. 7, IFI, Plenum.

Toyota, H., Vamada, T. and Nariai, Y. (1972). Improved radiation pyrometry for automatic emissivity compensation. *TMCSI*, **4**, Part 1, 611–618.

Ziman, J. M. (1972). "Principles of the Theory of Solids", 2nd Ed., Cambridge University Press.

8

Mercury-in-Glass
Thermometry

8-1   Introduction

While it is not appropriate here to enter in great detail into the design, construction and use of mercury-in-glass thermometers, it is, nevertheless, useful to outline the principles of modern mercury-in-glass thermometry and indicate to the interested reader where more detailed information can be found.

We have already mentioned mercury-in-glass thermometers: first, in Chapter 1, when outlining the developments that took place in thermometry during the seventeenth and eighteenth centuries; and second, in Chapter 2, when discussing the work of Chappuis who, at the end of the nineteenth century, used mercury-in-glass thermometers made by Tonnelot to record his "normal hydrogen" gas thermometer scale. The design and performance of mercury-in-glass thermometers at about that time were described in great detail by Guillaume in his *Traité pratique de la thermométrie de précision* published in 1889. Since then, the number of different types of mercury-in-glass thermometer has increased and a considerable amount of work has been done aimed at achieving greater precision and accuracy. This culminated in the work of Moreau *et al.* (1957) which led to the development of mercury-in-quartz thermometers. These thermometers had a zero stability of about 1 mK when used within the range 0 °C–100 °C. This was considerably better than that of the best mercury-in-glass thermometers which, as we shall see, always show long-term drifts as well as short-term reversible changes of zero following excursions to higher temperatures. The work of Moreau *et al.*, however, did not lead to commercial production of mercury-in-quartz thermometers. This was mainly due, at the time, to the technical problems encountered in the manufacture of quartz capillary tubing having a sufficiently uniform bore. The introduction, not long after, of the automatic a.c. resistance bridge and its subsequent development has rendered the high-precision mercury-in-glass or mercury-in-quartz thermometer obsolete. Not only do such thermometers require extremely skilled observers to obtain the best from them and are thus not amenable to automatic data recording, but they also lack the sensitivity of platinum resistance thermometers.

It now appears unlikely that any significant new developments will take place in the design and construction of mercury-in-glass thermometers. In most applications they are being replaced, slowly but surely, by resistance thermometers or other thermometers not requiring the active participation of a skilled observer. It will, however, be a very long time before the industrial and medical use of mercury-in-glass thermometers ceases to be of practical importance, and until this happens the basic principles of operation will continue to be important.

8-2 The mercury-in-glass thermometer

8-2-1 Types of thermometer

The mercury-in-glass thermometer, Fig. 8-1, is basically a very simple and straightforward instrument which allows relatively accurate measurements to be made over quite a wide range of temperature, Fig. 8-2. Its effectiveness, however, depends upon many things; above all upon the choice of glass, followed by the relative dimensions of the bulb and capillary, the thickness of the walls of the bulb compared to its diameter, and upon the way in which the thermometer is used and read.

The expansion of the mercury relative to the glass, which leads to the movement of the thread of mercury up the capillary, is largely determined by the temperature of the bulb, but it is also influenced by the temperature of the stem. For this reason, it is necessary to specify the immersion at which the thermometer is calibrated and to reproduce the same immersion during use or, if this is not possible, to correct for it. Mercury-in-glass thermometers are thus best suited to the measurement of the temperature of fluids, for which the level of immersion is well defined.

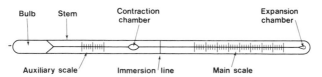

Fig. 8-1. The main features of a mercury-in-glass thermometer of the solid stem type. An auxiliary scale, to include the ice point, may be provided on those thermometers having a main scale that does not reach to the ice point. The immersion line is provided on partial-immersion thermometers only (see text). An expansion chamber is provided in gas-filled thermometers to prevent excessive pressures being developed when the thermometer is used at the top of its range and in other thermometers to avoid breakage in case of overheating.

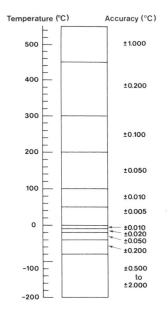

Fig. 8-2. This figure gives an indication of the accuracies that might be achieved using the best mercury-in-glass thermometers.

There are three categories of thermometer, distinguished by the different immersions at which they are used: first, total-immersion thermometers, in which the bulb and the stem up to the level of the top of the mercury thread are immersed. For these thermometers, the immersion varies with temperature and it must, therefore, be adjusted before each measurement. Second, partial-immersion thermometers, in which the bulb and a fixed, pre-determined, length of the stem is immersed. For these thermometers, an "emergent stem" correction must be applied. This is a function of the temperature of that part of the mercury thread above the immersion mark. Such a correction must also be applied to the readings of total-immersion thermometers if the mercury thread expands beyond the fluid level. Third, complete-immersion thermometers, these are thermometers designed to be used completely submerged below the level of the fluid.

Thermometers for use above about 100 °C are usually gas-filled, to suppress the distillation of the mercury to the expansion chamber which would otherwise occur in partial- and total-immersion thermometers used at high temperatures. The gas pressure must be quite high, of the order of 2 MPa, if the thermometer is intended for use up to 500 °C. The readings of a gas-filled thermometer used at complete immersion will be lower than those of the same thermometer used at partial or total immersion due to the effect of temperature on the pressure of the gas filling.

The relative expansion coefficients of glass and mercury are such that the volume of each degree-length of the mercury thread is about 1/6250 of that of the bulb. A stability of zero-reading of 5 mK, which can be achieved in the best thermometers, thus requires a stability in bulb volume of about one part per million. Seen in these terms, it is remarkable that thermometers are regularly made that show such stability. It is even more remarkable that mercury-in-quartz thermometers were made having a long-term zero stability and short-term reproducibility, on thermal cycling from 0 °C to 100 °C, of 1 mK; equivalent to a reproducibility of bulb volume of 2 parts in 10^7!

8-2-2   The maximum sensitivity attainable

The sensitivity of mercury-in-glass thermometers increases as the diameter of the capillary decreases, but a limit is reached beyond which the mercury thread no longer rises uniformly with increasing temperature, but in a series of jumps, Fig. 8-3. This phenomenon occurs because as the capillary diameter decreases, the surface tension forces become sufficiently large to cause significant variation in volume of the bulb as a function of the curvature of the mercury surface. On a rising temperature, for example, the mercury meniscus becomes increasingly convex, leading to an increase in capillary pressure which causes the bulb to expand. This continues until the internal pressure is sufficient to cause the mercury thread to move suddenly up the capillary, when the surface will flatten and the internal pressure will fall so that the mercury thread will once again remain stationary until the internal pressure leads to a further jump. The magnitude of these jumps depends upon the three parameters: capillary diameter, bulb diameter and wall thickness. This effect has been closely examined by Hall and Leaver (1959) who drew up design criteria which ensure that this jumping is minimized. They showed that the size of the jumps is proportional to the ratio of the external pressure coefficient of the bulb

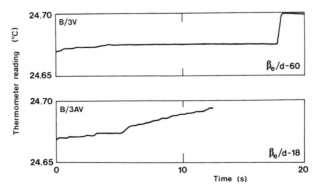

Fig. 8-3. Jumps observed in thermometers having small capillary bores. (a) the ratio of the pressure coefficient, β_e, to bore diameter, d, equals 60; (b) $\beta_e/d = 18$ for the same value of d (after Hall and Leaver, 1959).

volume to the diameter of the capillary bore. In a well designed thermometer the jumps should not exceed the equivalent of about 5 mK. The effect can be reduced even further, and made more reproducible, by tapping the thermometer stem regularly during the readings so that the mercury thread is released mechanically before the meniscus becomes sufficiently convex to lead to significant pressure rise.

8-2-3 Pressure coefficients

The external pressure obviously affects the position of the mercury thread. The external pressure coefficient β_e is defined as the change in scale reading, in millidegrees Celsius, for a change in external pressure of 1 cm of mercury. It is a function of the internal and external radii of the bulb R_i and R_e, respectively, and is given by

$$\beta_e = kR_e^2/(R_e^2 - R_i^2) \tag{8-1}$$

where k is a constant.

Guillaume found k to be 0.52 m °C/cm Hg, Hall and Leaver (1959) found a value of k a little lower than this, but since the effect is in any case small, the difference is not important. In any case, this coefficient is usually determined experimentally.

The change in reading of a mercury-in-glass thermometer with orientation, between the vertical and horizontal, say, is due to the change in bulb pressure resulting from the head of mercury in the thread. This internal pressure coefficient β_i is given by

$$\beta_i = \beta_e + 0.15 \text{ m °C/cm Hg} \tag{8-2}$$

The additional term of 0.15 m °C/cm Hg is due to the compressibility of mercury in glass.

8-2-4 Emergent-stem corrections

In total- and partial-immersion thermometers, the corrections that must be applied if the mercury thread emerges from the fluid more than is specified, are calculated on the basis of temperatures measured with supplementary thermometers. The usual way of making such measurements is by means of one of a series of special thermometers having long bulbs, known as faden thermometers. A faden thermometer having a bulb length approximately equal to that of the exposed mercury thread is placed alongside the thermometer being measured, Fig. 8-4. The tip of the faden thermometer is

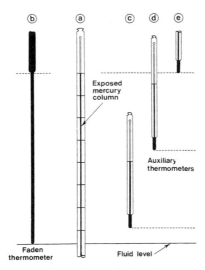

Fig. 8-4. The emergent stem correction to be applied to thermometer (a) is calculated using equation (8-3). The average temperature of the exposed mercury column, t, may be measured either with a single faden thermometer of appropriate length (b) or by means of three auxiliary thermometers (c), (d) and (e) placed as shown.

immersed in the fluid as shown in the figure. The correction due to the emergent stem C_{ems} is given by

$$C_{\text{ems}} = k_1 n(t_1 - t) \tag{8-3}$$

where k_1 is the differential coefficient of thermal expansion of mercury relative to the thermometer glass, n is the number of thermometer degrees exposed, t is the average temperature of the exposed stem measured by the faden thermometer, and t_1 is the temperature of the bulb of the thermometer. The value of k_1 is close to 16×10^{-5} for most thermometer glasses. This question is treated in more detail by Pemberton (1964).

8-3   Thermometric glasses

A great deal of experience has been built up concerning the behaviour of various thermometer glasses. There now exist more than a dozen well-established compositions for general use plus many others for special applications. The properties of glass in general are very complex and the particular requirements of a thermometric glass are that its volume and rigidity remain extremely stable on heating and on thermal cycling. In fact, these requirements are not very different from those demanded of pure platinum for use in a resistance thermometer, since in both cases, what we are looking for is a structural equilibrium that changes little or changes reproducibly on thermal cycling within the range of use of the thermometer.

Table 8-1. Various thermometer glasses and their temperature limits.[a]

	Strain point (°C)	Temperature limits (°C)	
		continuous	intermittent
Corning normal 7560	500	370	430
Kimble R6	490	360	420
Jena 16 III	495	365	425
Corning borosilicate 8800	529	400	460
Jena borosilicate 2954	548	420	480
Corning 1720	668	540	600
Jena Supremax 2955	665	535	595

[a] (after Thompson, 1962).

However, it is very much easier to approach structural equilibrium in a pure metal than it is in a glass. All that we can do with a glass is to find, empirically, an annealing procedure that leads to a state of metastable equilibrium at a temperature at which an annealed glass has a viscosity of $10^{13.5}$ Pa s Long called the strain point. The viscosity of the glass at the strain point should be sufficiently high for subsequent prolonged heating at lower temperatures to lead to only small structural changes. The strain point is defined as the temperature at which an annealed glass has a viscosity of $10^{13.5}$ Pa s. Long experience has shown that it is possible to achieve this using certain soda–lime glasses, which are very easy to use and which allow continuous operation up to about 350°C. These glasses are known as "normal" thermometric glasses and the most common types are listed in Table 8-1. For higher-temperature use, up to 460 °C, certain borosilicate glasses have been found suitable and at still higher temperatures, up to 600 °C, special proprietary glasses also listed in Table 8-1 are available. Thermometric glasses are usually identified by coloured stripes running the length of the stem or bulb.

8-4   Secular change and temporary depression of zero

Because of the nature of glass, small structural changes continue to take place below the strain point. These must be taken into account when using mercury-in-glass thermometers for precise measurements. The structural changes of the thermometer glass show themselves in the behaviour of the thermometer in two ways. First of all there is a very slow rise in the zero, called the secular change, which takes place at a decreasing rate over a period of many years. In the first year after manufacture, this will amount to a few hundredths of a degree Celsius. Obviously the secular change will be acclerated and will be larger in thermometers used at high temperatures.

The second way in which the structural changes of the glass effect the behaviour of the thermometer appears as a short-term reversible change in the zero on thermal cycling. It is found that the zero of the thermometer is depressed after use at high temperatures but that the zero recovers, more or less completely, if the thermometer is left at room temperature for a few days. The size of the so-called "temporary depression" of the zero is related to the maximum temperature reached before cooling and to the rate of cooling. It is about 0.05 °C after heating to 100 °C in normal glasses such as Jena 16III, or Whitefriar blue stripe, and about 0.02 °C in the high-temperature borosilicate glasses, such as Jena 2954 or Whitefriar borosilicate white stripe. If the thermometer is cooled over a period of 15 h or more then no temporary depression of the zero is observed (Van Dijk et al., 1958). There are two ways of dealing with the temporary depression of the zero: the first is to observe the zero immediately after the measurement of a higher temperature. In this way a different zero reading occurs for each measured temperature. Since the thermometer must, of course, be calibrated in the same way, this method leads to a very cumbersome measurement procedure. The second, and more usual, way of dealing with the temporary depression is to wait until the zero has recovered before making a zero reading. Although in many ways more convenient, this method leads to an important limitation on the use of the thermometer, namely that it can only be used to measure a series of increasing temperatures unless enough time is allowed to elapse for the zero to recover. However, in view of the smallness of the temporary depression of zero in a good thermometer, this is only a serious limitation in very precise work.

Both the secular change and the temporary depression of the zero were much smaller in the mercury-in-quartz thermometers described by Moreau et al. The secular change observed in quartz thermometers used up to 100 °C did not exceed 1 mK even after several years' use, and no temporary depression of the zero could be detected. The application of the calibration corrections is usually made on the basis of a linear change with change in zero.

8-5   Thermometers for special applications

8-5-1   Adjustable-range thermometers

Among the many thermometers for special applications, the one that calls for a description here is the thermometer designed to allow very accurate measurements to be made of small temperature differences over a relatively wide range of temperature. To avoid the need for an excessively long stem, with all the practical problems of construction and immersion that this would entail, thermometers are made which have adjustable ranges. The most common adjustable-range thermometer is the Beckmann thermometer Fig. 8.5(a), but there exists also the pipette type Fig. 8.5(b).

In these thermometers, a reservoir is provided at the top of the capillary into which a larger or smaller amount of mercury may be transferred from the bulb, thus changing the zero of the thermometer. In this way an expanded scale, covering a total of only five or six degrees Celsius, may be employed over a wide range of temperature. Such a scale would be divided at intervals of 0.01 °C, and an accuracy of temperature-difference measurement of 2 mK can be achieved. The use of such a thermometer is quite straightforward: to raise the zero of a Beckmann thermometer for example, the bulb is heated until a certain amount of mercury has run up round the top bend of the capillary and is hanging in a ball at the top of the reservoir. This may be detached and allowed to fall into the reservoir by a sharp tap given to the outside of the thermometer. To lower the zero, the thermometer is inverted and the bulb first heated to bring the mercury thread into the reservoir, then it is cooled to draw as much mercury as is required back into the bulb.

The pipette-type adjustable-range thermometer is one which can be set much more easily than a Beckmann to an exact temperature. The top of the capillary ends in a pear-shaped chamber, and the scale value of the top of the capillary is marked on the stem. To lower the zero to a particular value, the thermometer is placed in a bath having a temperature equal to the required zero plus the scale value of the top of the capillary. The excess mercury pushed out of the top of the capillary is shaken off into the reservoir by a sharp tap. To set the zero to a higher temperature, excess mercury is first drawn into the bulb by inverting the thermometer, as for the Beckmann, and then the procedure for lowering the zero to a predetermined value is carried out.

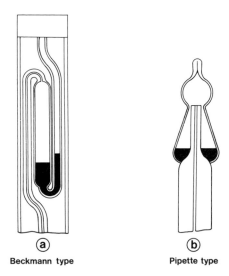

(a)	(b)
Beckmann type	Pipette type

Fig. 8-5. Portions of adjustable-range thermometers (a) the Beckmann type and (b) the pipette type.

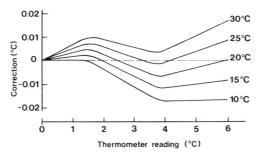

Fig. 8-6. The way in which scale corrections change with changes in zero setting of adjustable-range thermometers.

The length of scale representing a degree changes proportionately to the amount of mercury in the bulb. Hence all adjustable-range thermometers have scale corrections that vary with the setting. This is illustrated in Fig. 8.6. Calibrations of adjustable-range thermometers are usually given for a setting of 20 °C together with multiplying factors for settings above and below this temperature.

8-5-2   Other special-purpose thermometers

Among other special-purpose thermometers we find long-stem calorimeter thermometers and meteorological and clinical maximum thermometers as well as a very wide range of solid-stem and enclosed-scale laboratory and industrial thermometers. No mention has been made here of thermometers that use a liquid other than mercury, but for many applications such as those requiring measurements below the freezing point of mercury at −38.87 °C, various organic liquids such as ethyl alcohol (down to −80 °C), toluol (down to −100 °C) and pentane (down to −200 °C) may be used. Meteorological minimum thermometers also use alcohol as the thermometric liquid so that the glass index which indicates the minimum temperature reached can be maintained below the meniscus of the liquid thread in the capillary.

For details of these and other liquid in glass thermometers, their specifications, accuracies and uses, the reader is referred to the bibliography given below.

References

Guillaume, Ch-Ed. (1889). "Traité Pratique de la Thermométrie de Précision", Gauthier-Villars et Fils, Paris.

Hall, J. A. and Leaver, V. A. (1959). The design of mercury thermometers for calorimetry. *J. Sci. Instr.*, **36**, 183–187.

Moreau, H., Hall, J. A. and Leaver, V. M. (1957). Mercury-in-quartz thermometers for very high precision. *J. Sci. Instr.* **34**, 147–154.

Pemberton, L. H. (1964). Further considerations of emergent column correction in mercury thermometry. *J. Sci. Instr.* **41**, 234–236.

Van Dijk, S. J., Hall, J. A. and Leaver, V. M. (1958). The influence of rate of cooling on the zeros of mercury-in-glass thermometers. *J. Sci. Instr.* **35**, 334–338.

Bibliography of works related to mercury-in-glass thermometry additional to the articles referred to above.

Hall, J. A. (1966). "The Measurement of Temperature", Chapman and Hall, London.

Thompson, R. D. (1962). Recent developments in liquid-in-glass thermometry. *TMCSI*, **3**, Part 1, 201–218.

Wise, J. A. (1976). NBS Monograph 150, Liquid-in-Glass Thermometry.

The following British Standards:

    BS 1900 Secondary Reference Thermometers
    BS 593   Laboratory Thermometers
    BS 791   Calorimeter Thermometers
    BS 1365 Short-range, Short-stem Thermometers
    BS 1704 General Purpose Thermometers

The following International Standard:

    ISO-386 Liquid-in-glass laboratory thermometers — Principles of design, construction and use.

Appendixes

——————————————————————————————

I.   International Temperature Scales: dates and editions

II.   IPTS-68: Extracts from the text, Parts I (Introduction) and II (Definition)

III.   IEC Draft Table of Resistance Ratio/Temperature for industrial platinum resistance thermometers

IV.   Thermocouple Reference Tables: Skeleton tables for thermocouple Types B, E, J, K, R, S and T

V.   Polynomial expressions (forward and reverse) for the Thermocouple Reference Tables of Appendix IV

VI.   Nicrosil–Nisil thermocouple, skeleton reference table and reference functions

Appendix I

International Temperature Scales; dates and editions

The International Temperature Scale has gone through a number of editions beginning with the first, provisional, Scale adopted in 1927. The dates at which successive versions were adopted and the references to the French and English (where it exists) texts are as follows:

InternationalTemperature Scale of 1927, ITS-27.
    Comptes Rendus de la Septième Conférence Générale 1927, (BIPM, Paris).

International Temperature Scale of 1948, ITS-48.
    Comptes Rendus de la Neuvième Conférence Générale 1948, (BIPM, Paris).

International Practical Scale of Temperature of 1948 (amended edition of 1960), IPTS-48.
    Comptes Rendus de la Onzième Conférence Générale 1960, (BIPM, Paris).
    This version was published in English in *J. Res. NBS*, **65A**, 139–145, 1961.

International Practical Temperature Scale of 1968, IPTS-68.
    Comptes Rendus de la Treizième Conférence Générale 1968, (BIPM, Paris).
    This version was published in English in *Metrologia*, **5**, 35–44, 1969.

International Practical Temperature Scale of 1968 (amended edition of 1975), IPTS-68.
    Comptes Rendus de la Quinzième Conférence Générale 1975, (BIPM, Paris).
    This version was published in English in *Metrologia*, **13**, 7–17, 1976.

In addition to IPTS-68 there exists a provisional low temperature scale having the following title:
    The 1976 Provisional 0.5 K to 30 K Temperature Scale, EPT-76.
    The French version of this scale is available from BIPM and the English version is in *Metrologia*, **15**, 65–68, 1979.

Appendix II

IPTS-68: Extracts from the text. Parts I (Introduction) and II (Definition) (from Metrologia, **12**, 7–17, 1976)

The International Practical Temperature Scale of 1968 Amended Edition of 1975

The scale presented in this document was proposed by the International Committee of Weights and Measures in 1974 and adopted by the 15th General Conference of Weights and Measures in 1975. It constitutes *only* an amendment of the International Practical Temperature Scale of 1968 (IPTS-68), not a replacement. Any measured temperature, T_{68}, remains unchanged by this amendment of the IPTS-68.

I. Introduction

The unit of the fundamental physical quantity known as thermodynamic temperature, symbol T, is the kelvin, symbol K, defined as the fraction 1/273.16 of the thermodynamic temperature of the triple point of water.[1]

For historical reasons, connected with the way temperature scales were originally defined, it is common practice to express a temperature in terms of its difference from that of a thermal state 0.01 kelvins lower than the triple point of water. A thermodynamic temperature, T, expressed in this way is known as a Celsius temperature, symbol t, defined by

$$t = T - 273.15 \text{ K.} \qquad (1)$$

The unit of Celsius temperature is the degree Celsius, symbol °C, which is, by definition equal in magnitude to the kelvin. A difference of temperature may be expressed in kelvins or degrees Celsius.

The International Practical Temperature Scale of 1968 (IPTS-68) has been constructed in such a way that any temperature measured on it is a close approximation to the numerically corresponding thermodynamic temperature. Moreover, such measurements are easily made and are highly reproducible, whereas direct measurements of thermodynamic temperatures are both difficult to make and imprecise.

The IPTS-68 uses both International Practical Kelvin Temperatures, symbol T_{68}, and International Practical Celsius Temperatures, symbol t_{68}. The relation between T_{68} and t_{68} is the same as that between T and t, i.e.

---

[1] Comptes Rendus des Séances de la Treizième Conférence Générale des Poids et Mesures (1967–1968), Resolutions 3 and 4, p. 104.

$$t_{68} = T_{68} - 273.15 \text{ K}. \tag{2}$$

The units of T_{68} and t_{68} are the kelvin, symbol K, and the degree Celsius, symbol °C, as in the case of the thermodynamic temperature T and the Celsius temperature t.

The IPTS-68 was adopted by the International Committee of Weights and Measures as its meeting in 1968 in accordance with the power given to it by Resolution 8 of the 13th General Conference of Weights and Measures. That Scale replaced the International Practical Temperature Scale of 1948 (amended edition of 1960).

II.   Definition of the International Practical Temperature Scale of 1968 (IPTS-68)[2]

I.   Principle of the IPTS-68 and defining fixed points

The IPTS-68 is based on the assigned values of the temperatures of a number of reproducible equilibrium states (defining fixed points) and on standard instruments calibrated at those temperatures. These equilibrium states and the values of the International Practical Temperatures assigned to them are given in Table 1. Between the fixed point temperatures interpolation equations relate indications of the standard instruments to values of International Practical Temperature.

Table 1   Defining fixed points of the IPTS-68[a].

Equilibrium state	Assigned value of International Practical Temperature	
	T_{68} (K)	t_{68} (°C)
Equilibrium between the solid, liquid and vapour phases of equilibrium hydrogen (triple point of equilibrium hydrogen)[b]	13.81	−259.34
Equilibrium between the liquid and vapour phases of equilibrium hydrogen at a pressure of 33 330.6 Pa (25/76 standard atmosphere)[b,c]	17.042	−256.108
Equilibrium between the liquid and vapour phases of equilibrium hydrogen (boiling point of equilibrium hydrogen)[b,c]	20.28	−252.87

[2] In this document Kelvin temperatures are used, in general, below 0 °C and Celsius temperatures are used above 0 °C. This avoids the use of negative values and conforms with general usage.

Table 1 (continued)

Equilibrium state	Assigned value of International Practical Temperature	
	T_{68} (K)	t_{68} (°C)
Equilibrium between the liquid and vapour phases of neon (boiling point of neon)[c]	27.102	−246.048
Equilibrium between the solid, liquid and vapour phases of oxygen (triple point of oxygen)	54.361	−218.789
Equilibrium between the solid, liquid and vapour phases of argon (triple point of argon)[d]	83.798	−189.352
Equilibrium between the liquid and vapour phases of oxygen (condensation point of oxygen)[c,d]	90.188	−182.962
Equilibrium between the solid, liquid and vapour phases of water (triple point of water)	273.16	0.01
Equilibrium between the liquid and vapour phases of water (boiling point of water)[e]	373.15	100
Equilibrium between the solid and liquid phases of tin (freezing point of tin)[e]	505.1181	231.9681
Equilibrium between the solid and liquid phases of zinc (freezing point of zinc)	692.73	419.58
Equilibrium between the solid and liquid phase of silver (freezing point of silver)	1235.08	961.93
Equilibrium between the solid and liquid phases of gold (freezing point of gold)	1337.58	1064.43

[a] Except for the triple points and one equilibrium hydrogen point (17.042 K) the assigned values of temperature are for equilibrium states at a pressure $p_0 = 101\ 325$ Pa (1 standard atmosphere). The effects of small deviations from this pressure are shown in Table 5. In those cases where differing isotopic abundances could significantly affect the fixed point temperature, the abundances specified in section III must be used.
[b] The term equilibrium hydrogen is defined in section III, 5.
[c] Fractionation of isotopes or impurities dictates the use of boiling points (vanishingly small vapour fraction) for hydrogen and neon, and condensation point (vanishingly small liquid fraction) for oxygen (see section III).
[d] The triple point of argon may be used as an alternative to the condensation point of oxygen.
[e] The freezing point of tin ($t' = 231.9292$ °C, see equation (10)) may be used as an alternative to the boiling point of water.

The standard instrument used from 13.81 K to 630.74 °C is the platinum resistance thermometer. The thermometer resistor must be strain-free, annealed, pure platinum. Its resistance ratio $W(T_{68})$, defined by

$$W(T_{68}) = R(T_{68})/R(273.15 \text{ K}) \tag{3}$$

where R is the resistance, must not be less than 1.392 50 at $T_{68} = 373.15$ K. Below 0 °C the resistance-temperature relation of the thermometer is found from a reference function and specified deviation equations. From 0 °C to 630.74 °C two polynomial equations provide the resistance temperature relation.

The standard instrument used from 630.74 °C to 1064.43 °C is the platinum 10 % rhodium/platinum thermocouple, the electromotive force-temperature relation of which is represented by a quadratic equation.

Above 1064.43 °C the IPTS-68 is defined in terms of the Planck radiation law, using 1064.43 °C (1337.58 K) as a reference temperature and a value of 0.014 388 metre kelvin for c_2.

2. Definition of the IPTS-68 in different temperature ranges

(a) *The range from 13.81 K to 273.15 K*: From 14.81 K to 273.15 K the temperature T_{68} is defined by the relation

$$W(T_{68}) = W_{\text{CCT-68}}(T_{68}) + \Delta W_i(T_{68}) \tag{4}$$

where $_{\text{CCT-68}}(T_{68})$ is the resistance ratio given by the reference function[3]

$$T_{68} = \sum_{j=0}^{20} a_j \left(\frac{\ln W_{\text{CCT-68}}(T_{68}) + 3.28}{3.28} \right)^j \text{ K.} \tag{5}$$

The coefficients a_j of this reference function are set out in Table 2. The deviations $\Delta W_i(T_{68})$ at the temperatures of the defining fixed points are obtained from measured values of $W(T_{68})$ and the corresponding values of $W_{\text{CCT-68}}(T_{68})$, see Table 4. To find $\Delta W_i(T_{68})$ at intermediate temperatures, interpolation formulae are used. The range between 13.81 K and 273.15 K is divided into four parts in each of which $\Delta W_i(T_{68})$ is defined by a polynomial in T_{68}. The constants in the polynomials are determined from the values of $\Delta W_i(T_{68})$ at the fixed points and the condition that there shall be no discontinuity in $d\Delta W_i(T_{68})/dT_{68}$ across the junctions of the temperature ranges.[4]

---

[3] Equation (5) is a reformulation of equation (22) in the original version of the IPTS-68. Values of T_{68} derived from the two equations are virtually identical (within ± 10 μK); the original formulation is an acceptable alternative.

[4] Small discontinuities do exist in $d^2\Delta W_i(T_{68})/dT_{68}^2$ across the junctions at 20.28 K, 54.361 K, and 90.188 K; however, their magnitudes are such that they will be undetectable at 20.28 K and 54.361 K and barely detectable at 90.188 K in the most precise measurements of thermophysical quantities.

Table 2 Coefficients a_j of the reference function (both original and 1975 for platinum resistance thermometers for the range from 13.81 K–273.15 K.

j	$a_j(1968)$*	$a_j(1975)$†	c_j‡
0	0.273 150 000 000 000 0 E + 3	+ 38.59276	+ 161.82457
1	0.250 846 209 678 803 3 E + 3	+ 43.44837	+ 101.20451
2	0.135 099 869 964 999 7 E + 3	+ 39.10887	+ 51.52727
3	0.527 856 759 008 517 2 E + 2	+ 38.69352	+ 24.77384
4	0.276 768 548 854 105 2 E + 2	+ 32.56883	+ 10.11559
5	0.391 053 205 376 683 7 E + 2	+ 24.70158	+ 3.53282
6	0.655 613 230 578 069 3 E + 2	+ 53.03828	+ 0.91392
7	0.808 035 868 559 866 7 E + 2	+ 77.35767	+ 0.14666
8	0.705 242 118 234 052 0 E + 2	− 95.75103	− 0.00327
9	0.447 847 589 638 965 7 E + 2	− 223.52892	+ 0.00265
10	0.212 525 653 556 057 8 E + 2	+ 239.50285	+ 0.01172
11	0.767 976 358 170 845 8 E + 1	+ 524.64944	+ 0.01013
12	0.213 689 459 382 850 0 E + 1	− 319.79981	+ 0.00366
13	0.459 843 348 928 069 3 E + 0	− 787.60686	+ 0.00007
14	0.763 614 629 231 648 0 E − 1	+ 179.54782	− 0.00071
15	0.969 328 620 373 121 3 E − 2	+ 700.42832	− 0.00066
16	0.923 069 154 007 007 5 E − 3	+ 29.48666	− 0.00085
17	0.638 116 590 952 653 8 E − 4	− 335.24378	− 0.00028
18	0.302 293 237 874 619 2 E − 5	− 77.25660	+ 0.00034
19	0.877 551 391 303 760 2 E − 7	+ 66.76292	+ 0.00026
20	0.117 702 613 125 477 4 E − 8	+ 24.44911	+ 0.00005

The numbers in the second column of the above table are written in the form: decimal number E signed integer exponent, where E denotes the base 10.
* coefficients of equation (22) in the original formulation of IPTS-68.
† coefficients of equation (5) above.
‡ coefficients of a Chebyshev polynomial that gives the same values of T_{68} as do equation (5) and equation (22) of original formulation of IPTS-68.

From 13.81 K to 20.28 K the deviation function is

$$\Delta W_1(T_{68}) = A_1 + B_1 T_{68} + C_1 T_{68}^2 + D_1 T_{68}^3 \tag{6}$$

where the constants are determined by the measured deviations at the triple point of equilibrium hydrogen, the fixed point at 17.042 K, and the boiling point of equilibrium hydrogen, and by the first derivative of the deviation function at the boiling point of equilibrium hydrogen as derived from equation (7).

From 20.28 K to 54.361 K the deviation function is

$$\Delta W_2(T_{68}) = A_2 + B_2 T_{68} + C_2 T_{68}^2 + D_2 T_{68}^3 \tag{7}$$

where the constants are determined by the measured deviations at the boiling

Table 3 Values of $W_{CTT-68}(T_{68})$ according to equation (5) at integral values of T_{68}.
[This table is not reproduced here.]

Table 4 Values of $W_{CCT-68}(T_{68})$, according to the data given in Table 2, at the fixed-point temperatures.

Fixed point	T_{68} (K)	t_{68} (°C)	W_{CCT-68}
e-H$_2$ triple	13.81	−259.34	0.001 412 08
e-H$_2$ 17.042 K	17.042	−256.108	0.002 534 45
e-H$_2$ boiling	20.28	−252.87	0.004 485 17
Ne boiling	27.102	−246.048	0.012 212 72
O$_2$ triple	54.361	−218.789	0.091 972 53
Ar triple	83.798	−189.352	0.216 057 05
O$_2$ condensation	90.188	−182.962	0.243 799 12
	273.15	0	1
H$_2$O boiling	373.15	100	1.392 596 68

points of equilibrium hydrogen and neon and at the triple point of oxygen, and by the first derivative of the deviation function at the triple point of oxygen as derived from equation (8).

From 54.361 K to 90.188 K the deviation function is

$$\Delta W_3(T_{68}) = A_3 + B_3 T_{68} + C_3 T_{68}^2, \tag{8}$$

where the constants are determined by the measured deviations at the triple point and the condensation point of oxygen (or the triple point of argon, see Note d, Table 1) and by the first derivative of the deviation function at the condensation point of oxygen as derived from equation (9).

From 90.188 K to 273.15 K the deviation function is

$$\Delta W_4(T_{68}) = b_4(T_{68} - 273.15 \text{ K}) + e_4(T_{68} - 273.15 \text{ K})^3 \times (T_{68} - 373.15 \text{ K}) \tag{9}$$

where the constants are determined by the measured deviations at the condensation point of oxygen (or the triple point of argon, see Note d, Table 1) and the boiling point of water.[5]

(b) *The range from 0 °C (273.15 K) to 630.74 °C*: From 0 °C to 630.74 °C the temperature t_{68} is defined by

---

[5] If the freezing point of tin (see Note e, Table 1) is used as a fixed point instead of the boiling point of water, $W(100\ °C)$ for the platinum thermometer should be calculated from equations (10) and (11).

$$t_{68} = t' + 0.045 \left(\frac{t'}{100\ °C}\right)\left(\frac{t'}{100\ °C} - 1\right)\left(\frac{t'}{419.58\ °C} - 1\right)\left(\frac{t'}{630.74\ °C} - 1\right) °C \tag{10}$$

where t' is defined by the equation

$$t' = \left\{\frac{1}{\alpha}[W(t') - 1] + \delta\left(\frac{t'}{100\ °C}\right)\left(\frac{t'}{100\ °C} - 1\right)\right\} °C \tag{11a}$$

with $W(t') = R(t')/R(0\ °C)$. The constants $R(0\ °C)$, α and δ are determined from measurements of the resistances at the triple point of water, the boiling point of water (or the freezing point of tin, see Note e, Table 1), and the freezing point of zinc.

Equation (11a) is equivalent to the equation

$$W(t') = 1 + At' + Bt'^2 \tag{11b}$$

where $A = \alpha(1 + \delta/100\ °C)$ and $B = -10^{-4}\ \alpha\delta\ °C^{-2}$.

(c) *The range from 630.74 °C to 1064.43 °C*: From 630.74 °C to 1064.43 °C the temperature t_{68} is defined by the equation

$$E(t_{68}) = a + bt_{68} + ct_{68}^2 \tag{12}$$

where $E(t_{68})$ is the electromotive force of a standard platinum 10 % rhodium/platinum thermocouple when one junction is at 0 °C and the other is at t_{68}. The constants a, b, and c are calculated from the values of $E(t_{68})$ at 630.74 °C ± 0.2 °C, as determined by a platinum resistance thermometer, and at the freezing points of silver ($t_{68}(Ag)$) and gold ($t_{68}(Au)$).

The purity of the platinum wire of the standard thermocouple shall be such that $W(100\ °C)$ is not less than 1.3920. The platinum–rhodium wire shall contain nominally 10 % rhodium and 90 % platinum by mass. The thermocouple shall be such that the electromotive forces $E(630.74\ °C)$, $E(t_{68}(Ag))$ and $E(t_{68}(Au))$ satisfy the following relations:

$$E(t_{68}(Au)) = 10\ 334\ \mu V \pm 30\ \mu V \tag{13}$$

$$E(t_{68}(Au)) - E(t_{68}(Ag)) = 1186\ \mu V$$

$$+ 0.17[E(t_{68}(Au)) - 10\ 334\ \mu V] \pm 3\ \mu V \tag{14}$$

$$E(t_{68}(Au)) - E(630.74\ °C) = 4782\ \mu V$$

$$+ 0.63[E(t_{68}(Au)) - 10\ 334\ \mu V] \pm 5\ \mu V \tag{15}$$

Part III Supplementary Information — including Tables 5 and 6 — is not reproduced here.

Table 7 The approximate numerical differences $t_{68} - t_{48}$ in kelvins.

(a) *For the range* $-180\,^{\circ}C$ *to* $0\,^{\circ}C$

t_{68} (°C)	0	-10	-20	-30	-40	-50	-60	-70	-80	-90	-100
-100	0.022	0.013	0.003	-0.006	-0.013	-0.013	-0.005	0.007	0.012	0.029	0.022
0	0.000	0.006	0.012	0.018	0.024	0.029	0.032	0.034	0.033		

(b) *For the range* $0\,^{\circ}C$ *to* $1070\,^{\circ}C$

t_{68} (°C)	0	10	20	30	40	50	60	70	80	90	100
0	0.000	-0.004	-0.007	-0.009	-0.010	-0.010	-0.010	-0.008	-0.006	-0.003	0.000
100	0.000	0.004	0.007	0.012	0.016	0.020	0.025	0.029	0.034	0.038	0.043
200	0.043	0.047	0.051	0.054	0.058	0.061	0.064	0.067	0.069	0.071	0.073
300	0.073	0.074	0.075	0.076	0.077	0.077	0.077	0.077	0.077	0.076	0.076
400	0.076	0.075	0.075	0.075	0.074	0.074	0.074	0.075	0.076	0.077	0.079
500	0.079	0.082	0.085	0.089	0.094	0.100	0.108	0.116	0.126	0.137	0.150
600	0.150	0.165	0.182	0.200	0.23	0.25	0.28	0.31	0.34	0.36	0.39
700	0.39	0.42	0.45	0.47	0.50	0.53	0.56	0.58	0.61	0.64	0.67
800	0.67	0.70	0.72	0.75	0.78	0.81	0.84	0.87	0.89	0.92	0.95
900	0.95	0.98	1.01	1.04	1.07	1.10	1.12	1.15	1.18	1.21	1.24
1000	1.24	1.27	1.30	1.33	1.36	1.39	1.42	1.44			

(c) *For the range* $1100\,^{\circ}C$ *to* $4000\,^{\circ}C$

t_{68} (°C)	0	100	200	300	400	500	600	700	800	900	1000
1000		1.5	1.7	1.8	2.0	2.2	2.4	2.6	2.8	3.0	3.2
2000	3.2	3.5	3.7	4.0	4.2	4.5	4.8	5.0	5.3	5.6	5.9
3000	5.9	6.2	6.5	6.9	7.2	7.5	7.9	8.2	8.6	9.0	9.3

(d) *The range above 1064.43 °C:* Above 1064.43 °C (1337.58 K) the temperature t_{68} ($=T_{68}-273.15$ K) is defined by means of the equation:

$$\frac{L_\lambda(T_{68})}{L_\lambda(T_{68}(\text{Au}))}=\frac{\exp\left[\dfrac{c_2}{\lambda T_{68}(\text{Au})}\right]-1}{\exp\left[\dfrac{c_2}{\lambda T_{68}}\right]-1} \tag{16}$$

in which $L_\lambda(T_{68})$ and $L_\lambda(T_{68}(\text{Au}))$ are the spectral concentrations of the radiance of a blackbody at the wavelength (in vacuo) λ at T_{68} and at $T_{68}(\text{Au})$ respectively, and $c_2=0.014\,338$ m · K. It is not necessary in practice to specify the value of the wavelength to be employed in the measurements because $T_{68}(\text{Au})$ is sufficiently close to the thermodynamic temperature of the freezing point of gold and the specified value of c_2 is sufficiently close to the true value of the second radiation constant in the Planck equation for any wavelength dependence of T_{68} to be negligible.

Appendix III

IEC draft Table of Resistance Ratio/Temperature for industrial platinum resistance thermometers

The following table is based upon the equations:

-200 to $0\,°C$: $W_t=1+At+Bt^2+C(t-100)t^3$

0 to $850\,°C$: $W_t=1+At+Bt^2$

where $W_t=R_t/R_0$

$A=\alpha+\alpha\delta/100$

$B=-\alpha\delta/10^4$

$C=-\alpha\beta/10^8$

$\alpha=0.00385$

$\delta=1.507$

$\beta=0.111$

For comparison, the present tables from BS 1904: 1960 (amended in 1970) and from DIN 43760 are also listed.

Temp °C	Proposed table	BS 1904	DIN 43760
−200	0.1849	0.1856	0.1853
−150	0.3971	0.3972	0.3965
−100	0.6025	0.6028	0.6020
−50	0.8031	0.8032	0.8025
0	1.0000	1.0000	1.0000
50	1.1940	1.1940	1.1940
100	1.3850	1.3850	1.3850
150	1.5731	1.5731	1.5732
200	1.7584	1.7583	1.7584
250	1.9407	1.9407	1.9408
300	2.1202	2.1202	2.1203
350	2.2967	2.2969	2.2969
400	2.4704	2.4708	2.4706
450	2.6411	2.6417	2.6414
500	2.8090	2.8098	2.8093
550	2.9739	2.9750	2.9743
600	3.1359	3.1372	3.1365
650	3.2951	3.296	3.2957
700	3.4513	3.454	3.4521
750	3.6047	3.606	3.6055
800	3.7551	3.756	3.7561
850	3.9026	3.904	3.9038

Appendix IV

Thermocouple Reference Tables: skeleton Tables for thermocouple Types B, E, J, K, R, S and T

These Tables are extracts from NBS Monograph 125, Reference Tables for Thermocouples, 1974 and, for types R and S, upon Bedford *et al.*, 1972 *TMCSI*, 1585–1602. The values are identical to those appearing in IEC 584-1, BS4937 and ASTM-E-230, which are based upon the same sources.

Platinum–30 % rhodium/platinum–6 % rhodium thermocouple table — Type B

Temperatures °C (IPTS-68)

Reference junction at 0 °C

e.m.f. µV

Temp.	0	10	20	30	40	50	60	70	80	90	Temp.
0	0	−2	−3	−2	−0	2	6	11	17	25	0
100	33	43	53	65	78	92	107	123	140	159	100
200	178	199	220	243	266	291	317	344	372	401	200
300	431	462	494	527	561	596	632	669	707	746	300
400	786	827	870	913	957	1002	1048	1095	1143	1192	400
500	1241	1292	1344	1397	1450	1505	1560	1617	1674	1732	500
600	1791	1851	1912	1974	2036	2100	2164	2230	2296	2363	600
700	2430	2499	2569	2639	2710	2782	2855	2928	3003	3078	700
800	3154	3231	3308	3387	3466	3546	3626	3708	3790	3873	800
900	3957	4041	4126	4212	4298	4386	4474	4562	4652	4742	900
1000	4833	4924	5016	5109	5202	5297	5391	5487	5583	5680	1000
1100	5777	5875	5973	6073	6172	6273	6374	6475	6577	6680	1100
1200	6783	6887	6991	7096	7202	7308	7414	7521	7628	7736	1200
1300	7845	7953	8063	8172	8283	8393	8504	8616	8727	8839	1300
1400	8952	9065	9178	9291	9405	9519	9634	9748	9863	9979	1400
1500	10094	10210	10325	10441	10558	10674	10790	10907	11024	11141	1500
1600	11257	11374	11491	11608	11725	11842	11959	12076	12193	12310	1600
1700	12426	12543	12659	12776	12892	13008	13124	13239	13354	13470	1700
1800	13585	13699	13814								1800

Nickel–chromium/copper–nickel thermocouple tables — Type E

Temperatures °C (IPTS-68)

Temp.	0	10	20	30	40	50	60	70	80	90	Temp.
					e.m.f. μV				Reference junction at 0 °C		
−200	−8 824	−9 063	−9 274	−9 455	−9 604	−9 719	−9 797	−9 835			−200
−100	−5 237	−5 680	−6 107	−6 516	−6 907	−7 279	−7 631	−7 963	−8 273	−8 561	−100
0	0	−581	−1 151	−1 709	−2 254	−2 787	−3 306	−3 811	−4 301	−4 777	0
0	0	591	1 192	1 801	2 419	3 047	3 683	4 329	4 983	5 646	0
100	6 317	6 996	7 683	8 377	9 078	9 787	10 501	11 222	11 949	12 681	100
200	13 419	14 161	14 909	15 661	16 417	17 178	17 942	18 710	19 481	20 256	200
300	21 033	21 814	22 597	23 383	24 171	24 961	25 754	26 549	27 345	28 143	300
400	28 943	29 744	30 546	31 350	32 155	32 960	33 767	34 574	35 382	36 190	400
500	36 999	37 808	38 617	39 426	40 236	41 045	41 853	42 662	43 470	44 278	500
600	45 085	45 891	46 697	47 502	48 306	49 109	49 911	50 713	51 513	52 312	600
700	53 110	53 907	54 703	55 498	56 291	57 083	57 873	58 663	59 451	60 237	700
800	61 022	61 806	62 588	63 368	64 147	64 924	65 700	66 473	67 245	68 015	800
900	68 783	69 549	70 313	71 075	71 835	72 593	73 350	74 104	74 857	75 608	900
1000	76 358										

Iron/copper–nickel thermocouple tables — Type J

Temperatures °C (IPTS-68) — Reference junction at 0 °C

e.m.f. µV

Temp.	0	10	20	30	40	50	60	70	80	90	Temp.
−200	−7890	−8096									−200
−100	−4632	−5036	−5426	−5801	−6159	−6499	−6821	−7122	−7402	−7659	−100
0	0	−501	−995	−1481	−1960	−2431	−2892	−3344	−3785	−4215	0
0	0	507	1019	1536	2058	2585	3115	3649	4186	4725	0
100	5268	5812	6359	6907	7457	8008	8560	9113	9667	10222	100
200	10777	11332	11887	12442	12998	13553	14108	14663	15217	15771	200
300	16325	16879	17432	17984	18537	19089	19640	20192	20743	21295	300
400	21846	22397	22949	23501	24054	24607	25161	25716	26272	26829	400
500	27388	27949	28511	29075	29642	30210	30782	31356	31933	32513	500
600	33096	33683	34273	34867	35464	36066	36671	37280	37893	38510	600
700	39130	39754	40382	41013	41647	42283	42922	43563	44207	44852	700
800	45498	46144	46790	47434	48076	48716	49354	49989	50621	51249	800
900	51875	52496	53115	53729	54341	54948	55553	56155	56753	57349	900
1000	57942	58533	59121	59708	60293	60876	61459	62039	62619	63199	1000
1100	63777	64355	64933	65510	66087	66664	67240	67815	68390	68964	1100
1200	69536										

Nickel–chromium/nickel–aluminium thermocouple tables — Type K

Temperatures °C (IPTS-68)

Reference junction at 0 °C

e.m.f. μV

Temp.	0	10	20	30	40	50	60	70	80	90	Temp.
-200	-5 891	-6 035	-6 158	-6 262	-6 344	-6 404	-6 441	-6 458			-200
-100	-3 553	-3 852	-4 138	-4 410	-4 669	-4 912	-5 141	-5 354	-5 550	-5 730	-100
0	0	-392	-777	-1 156	-1 527	-1 889	-2 243	-2 586	-2 920	-3 242	0
0	0	397	798	1 203	1 611	2 022	2 436	2 850	3 266	3 681	0
100	4 095	4 508	4 919	5 327	5 733	6 137	6 539	6 939	7 338	7 737	100
200	8 137	8 537	8 938	9 341	9 745	10 151	10 560	10 969	11 381	11 793	200
300	12 207	12 623	13 039	13 456	13 874	14 292	14 712	15 132	15 552	15 974	300
400	16 395	16 818	17 241	17 664	18 088	18 513	18 938	19 363	19 788	20 214	400
500	20 640	21 066	21 493	21 919	22 346	22 772	23 198	23 624	24 050	24 476	500
600	24 902	25 327	25 751	26 176	26 599	27 022	27 445	27 867	28 288	28 709	600
700	29 128	29 547	29 965	30 383	30 799	31 214	31 629	32 042	32 455	32 866	700
800	33 277	33 686	34 095	34 502	34 909	35 314	35 718	36 121	36 524	36 925	800
900	37 325	37 724	38 122	38 519	38 915	39 310	39 703	40 096	40 488	40 879	900
1000	41 269	41 657	42 045	42 432	42 817	43 202	43 585	43 968	44 349	44 729	1000
1100	45 108	45 486	45 863	46 238	46 612	46 985	47 356	47 726	48 095	48 462	1100
1200	48 828	49 192	49 555	49 916	50 276	50 633	50 990	51 344	51 697	52 049	1200
1300	52 398	52 747	53 093	53 439	53 782	54 125	54 466	54 807			1300

Platinum-13 % rhodium/platinum thermocouple tables — Type R

| Temperature °C (IPTS-68) | | | | | | Reference junction at 0 °C | | | | | |
Temp.	0	10	20	30	40	50	60	70	80	90	Temp.
					e.m.f. µV						
0	0	−51	−100	−145	−188	−226					0
100	647	723	800	879	959	1 041	1 124	1 208	1 294	1 380	100
200	1 468	1 557	1 647	1 738	1 830	1 923	2 017	2 111	2 207	2 303	200
300	2 400	2 498	2 596	2 695	2 795	2 896	2 997	3 099	3 201	3 304	300
400	3 407	3 511	3 616	3 721	3 826	3 933	4 039	4 146	4 254	4 362	400
500	4 471	4 580	4 689	4 799	4 910	5 021	5 132	5 244	5 356	5 469	500
600	5 582	5 696	5 810	5 925	6 040	6 155	6 272	6 388	6 505	6 623	600
700	6 741	6 860	6 979	7 093	7 218	7 339	7 460	7 582	7 703	7 826	700
800	7 949	8 072	8 196	8 320	8 445	8 570	8 696	8 822	8 949	9 076	800
900	9 203	9 331	9 460	9 589	9 718	9 848	9 978	10 109	10 240	10 371	900
1000	10 503	10 636	10 768	10 902	11 035	11 170	11 304	11 439	11 574	11 710	1000
1100	11 846	11 983	12 119	12 257	12 394	12 532	12 669	12 808	12 946	13 085	1100
1200	13 224	13 363	13 502	13 642	13 782	13 922	14 062	14 202	14 343	14 483	1200
1300	14 624	14 765	14 906	15 047	15 188	15 329	15 470	15 611	15 752	15 893	1300
1400	16 035	16 176	16 317	16 458	16 599	16 741	16 882	17 022	17 163	17 304	1400
1500	17 445	17 585	17 726	17 866	18 006	18 146	18 286	18 425	18 564	18 703	1500
1600	18 842	18 981	19 119	19 257	19 395	19 533	19 670	19 807	19 944	20 080	1600
1700	20 215	20 350	20 483	20 616	20 748	20 878	21 006				1700

Platinum-10 % rhodium/platinum thermocouple tables — Type S

Temperature °C (IPTS-68)

Reference junction at 0 °C.

e.m.f. µV

Temp.	0	10	20	30	40	50	60	70	80	90	Temp.
0	0	−53	−103	−150	−194	−236				0	0
0	0	55	113	173	235	299	365	432	502	573	0
100	645	719	795	872	950	1 029	1 109	1 190	1 273	1 356	100
200	1 440	1 525	1 611	1 698	1 785	1 873	1 962	2 051	2 141	2 232	200
300	2 323	2 414	2 506	2 599	2 692	2 786	2 880	2 974	3 069	3 164	300
400	3 260	3 356	3 432	3 549	3 645	3 743	3 840	3 938	4 036	4 135	400
500	4 234	4 333	4 432	4 532	4 632	4 732	4 832	4 933	5 034	5 136	500
600	5 237	5 339	5 442	5 544	5 648	5 751	5 855	5 960	6 064	6 169	600
700	6 274	6 380	6 486	6 592	6 699	6 805	6 913	7 020	7 128	7 236	700
800	7 345	7 454	7 563	7 672	7 782	7 892	8 003	8 114	8 225	8 336	800
900	8 448	8 560	8 673	8 786	8 899	9 012	9 126	9 240	9 355	9 470	900
1000	9 585	9 700	9 816	9 932	10 048	10 165	10 282	10 400	10 517	10 635	1000
1100	10 754	10 872	10 991	11 110	11 229	11 348	11 467	11 587	11 707	11 827	1100
1200	11 947	12 067	12 188	12 308	12 429	12 550	12 671	12 792	12 913	13 034	1200
1300	13 155	13 276	13 397	13 519	13 640	13 761	13 883	14 004	14 125	14 247	1300
1400	14 368	14 489	14 610	14 731	14 852	14 973	15 094	15 215	15 336	15 456	1400
1500	15 576	15 697	15 817	15 937	16 057	16 176	16 296	16 415	16 534	16 653	1500
1600	16 771	16 890	17 008	17 125	17 243	17 360	17 477	17 594	17 711	17 826	1600
1700	17 942	18 056	18 170	18 282	18 394	18 504	18 612				

Copper/copper–nickel thermocouple — Type T

Temperatures °C (IPTS-68)

Reference junction at 0 °C

e.m.f. μV

Temp.	0	10	20	30	40	50	60	70	80	90	Temp.
-200	-5603	-5753	-5889	-6007	-6105	-6181	-6232	-6258			-200
-100	-3378	-3656	-3923	-4177	-4419	-4648	-4865	-5069	-5261	-5439	-100
0	0	-383	-757	-1121	-1475	-1819	-2152	-2475	-2788	-3089	0
0	0	391	789	1196	1611	2035	2467	2908	3357	3813	0
100	4277	4749	5227	5712	6204	6702	7207	7718	8235	8757	100
200	9286	9820	10360	10905	11456	12011	12572	13137	13707	14281	200
300	14860	15443	16030	16621	17217	17816	18420	19027	19638	20252	300
400	20869										400

Appendix V

In this Appendix are to be found the polynomials which describe the IEC*
reference tables for thermocouples.

The polynomials (A) are of the form $E = f(T)$, where $E =$ emf/microvolts;
$T =$ temperature/degrees Celsius.

In the first column are listed the numbered terms of the polynomials, in the
second the coefficients exactly as in IEC (also BS 4937 and ASTM-E-230), in
the third are the powers of ten by which each coefficient must be multiplied, in
the fourth the truncated coefficients[†] (to be multiplied by the same power of
ten) which give the IEC tables with errors not exceeding 1 μV, and in the last
column the coefficients of the Chebyshev polynomials[†] which give the IEC
tables with full precision.

The polynomials (B)[‡] are of the form $T = f(E)$.

In the first column are given the numbered terms of the polynomial, in the
second the coefficients (with sufficient digits to give errors of less than about 1
μV), and in the last column the powers of ten by which each coefficient must be
multiplied. These polynomials, B, are approximations which do not lead to
errors exceeding about 0.1 °C or 2 μV.

For a discussion on the derivation of these approximations, the reader is
referred to Coates (1978).

Type B thermocouple

A. 0 °C to 1820 °C $\qquad\qquad E = \sum_{j=1}^{j=8} a_j T^j$

Term	Polynomial coefficients IEC		IEC truncated	Chebyshev
0	–			10 997.210
1	− 2.467 460 1620	− 1	− 2.467 46	7 122.783
2	5.910 211 1169	− 3	5.910 21	1 434.472
3	− 1.430 712 3430	− 6	− 1.430 71	− 209.452
4	2.150 914 9750	− 9	2.150 91	− 25.267
5	− 3.175 780 0720	− 12	− 3.175 78	− 7.052
6	2.401 036 7459	− 15	2.401 03	− 1.544
7	− 9.092 814 8159	− 19	− 9.092 81	0.476
8	1.329 950 5137	− 22	1.329 95	0.489

* IEC 584-1, 1977.
† These are taken from Coates, P. B. and Smith, A. C. K., 1977 Polynomial representations of
standard thermocouple reference tables, NPL Report QU 36.
‡ These are taken from Coates, P. B., 1978, Functional approximations to the standard
thermocouple reference tables.

B. 250 ° C to 1820 °C $T = l_0 + \sum_{j=1}^{j=4} l_j I^j + c_0 E^{1/2}$

Term	Polynomial coefficient
l_0	$2.1788 + 1$
l_1	$9.3164 - 3$
l_2	$8.7507 - 7$
l_3	$-6.4437 - 11$
l_4	$3.2941 - 15$
c_0	$1.3208 + 1$

Type E thermocouple

A. (i) -270 °C to 0 °C $E = \sum_{j=1}^{j=13} a_j T^j$

Term	Polynomial coefficients IEC	IEC truncated	Chebyshev
0			$-11\,661.290$
1	$5.869\,585\,7799$	1	5015.620
2	$5.166\,751\,7705$	-2	897.483
3	$-4.465\,268\,3347$	-4	-94.451
4	$-1.734\,627\,0905$	-5	15.190
5	$-4.871\,936\,8427$	-7	-3.898
6	$-8.889\,655\,0447$	-9	0.563
7	$-1.093\,076\,7375$	-10	0.360
8	$-9.178\,453\,5039$	-13	-0.265
9	$-5.257\,515\,8521$	-15	0.064
10	$-2.016\,960\,1996$	-17	0.045
11	$-4.950\,213\,8782$	-20	-0.128
12	$-7.017\,798\,0633$	-23	0.116
13	$-4.367\,180\,8488$	-26	-0.053

NOT SIMPLY TRUNCATED

(ii) 0 °C to 1000 °C $\qquad E = \sum_{j=1}^{j=9} a_j T^j$

0	–			74 958.67
1	5.869 585 7799	1	5.869 586	38 767.70
2	4.311 094 5462	−2	4.311 094	589.54
3	5.722 035 8202	−5	5.722 035	−577.44
4	−5.402 066 8085	−7	−5.402 066	110.89
5	1.542 592 2111	−9	1.542 592	−14.04
6	−2.485 008 9136	−12	−2.485 009	0.35
7	2.338 972 1459	−15	2.338 972	0.58
8	−1.194 629 6815	−18	−1.194 629	−1.35
9	2.556 112 7497	−22	2.556 112	1.95

B. 0 °C to 1000 °C $\qquad T = \sum_{j=1}^{j=3} l_j E^j + c_0 E \exp(c_1 E)$

Term	Polynomial coefficient
l_1	1.43054 −2
l_2	−3.6975 −8
l_3	2.749 −13
c_0	2.7126 −3
c_1	−7.0571 −5

Type J thermocouple

A. −210 °C to 760 °C $\qquad E = \sum_{j=1}^{j=7} a_j T^j$

Term	Polynomial coefficients IEC		IEC truncated	Chebyshev
0	–			31 626.2
1	5.037 275 3027	1	5.037 28	25 988.1
2	3.042 549 1284	−2	3.042 55	1 233.1
3	−8.566 975 0464	−5	−8.566 98	−398.3
4	1.334 882 5735	−7	1.334 88	363.5
5	−1.702 240 5966	−10	−1.702 24	−71.5
6	1.941 609 1001	−13	1.941 61	3.5
7	−9.639 184 4859	−17	−9.639 18	−9.5

B. 0 °C to 760 °C $\qquad T = \sum_{j=1}^{j=5} l_j E^j$

Term	Polynomial coefficient
l_1	1.98545 −2
l_2	−2.1634 −7
l_3	1.14521 −11
l_4	−2.62246 −16
l_5	1.9961 −21

Type K thermocouple

A. (i) −270 °C to 0 °C $\qquad E = \sum_{j=1}^{j=10} a_j T^j$

Term	Polynomial coefficients			Chebyshev
	IEC		IEC truncated	
0	−			−7772.376
1	3.947 543 3139	1	3.947 54	3 295.457
2	2.746 525 1138	−2	2.746 52	656.432
3	−1.656 540 6716	−4	−1.656 54	−66.404
4	−1.519 091 2392	−6	−1.519 09	1.439
5	−2.458 167 0924	−8	−2.458 16	−0.783
6	−2.475 791 7816	−10	−2.475 79	−0.193
7	−1.558 527 6173	−12	−1.558 52	0.481
8	−5.972 992 1255	−15	−5.972 99	0.353
9	−1.268 880 1216	−17	−1.268 88	0.156
10	−1.138 279 7374	−20	−1.138 27	−0.045

(ii) 0 °C to 1372 °C $\qquad E = a_0 + \sum_{j=1}^{j=8} a_j T^j + 125 \exp\left[-\frac{1}{2}\left(\frac{T-127}{65}\right)^2 \right]$

0	−1.853 306 3273	1	−1.853 3	55 904.37
1	3.891 834 4612	1	3.891 834	27 751.81
2	1.664 515 4356	−2	1.644 515	−543.98
3	−7.870 237 4448	−5	−7.870 237	−315.96
4	2.283 578 5557	−7	2.283 578	23.63
5	−3.570 023 1258	−10	−3.570 023	18.07
6	2.993 290 9136	−13	2.993 290	−12.18
7	−1.284 984 8798	−16	−1.284 984	−7.20
8	2.223 997 4336	−20	2.233 997	8.52

B. 200 °C to 1200 °C $\qquad T = l_0 + \sum_{j=1}^{j=4} l_j E^j$

Term	Polynomial coefficient
l_0	$-1.4185 +1$
l_2	$2.7906 -2$
l_2	$-2.3042 -7$
l_3	$4.6265 -12$
l_4	$-2.4231 -17$

Type R thermocouple

A. (i) -50 °C to 630.74 °C $\qquad E = \sum_{j=1}^{j-7} a_j T^j$

Term	Polynomial coefficients IEC		IEC truncated	Chebyshev
0	–			5 132.55
1	5.289 139 5059		5.289	3 136.75
2	1.391 110 9947	-2	1.391	273.00
3	$-2.400\,523\,8430$	-5	-2.400	-54.61
4	3.620 141 0595	-8	3.620	13.69
5	$-4.464\,501\,9036$	-11	-4.464	-2.26
6	3.849 769 1865	-14	3.849	0.35
7	$-1.537\,264\,1559$	-17	-1.537	-0.13

(ii) 630.74 °C to 1064.43 °C $\qquad E = a_0 + \sum_{j=1}^{j=3} a_j T^j$

0	$-2.641\,800\,7025$	2	-2.641	17 188.50
1	8.046 868 6747		8.046	2 716.07
2	2.989 229 3723	-3	2.989	54.21
3	$-2.687\,605\,8617$	-7	-2.687	-0.68

(iii) 1064.43 °C to 1665 °C $\qquad E = a_0 + \sum\limits_{j=1}^{j=3} a_j T^j$

0	1.490 170 2702	3	1.490	31 087.65
1	2.863 986 7552		2.863	4 200.50
2	8.082 363 1189	−3	8.082	7.43
3	−1.933 847 7638	−6	−1.933	−13.09

(iv) 1665 °C to 1769 °C $\qquad E = a_0 + \sum\limits_{j=1}^{j=3} a_j T^j$

0	9.544 555 9910	4	9.544	
1	−1.664 250 0359	2	−1.664	691.86
2	1.097 574 3239	−1	1.097	−6.83
3	−2.228 921 6980	−5	−2.228	−0.78

B. (i) 0 °C to 500 °C $\qquad T = \sum\limits_{j=1}^{j=3} l_j E^j + c_0 E(1 + c_1 E)^{-1}$

Term	Polynomial coefficient
l_1	1.09713 −1
l_2	−3.4698 −6
l_3	1.837 −10
c_0	7.8372 −2
c_1	1.03044 −3

(ii) 500 °C to 1500 °C $\qquad T = l_0 + \sum\limits_{j=1}^{j=2} l_j E^j + c_0 \exp(c_1 E)$

l_0	−4.86246 +3
l_1	−6.3783 −2
l_2	−5.3573 −6
c_0	4.91307 +3
c_1	3.5368 5

(iii) 1000 °C to 1769 °C $\qquad T = \sum_{j=1}^{j=4} l_j E^j + E \exp[c_0(E - c_1)]$

l_1	1.25615 − 1
l_2	−4.0579 −6
l_3	1.23343 − 10
l_4	−1.20083 − 15
c_0	1.7631 −3
c_1	2.6155 +4

Type S thermocouple

A. (i) −50 °C to 630.74 °C $\qquad E = \sum_{j=1}^{j=6} a_j T^j$

Term	Polynomial coefficients			
	IEC		IEC truncated	Chebyshev
0	−			4 865.93
1	5.399 578 2346		5.399 5	2 947.94
2	1.251 977 0000	−2	1.251 9	211.22
3	−2.244 821 7997	−5	−2.244 8	−51.87
4	2.845 216 4949	−8	2.845 2	13.60
5	−2.244 058 4544	−11	−2.244 0	−2.18
6	8.505 416 6936	−15	8.505 4	0.41

(ii) 630.74 °C to 1064.43 °C $\qquad E = a_0 + \sum_{j=1}^{j=2} a_j T^j$

0	−2.982 448 1615	2	−2.982	15 809.03
1	8.237 552 8221		8.237	2 391.10
2	1.645 390 9942	−3	1.645	38.68

(iii) 1064.43 °C to 1665 °C $E = a_0 + \sum\limits_{j=1}^{j=3} a_j T^j$

0	1.276 629 2175	3	1.276	27 875.04
1	3.497 090 8041		3.497	3 611.41
2	6.382 464 8666	−3	6.382	−2.36
3	−1.572 242 4599	−6	−1.572	−10.64

(iv) 1665 °C to 1769 °C $E = a_0 + \sum\limits_{j=1}^{j=3} a_j T^j$

0	9.784 665 5361	4	9.784 66	36 258.10
1	−1.705 029 5632	2	−1.705 02	587.28
2	1.108 869 9768	−1	1.108 87	−6.73
3	−2.249 407 0849	−5	−2.249 40	−0.79

B. (i) 0 °C to 600 °C $T = \sum\limits_{j=1}^{j=3} l_j E^j + c_0 E (1 + c_1 E)^{-1}$

Term	Polynomial coefficient
l_1	1.06735 − 1
l_2	−1.3338 − 6
l_3	3.96 − 11
c_0	7.7043 − 2
c_1	8.8025 − 4

(ii) 500 °C to 1500 °C $T = l_0 + \sum\limits_{j=1}^{j=4} l_j E^j$

l_0	4.3662 + 1
l_1	1.14259 − 1
l_2	−1.4436 − 6
l_3	−2.9938 − 11
l_4	2.379 − 15

(iii) 1000 °C to 1769 °C $T = \sum_{j=1}^{j=4} l_j E^j + E \exp[c_0(E - c_1)]$

l_1	1.30728 −1	
l_2	−3.6995 −6	
l_3	1.0367 −10	Below $E \approx 16\,500\ \mu V$
l_4	−5.177 −16	($\sim$1580 °C) the exponential
c_0	2.5827 −3	term is insignificant.
c_1	2.2131 +4	

Type S thermocouple

A. (i) −270 °C to 0 °C $E = \sum_{j=1}^{j=14} a_j T^j$

Term	Polynomial coefficients		IEC truncated	Chebyshev
	IEC			
0	–			−7 448.163
1	3.874 077 3840	1		3 173.673
2	4.412 393 2482	−2		583.495
3	1.140 523 8498	−4		−39.522
4	1.997 440 6568	−5		10.474
5	9.044 540 1187	−7	NOT	−5.651
6	2.276 601 8504	−8		1.855
7	3.624 740 9380	−10	SIMPLY	0.080
8	3.864 892 4201	−12		−0.558
9	2.829 867 8519	−14	TRUNCATED	0.404
10	1.428 138 3349	−16		−0.158
11	4.883 325 4364	−19		−0.055
12	1.080 347 4683	−21		0.117
13	1.394 929 1026	−24		−0.137
14	7.979 589 3156	−28		0.065

(ii) 0 °C to 400 °C $\displaystyle E=\sum_{j=1}^{j=8} a_j T^j$

0	–			19 715.11
1	3.874 077 3840	1	3.874 08	10 483.64
2	3.319 019 8092	−2	3.319 02	574.57
3	2.071 418 3645	−4	2.071 42	−49.29
4	−2.194 583 4823	−6	−2.194 58	3.33
5	1.103 190 0550	−8	1.103 19	−0.04
6	−3.092 758 1898	−11	−3.092 75	−0.31
7	4.565 333 7165	−14	4.565 33	0.29
8	−2.761 687 8040	−17	−2.761 68	−0.55

B*. −200 °C to 400 °C $T = l_1 E + c_0 E(1 + c_1 E)^{-1} + c_2 E(1 + c_3 E)$

l_1	1.3332 − 2
c_0	1.0091 − 2
c_1	4.567 − 5
c_2	2.461 − 3
c_3	1.2859 − 4

* From Boudry, M. R. (1976). *J. Phys. E.* **9**, 1064.

Appendix VI

Nicrosil–Nisil thermocouple, reference functions

The emf/temperature relation for Nicrosil–Nisil is given by three polynomial expressions. The first covers the range −270 °C to 0 °C and the second 0 °C to 400 °C, both for thin wires (AWG 28 or 0.32 mm diameter). For thick wires (AWG 14 or 1.63 mm diameter), the third polynomial covers the range 0 °C to 1300 °C. The emf/temperature values in the range 0 °C to 400 °C are slightly different for the two sizes of wire. The coefficients of these polynomials are given in the Table below, taken from "The Nicrosil versus Nisil thermocouple: properties and thermoelectric reference data" by N. A. Burley, R. L. Powell, G. W. Burns and M. G. Scroger, 1978, NBS Monograph 161.

Wire gauge	Temperature range	Degree	Coefficients	Term
AWG 28	−270 to 0 °C	8	$+2.6153540164 \times 10^1$	T
			$+1.0933114132 \times 10^{-2}$	T^2
			$-9.3917128470 \times 10^{-5}$	T^3
			$-5.3592739285 \times 10^{-8}$	T^4
			$-2.7406835184 \times 10^{-9}$	T^5
			$-2.3370710645 \times 10^{-11}$	T^6
			$-7.8250681060 \times 10^{-14}$	T^7
			$-9.5885491371 \times 10^{-17}$	T^8
AWG 28	0 to 400 °C	7	$+2.6153540164 \times 10^1$	T
			$+9.3169626960 \times 10^{-3}$	T^2
			$+1.3507720863 \times 10^{-4}$	T^3
			$-8.5131026625 \times 10^{-7}$	T^4
			$+2.5853558632 \times 10^{-9}$	T^5
			$-3.9887895408 \times 10^{-12}$	T^6
			$+2.4633802582 \times 10^{-15}$	T^7

Wire gauge	Temperature range	Degree	Coefficients	Term
AWG 14	0 to 1300 °C	9	$+2.5897798582 \times 10^1$	T
			$+1.6656127713 \times 10^{-2}$	T^2
			$+3.1234962101 \times 10^{-5}$	T^2
			$-1.7248130773 \times 10^{-7}$	T^4
			$+3.6526665920 \times 10^{-10}$	T^5
			$-4.4390833504 \times 10^{-13}$	T^6
			$+3.1553382729 \times 10^{-16}$	T^7
			$-1.2150879468 \times 10^{-19}$	T^8
			$+1.9557197559 \times 10^{-23}$	T^9

In NBS Monograph 161 various approximate functions are given which are low-order polynomials and which adequately describe the emf/temperature relation over restricted ranges. These low-order polynomials are designed for use with small calculators.

Skeleton reference table for Nicrosil–Nisil thermocouples.
(a) thin wire (0.32 mm dia.), (b) thick wire (1.63 mm dia.)

t_{68} (°C)	E (μV)	E (μV)	dE/dt_{68} (μV/°C)
−270	−4345		0.3
−250	−4313		2.9
−200	−3990		9.9
−150	−3336		16.0
−100	−2407		20.9
−50	−1268		24.3
0	0	0	25.9
50	1343	1339	27.7
100	2781	2774	29.6
150	4313	4301	31.6
200	5925	5912	33.0
250	7607	7596	34.3
300	9349	9340	35.4
350	11143	11135	36.4
400	12975	12972	37.1
500		16744	38.3
600		20609	39.0
700		24526	39.3
800		28456	39.3
900		32370	39.0
1000		36248	38.6
1100		40076	38.0
1200		43836	37.2
1300		47502	36.1

Index

α-coefficient
  definition of, 183
  ideal, 184
Absolute zero, 37
Accessible state, 11
Acoustic thermometry, 84–98
  at high frequencies, 95
  at low frequencies, 97
  boundary layer in, 90–92
  higher modes in, 93–95
  non-plane waves in, 93–95
  spherical resonators in, 97
  theory of, 86–96
Adiabatic wall, 5
Adsorption (see Sorption)
Aerostatic head correction
  in boiling points, 141
  in gas thermometry, 80
Amontons, 18
Annealing
  of resistance thermometers, 194, 197, 202
  of thermocouples, 259
Average mode density (see Jeans number)
Avogadro constant, 15

BIPM (Bureau International des Poids et Mesures), 26
Bireciprocal potential, 65
Blackbody lamp, 332
Blackbody radiation, 286–296
Blackbody radiator
  emissivity of, 300–315
    integral equation method, 302
    series reflectivity method, 307
    for specularly reflecting cavities, 313
    for non-isothermal cavities, 315
  practical design of, 316

Blackbody radiator (continued)
  radiation emitted by, 292 et seq.
  spectral exitance of, 292
  spectral radiance of, 293
  total exitance of, 293
Boiling points
  aerostatic head correction in, 141
  helium (see also Helium), 57
  hydrogen (see also Hydrogen), 139
  neon, 142
  oxygen, 143
  sulphur, 133
  water, 133
Boltzmann constant, 15, 11
Boundary layer, acoustic, 90
Boyle temperature, 66
Boyle's law, 10
British Association for the Advancement of Science (BAAS), 25, 29

Callendar, 28, 29
Carnot cycle, 8
Canonical ensemble, 11
Caratheodory, 7
Carbon-glass thermometer, 225
Carbon thermometer
  Allen-Bradley type, 223
  calibration of, 224
  Ohmite type, 223
  Speer type, 224
  thermal anchoring of, 225
Cavity (see Blackbody radiator)
CCT (Comité Consultatif de Thermométrie), 31
CCT-64, 39
Celsius, 21
CGPM (Conférence Générale des Poids et Mesures), 26
Chappuis, 27

Charles, 21
CIPM (Comité International des Poids et Mesures), 26
Clausius, 7
Clausius-Mossotti relation, 113
Cold spot, in v.p. measurement, 40
Comparison-baths, 122–124
    fluidized bed, 125
    salt, safety precautions in use of, 124
    stirred water and oil, 122
Comparison furnaces, 125–128
Cone, emissivity of specular, 314
Convention du Mètre, 26
Corresponding states, principle of, 66
Curie constant, 108
Curie law, 108
Cylinder, emissivity of diffuse, 302 *et seq.*

Dalton, 21
Desorption (*see* Sorption)
Detailed balance, principle of, 296
Diathermal wall, 5
Dielectric constant thermometry, 113–115

Emittance of surfaces, 295 *et seq.*
Entropy, 8
EPT-76, 49–53
Equation of state, 6
Eutectic alloys, melting and freezing of, 160

Fahrenheit, 19, 20
Ferdinand II, 19
Ferric hydroxide, 136
Fixed points, 121–163
    (*see also* Comparison baths), Heat pipes, Freezing points of metals, Superconducting transition points)
Freezing points of metals, 150–160
    effect of radius of curvature of crystals on, 159
    effects of impurities on, 153
    hydrostatic head correction in, 193
    nucleation of the freeze in, 157

Freezing points of metals (*continued*)
    oxygen contamination in, 159
    phase equilibrium in, 152
    practical realization of, 154

Galen, 18
Galileo, 18
Gas constant, 16
Gas thermometry, 61–83
    aerostatic head correction in, 80
    at high temperatures, 78
    at low temperatures, 76
    constant volume, 73
    deadspace correction in, 73, 78
    isotherm, 72
    practice of, 72–74
    sorption in, 74
    thermomolecular pressure in, 81
Gay Lussac, 21
Germanium resistance thermometer, a.c./d.c. effects in, 215
    construction of, 213
    interpolation equations for, 218
    long-term stability of, 216
    rf pick-up in, 216
    self-heating in, 214
    thermal anchoring of, 214
    thermal cycling of, 216
Gibbs, 11
Gibbs ensemble, 11
Gladstone and Dale law, 116

Harker, 28
Hartshorn bridge, 109
Heat, 6
Heat pipes, 129–131
    materials for use in, 132
Helium
    boiling point, 57
    critical point, 57
    lambda point, 56
    saturated vapour pressure, 53 *et seq.*
Hooke, Robert, 19
Hydrogen (*see also* Boiling points, Triple points)
    ortho-, para- states, 134
    conversion catalyst, 136
    normal scale, 27
    safety in the use of, 141

Immersion characteristics of Pt
 thermometer, 190
Inductive voltage divider, 234
Industrial platinum resistance
 thermometer
 (*see* Platinum resistance
 thermometer)
Intermolecular forces, 65
International Practical Temperature
 Scale of 1968
 (*see* IPTS-68)
International system of Units (SI), 1
International Temperature scales
 dates and editions, 380
IPTS-68 (*see also* Temperature scales)
 departure from thermodynamic
 temperature, 46
 extracts from text, 381
 outline of, 37–43
 uniqueness at low temperature, 43

Jeans number, 287

Kelvin (Lord), 7, 8
Kelvin, definition, of, 36
 degree, 36
 international practical degree, 36
Kirchhoff's law, 296

Landé factor, 108
Latent heat of melting
 Table of, for various gases, 145
Laws of Thermodynamics (*see*
 Thermodynamics, laws of)
Lead, absolute thermopower of, 254
Leads, problems of, 236
Lennard-Jones potential, 65
Limiting density, method of, 16
Liquid-in-glass (*see* Mercury-in-glass)
Lorentz-Lorenz equation, 116

Magnetic thermometry, 107–112
 accuracy of, 112
 Hartshorn bridge in, 109
 practice of, 109

Magnetic thermometry (*continued*)
 salts used in, 110
 theory of, 107–108
Magnetoresistance effects
 carbon, 227
 germanium, 230
 glass ceramic, 232
 platinum, 233
 p-n junction diode, 230
 rhodium-iron, 233
 Table of, 228
 thermistors, 229
Manometry (*see* Pressure measurement)
Maxwell-Boltzmann distribution, 10
Mercury-in-glass thermometers,
 368–378
 accuracy of, 370
 adjustable range, 375
 Beckmann, 375
 emergent stem correction, 372
 glasses for, 373
 pressure coefficient of, 372
 secular change of zero in, 374
 sensitivity of, 371
 temporary depression of zero in, 374
 types of, 369
Metric system (*see* Convention du
 Mètre)
Molar volume
 measurement of, 16
Moser wobble, 41
Mutual inductance, 109

Noise thermometry, 98–107
 conditions for accurate, 101
 correlation method, 102
 equal resistor method, 101
 noise power method, 103
 resistive SQUID method, 104
 statistical limitations in, 99
 unequal resistor method, 100
NPL-75, 49

Optical properties of surfaces (*see also*
 Emittance of surfaces), 299
 nomenclature for, 297
Ortho-para conversion (*see* Hydrogen)

Paramagnetism, 107
Phase-shift sums, 68
Phase, space, 10
Photon tunnelling, 292
Planck's law
  derivation of, 287–288
Platinum resistance thermometers
  a.c./d.c. effects, 189
  Callendar equation for, 183
  Callendar-van Dusen equation for,
   185
  capsule type, 187
  causes of failure, 188
  early history of, 28
  effects of impurities on, 184
  effects of oxidation on, 199
  high temperature type, 193
    lattice defects in, 194
    quenching of, 196
    interpolating equations for, 198
  industrial types, 200–209
    fine-wire, 201
    screen-printed, 202
    calibration Table for, 389
    in steam and turbine temperature
      measurement, 204–205
    in aeronautics, 206–208
  IPTS-68 reference function, 185
    alternative formulations for, 186
  long-stem type
    immersion effects, 192
    radiation losses in, 192
  resistance/temperature relation for,
    182
  self-heating effect, 189
  thermal anchoring of, 188

Pressure measurement, 82
Primary thermometry, 20
Process
  infinitesimal, 9
  isentropic, 9
  reversible, 9

Radiation laws (see under names of
    each law)
Radiation pyrometer, 334–349
  auxiliary sources, 355

Radiation pyrometer (continued)
  detectors for use in, 345
  disappearing filament, 335
  effective wavelength of, 338
  filters for, 346
  multi-wavelength, 359
  photoelectric, 341
  photon-counting, 345
  polarization, 356
  sectored discs, use of, 343
  semi-transparent media, 359
  size-of-source effect in, 347
  thermodynamic temperature,
    determination of, 349
  two-colour, 351
    laser absorption, method of, 354
  use of, to set up IPTS-68, 343
Radiation thermometry, 284–367
Ratio transformer (see Inductive
    voltage divider)
Rayleigh-Jeans law, 288
Reaumur, 21
Recovery temperature, 206
Refractive index thermometry, 115
Regnault, 21
Resistance measurement, 233–237
  Hill and Miller bridge, 234
  multi-stage transformer bridges, 235
  current comparator bridge, 236
Resistance of alloys, 172
Resistance of metals,
  band theory of, 171
  effect of lattice vacancies on, 176
  Fermi surface, 169
  free-electron theory of, 170
  temperature dependence of, 174–177
  theory of, 168
Resistance ratio, 38
Resistance thermometer (see under
    individual types)
Resistance thermometry, 167–237
Rhodium-iron thermometer, 209–213
  sensitivity of, 211
  reproducibility of, 212
Römer, 20
RTD, 200

Saturated vapour pressure, 53
Secondary thermometry, 23
Self-heating (see under individual
    thermometers)

Semiconductors
  resistance/temperature
    characteristics, 177 *et seq.*
Sorption
  in gas thermometry, 74
Spectral radiance temperature, 321
Spectral volume radiance, 361
Speed of sound (*see* Acoustic
    thermometry)
Spherical cavity, emissivity of, 310
SQUID (*see* Noise thermometry)
State parameter, 5
Static air temperature, 206
Statistical mechanics, 11
Steam point (*see* Boiling points, water)
Stefan-Boltzmann constant, 17
Stefan-Boltzmann law
  derivation of 288–289
Superconducting transition points, 148
  reproducibility of, 150
  SRM 767, 149
  SRM 768, 150
Susceptibility, paramagnetic, 107
System
  closed, 4
  isolated, 4

Temperature
  definition of, 6, 8, 12
  empirical, 6
  in classical thermodynamics, 5
  in non-equilibrium conditions, 11
  in statistical mechanics, 10
  negative, 14
  thermodynamic, definition of, 8
Temperature scales
  differences between, 35, 36
  EPT-76, 49–53
  Helium vapour pressure, 53–58
  IPTS-48, 34
  IPTS-68 (*see also* IPTS-68), 37 *et
    seq.*
  ITS-27, 31, 33
  NBS-55, 38
  normal hydrogen, 27
  NPL-75, 49
  practical scales, principles of, 31
  smoothness of, 32
  $T_{X_{AC}}$, 51
  T_{58}, 54
  T_{62}, 54

Temperature scales (*continued*)
  uniqueness of, 32
  wire scale, 33
Thermal contact, 13
Thermal equilibrium
  definition of, 4
  local, 14
Thermal radiation (*see also* Blackbody
    radiation), 286 *et seq.*
  in small cavities, 289 *et seq.*
Thermal transpiration (*see*
    Thermomolecular pressure)
Thermistors, 220–223
  conduction processes in, 220
  interpolation equations for, 221
  stability of, 221
  types of, 220
Thermocouple types, 250
  chemical composition, 251
  Cu/Cu-Ni, Type T, 264
  Fe/Cu-Ni, Type J, 264
  gold/iron, 269
  Ni-Cr/Ni-Al, Type E, 263
  Ni-Cr/Ni-Al, Type K, 264
  Nicrosil/Nisil, 266
  other noble metal couples, 258
  Platinum-67, 251
  Pt-10% Rh/Pt, Type S, 253
  Pt-13% Rh/Pt, Type R, 254
  Pt-30% Rh/Pt-6% Rh, Type B, 257
  standard reference tables for
    (*see* Thermocouples)
  tungsten-rhenium, 267, 271
  various Pt-Rh alloys, 255
Thermocouples, 241–283
  absolute thermopower of, 253
    (*see also* Lead, absolute
    thermopower of)
  calibration of, 276
  elementary theory of, 243–249
  extension and compensation cables,
    272
  for high temperatures, 267
  for low temperatures, 268
  for nuclear reactors, 270
  insulator and sheaths for Pt-Rh
    couples, 258
  IPTS-68, criteria for Type S, 256
  Kelvin relations, 247
  Onsager relations, 247
  Peltier effect, 246
  polynomial equations for, 398–407
  pressure effects in, 280

Thermocouples (*continued*)
  Pt alloy sheaths for Types S and R,
    262
  reference junctions for, 279
  Rh migration in Types S and R, 260
  Seebeck effect, 244
  standard reference tables for, 274,
    390–397, 408
  thermopower, expression for, 248
  Thomson effect, 247
Thermodynamic parameters, 5
Thermodynamic temperature (*see*
    Temperature)
Thermodynamics, laws of
  first law, 6
  second law, 7
  zeroth law, 5
Thermometer (*see also* under names of
    individual thermometers)
  early types
    alcohol-in-glass, 17
    Florentine, 17
  invention of, 16
Thermometry
  origins and early history of, 15
  primary, definition of, 23
  secondary, definition of, 23
  semi-primary, 23
Thermomolecular pressure, 80
Thomson, William (*see* Kelvin)
Tonnelot, 26
Total air temperature, 206
Total radiation thermometry, 349
Triple point
  gallium, 162
  hydrogen (*see also* Hydrogen), 137
  of other gases, 144–146
  sealed cells for, 146–148
  water, 160–161
Tunnel-diode, use in gas thermometry,
    83

Tungsten
  emissivity of, 321
  evaporation of, 324
  grain boundary grooves in, 325
  grain surface facets, 325
  outgassing of, 323
  recrystallization of, 325
Tungsten ribbon lamp, 320–332 (*see
    also* Blackbody lamp)
  ambient-temperature coefficient of,
    330
  design of, 326
  effects of orientation of, 329
  pin-temperature coefficient of, 330
  reproducibility of, 328
  temperature gradients in, 329
  water-cycle effect in, 324
  wavelength corrections for, 339
  window transmission of, 332

"V"-groove
  emissivity of, 317
Vapour pressure thermometry, 53 *et
    seq.*
Virial coefficients, 66–72
  function of temperature, 69–71
  2nd, for helium, 71
  theoretical calculation of, 67
Virial equation of state, 62
  theoretical basis of, 64

W(T) (*see* Resistance ratio)
Weber-Schmidt equation, 81
Weyl, 290
Wien displacement law, 293
Wire-scale (*see* Temperature scales)

Z-function, 37